create™

Annual Editions:
Global Issues, 32e

Robert Weiner

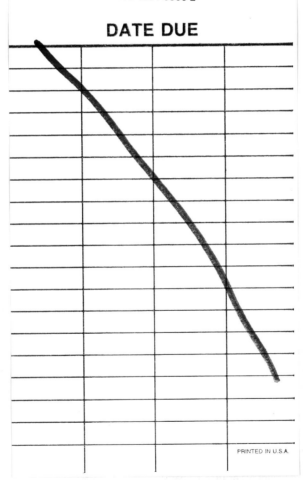

DATE DUE

PRINTED IN U.S.A.

http://create.mheducation.com

ISBN-10: 1259676005   ISBN-13: 9781259676000

Contents

Detailed Table of Contents

Preface

This book engages in an analysis of contemporary global issues based on a careful reading of the major elite newspapers and magazines, as well as the issues that have been emphasized at recent meetings of the International Studies Association and the International Political Science Association. An effort to identify important global issues has also been culled from an analysis of major US governmental reports such as the National intelligence Council, the Pentagon Quadrennial Defense Review, and the US State Department's Quadrennial Diplomatic report.

The ability of the international community to deal with global issues takes place within the framework of the forces of globalization, in a mixed international system, whose structure consists of both state and nonstate actors. Consequently, globalization is occurring in a multipolar system that is characterized by a diffusion of power. Although states, especially the "Great Powers," are still the primary actors in the international system, their power has been eroded somewhat by the phenomenal growth of such nonstate actors as international governmental organizations, nongovernmental organizations, multinational corporations, and terrorist organizations such as al Qaeda and the Islamic State.

In publishing Annual Editions we recognize the enormous role played by magazines, newspapers, and journals of the public press in a broad spectrum of areas. A number of articles are drawn from such influential journals as *Foreign Affairs, Foreign Policy,* and *The National Interest* as well, which deal with the most important global issues of the day.? Many of these articles are appropriate for students, researchers, and professionals seeking accurate, current information to help bridge the gap between theories and the real world. These articles, however, become more useful for study when those of lasting value are carefully collected, organized, indexed, and reproduced in a low-cost format that provides easy and permanent access when the material is needed. That is the role of Annual Editions.

A number of learning tools are also included in the book. Each article is followed by a set of *Critical Thinking* questions designed to allow the student to engage in further research and to stimulate classroom discussion, and valuable *Internet References* that provide the reader with more information about the themes addressed in each article. Each article is also preceded by *Learning*

Outcomes, which helps the student to focus on the major themes of each article.

I would like to express my thanks to McGraw-Hill's Senior Product Developer, Jill Meloy, without whose guidance this project would not have been completed. Special thanks are also due to Dan Torres, whose research assistance was invaluable in selecting articles that appear in this book.

Robert Weiner
Editor

Editor

Robert Weiner is a Center Associate at the Davis Center for Russian and Eurasian Studies at Harvard University and a Fellow at the Center for Peace, Democracy, and Development at the McCormick Graduate School of Policy and Global Studies, University of Massachusetts/Boston. He has worked as a consultant for Global Integrity, a Washington-based nongovernmental organization that investigates corruption in countries around the world. He is the author of *Romanian Foreign Policy at the United Nations* and *Change in Eastern Europe.* He is also the author of more than 20 book chapters and articles and book reviews. Most recently, he has published chapters entitled "The European Union and Democratization in Moldova" and "The Failure to Prevent and Punish Genocide." He has published articles and book reviews in such journals as *The Slavic Review, Sudost-Europa, The East European Quarterly, The International and Comparative Law Quarterly, Orbis, The Journal of Cold War Studies,* and *The International Studies Encyclopedia.* Between 2001 and 2011, he was the Graduate Program Director of the Master's program in International Relations at the McCormack Graduate School of Policy and Global Studies at the University of Massachusetts/Boston. He is currently a lecturer in the online BA Program in Global Affairs at the University of Massachusetts/Boston.

Academic Advisory Board

Members of the Academic Advisory Board are instrumental in the final selection of articles for Annual Editions books. Their review of the articles for content, level, and appropriateness provides critical direction to the editor(s) and staff. We think that you will find their careful consideration reflected here.

Tahereh Alavi Hojjat
Desales University

Chi Anyansi-Archibong
North Carolina A&T State University

Augustine Ayuk
Clayton State University

Dilchoda Berdieva
Miami University of Ohio

Karl Buschman
Harper College

Steven J. Campbell
University of South Carolina, Lancaster

Jianyue Chen
Northeast Lakeview College

Ravi Dhangria
Sacred Heart University

Charles Fenner
SUNY Canton

Dorith Grant-Wisdom
University of Maryland, College Park

Heather Hawn
Mars Hill University

John Patrick Ifedi
Howard University

Richard Katz
Antioch University

Steven L. Lamy
University of Southern California

Allan Mooney
Strayer University/Ashford University

Derek Mosley
Meridian Community College

Vanja Petricevic
Florida Gulf Coast University

Nathan Phelps
Western Kentucky University

Amanda Rees
Columbus State University

Kanishkan Sathasivam
Salem State University

Thomas Schunk
SUNY Orange County Community College

James C. Sperling
University of Akron

Uma Tripathi
St. John's University

Unit 1

Prepared by: Robert Weiner, *University of Massachusetts, Boston*

UNIT

Global Issues in the Twenty-First Century: An Overview

As the various units that follow indicate, globalization has not necessarily meant the institutionalization of a more effective system of global governance. This was evident in 2014 by the rather slow reaction of the World Health Organization to the outbreak of the Ebola virus in the West African states of Liberia, Sierra Leone, and Guinea. In 2015, the Ebola epidemic was eventually contained, but its lingering effects continued to be felt in post-conflict societies like Liberia.

Globalization has also been accompanied by the growth of the world's population to over 7 billion, putting a strain on global natural resources. As the Millennium State of the Future Report for 2014 indicates, humanity is becoming healthier, wealthier, and better educated. However, the growth in the world's population has been accompanied by increased competition for scarce resources, such as fresh water. For example, about 60% of the world's fresh water is located in a small number of states. Central Asian states, such as Kazakhstan, find themselves facing the dilemma of seeking access to safe, clean fresh water.

China imports via the oceans, much of the resources, such as oil and iron ore, which it needs to maintain its economic development. On the other hand, in 2015, there was a glut of energy in the world market, as the benchmark price of Brent crude oil fell below $50 a barrel. Advanced economies such as the United States have developed new energy technologies that allow them to significantly increase the extraction of natural gas and oil domestically. This has profound implications for the geopolitics of energy, as the United States is poised to become the world's leading energy producer.

Globalization has not eliminated conflict. Civil conflicts have resulted in the displacement of millions of people, with the number of migrants, internally displaced persons, and refugees standing at one of the highest marks since the end of the Second World War, straining the capacity of the international community to care for them. New masses of "boat people" fleeing the wars in the Middle East, especially in Syria, have sought sanctuary and safety in Europe, with several thousand drowning making the dangerous crossing across the Mediterranean, or putting their lives in the hands of criminal gangs of human traffickers. The European Union, faced with hundreds of thousands of migrants flooding into Eastern and Western Europe, has lacked a coherent policy to deal with this latest crisis. Immigration also emerged as a major issue in the 2016 Presidential campaign. The U.S. presidential debate revolved around immigration reform and the question of the deportation of illegal immigrants and their children, especially to Mexico. The United States also recently became a sanctuary for thousands of children from Central America fleeing poverty and criminal gangs.

Globalization is also occurring in an international system which is marked by a diffusion of power, along with the rise and decline of "Great Powers," and the emergence of new centers of power. Change and power transitions in the international system have resulted in what might be described as an emerging multipolar system. China's rise, for example, has drawn international attention to its Grand Strategy and the dangers of maritime conflict in the South and East China Seas, as Beijing seeks to consolidate its position as a regional hegemon. China seeks to deny access to external powers to what it sees as its sphere of influence, as Beijing emphasizes its sovereignty over various disputed islands in the South China and East China Seas. China is projecting its power into the Indian Ocean and, therefore, can also be viewed as an Indo-Pacific power as well. The construction of artificial islands in the Seas also contributes to a second line of defense for China. China identifies itself as a Eurasian power and seeks to expand its influence through the reconstruction of ancient trade routes through Central Asia.

The United States, on the other hand, has been categorized by a number of analysts as a declining power. However, China's "peaceful" rise has been countered by U.S. "rebalancing" with the creation of the TransPacific Partnership, and U.S. acquiescence in Japanese development of more offensive military strategy.

The emerging multipolar system has been characterized by the rise of the BRICS (Brazil, Russia, India, China, South Africa) as new centers of power. However, by 2015, most of the BRICS states were experiencing serious economic difficulties. Older blocs of states, like the G-77 (Group of 77) continued to experience a deterioration of their internal cohesion. Nevertheless, several African states have undergone impressive amounts of economic growth within the framework of a western dominated international and economic financial system.

The emerging multipolar system has also been marked by regional instability, especially in the Middle East, as ISIL (the Islamic State in Iraq and the Levant) continued to function as a major force in the civil conflicts in Syria and Iraq, against a backdrop of unsuccessful U.S. efforts to extricate itself from the region. In 2015 the world witnessed the deployment of Russian military assets in Syria, in support of President Assad. Critics argued that the United States needed a Grand Strategy to deal with the regional conflicts in the Middle East, raging in Syria, Iraq, Yemen, and Libya. A nuclear agreement, reached between Iran and the United States in the summer of 2015, also had an effect on the regional security architecture in the Middle East, given the opposition of such states as Saudi Arabia and Israel to the agreement.

Terrorism continued as another feature of globalization as underscored by the murder by jihadists of the staff of a French satirical magazine in Paris.

The United States in 2015 found it difficult to disentangle itself from Afghanistan as promised by President Obama, given a resurgent Taliban, as well as the inroads being made into Afghanistan by ISIL.

The globalized world of the 21st century has witnessed the growth of new digital technologies, such as "the Internet of things." On the darker digital side of globalization, the revolution in global communications and technology has increased the vulnerability of advanced information societies to cyberthreats, hacking, and cyberwar in cyberspace. Moreover, as Edward Snowden revealed, information technology has also increased the surveillance capacity of the state both globally and at the domestic level.

Globalization has taken place within the framework of increasing threats to the environment, such as climate warming. Most scientists agree that climate warming is here now, due to the release into the atmosphere of greenhouse gases, which has the result of raising the temperature of the earth. This has a number of effects, such as the melting of the ice in the Arctic, and the melting of the ice sheets in Antarctica. The warmer climate also causes extreme weather events, such as killer typhoons and hurricanes, as well as droughts that compound the problem of dealing with water scarcity.

Finally, from a liberal international point of view, there is a need to effectively implement a set of norms that can serve as a benchmark for the behavior of states in the international system. States have a responsibility not only to protect their own populations, but also vulnerable populations of other states from gross and mass violations of human rights. Human security needs to be given at least as much weight as the traditional concern of national security.

Article   Prepared by: Robert Weiner, *University of Massachusetts, Boston*

Our Global Situation and Prospects for the Future

Humanity is making momentous strides forward in health, literacy, and many other critical areas, but also stalling or moving backward on many others, warns The Millennium Project in its latest *State of the Future* report.

JEROME C. GLENN

Learning Outcomes

After reading this article, you will be able to:

- Identify the critical areas in which humanity is not moving forward.

- Discuss the effects of an increasingly interconnected world.

The global situation for humanity continues to improve in general, but at the expense of the environment. Massive transitions from isolated subsistence agriculture and industry to a global, Internet-connected, pluralistic civilization are occurring at unprecedented speed and with never-before-seen levels of uncertainty.

The indicators of progress, from health and education to water and energy, show that we are winning more than we are losing—but where we are losing is very serious. As The Millennium Project has documented over the past 17 years in its annual *State of the Future* reports, humanity clearly has the ideas and resources to address its global challenges, but it has not yet shown the leadership, policies, and management on the scale necessary.

On one hand, people around the world are becoming healthier, wealthier, better educated, more peaceful, and increasingly connected, and they are living longer. The child mortality rate has dropped 47% since 1990, while life expectancy has risen by 10 years to reach 70.5 years today. Extreme poverty in the developing world fell from 50% in 1981 to 21% in 2010, primary-school completion rates grew from 81% in 1990 to 91% in 2011, and only one transborder war occurred in 2013. Furthermore, nearly 40% of humanity is now connected via the Internet.

However, water tables on all continents are falling, glaciers are melting, coral reefs are dying, ocean acidity is increasing, ocean dead zones have doubled every decade since the 1960s, and half the world's topsoil has been destroyed.

Some critical socioeconomic fault lines are worsening, as well: Intrastate conflicts and refugee numbers are rising, income gaps are increasingly obscene, and youth unemployment has reached dangerous proportions. Meanwhile, traffic jams and air pollution are strangling cities. In addition, between $1 trillion and $1.6 trillion is paid in bribes, organized crime takes in twice as much money per year as all military budgets combined, civil liberties are increasingly threatened, and half the world is potentially unstable.

The International Monetary Fund expects the global economy to grow from 3% in 2013 to 3.7% during 2014 and possibly 3.9% in 2015. The world population having grown 1.1% in 2013, global per capita income will be increasing by 2.6% or more a year. Our world is reducing poverty faster than many thought was possible.

Nevertheless, the divide between the rich and poor is growing fast: According to Oxfam, the total wealth of the richest 85 people equals that of 3.6 billion people in the bottom half of the world's economy, and half of the world's wealth is owned by just 1% of the population. We need to continue the successful efforts that are reducing poverty, but we also need to focus far more seriously on reducing income inequality in order to avoid long-term instability.

Instability has already been erupting and expanding in many parts of the world over the last five years, due to a confluence of rising food and energy prices, failing states, falling water tables, climate change, desertification, and increasing migrations resulting from political, environmental, and economic conditions. And, because the world is better educated and increasingly connected, people are becoming less tolerant of the abuse of elite power than in the past. Unless these elites open the conversation about the future with the rest of their populations, unrest and revolutions are likely to continue and increase.

Although wars between states are becoming fewer and fewer, and the numbers of both nuclear weapons and battle-related deaths have been decreasing, conflicts within countries are increasing: A third of Syria's 21 million people are displaced or live as refugees, and the world ignores 6 million war-related deaths in the Congo.

Other fault lines are emerging worldwide in the form of rapidly rising frequency of cyberattacks and espionage, an escalation in territorial tensions among Asian countries, and overlapping jurisdictions for energy access to the melting Arctic. It will be a test of humanity's maturity to resolve all these conflicts peacefully.

Meanwhile, the world is automating jobs far more broadly and quickly than it did in earlier eras. How many truck and taxi-cab drivers will future self-driving vehicles replace? How many industrial laborers will lose their jobs to robotic manufacturing? How many telephone support personnel will be supplanted by AI telephone systems?

In every industry and sector, the number of employees per business revenue is falling, giving rise to employment-less economic growth. Job seekers will need more opportunities for one-person Internet-based self-employment and for markets for their interests and abilities in other job markets worldwide. Successfully leapfrogging slower linear development processes in lower-income countries is likely to require implementing futuristic possibilities—from 3-D printing to seawater agriculture—and making increasing individual and collective intelligence a national objective of each country.

> **"Because the world is better educated and increasingly connected, people are becoming less tolerant of the abuse of elite power than in the past."**

The explosive, accelerating growth of knowledge in a rapidly changing and increasingly interdependent world gives us so much to know about so many things. Unfortunately, we are also flooded with so much trivial news that serious issues get little attention or interest, and too much time is wasted going through useless information.

At the same time, the world is increasingly engaged in diverse conversations about how to relate to the environment and to our fellow humans, and about what technologies, economics, and laws are right for our common future. These conversations are emerging from countless international negotiations, UN gatherings, and thousands of Internet discussion groups and big-data analyses. Humanity is slowly but surely becoming aware of itself as an integrated system of cultures, economies, technologies, natural and built environments, and governance systems.

Collecting Our Intelligence

These great conversations will be better informed if we realize that the world is improving more than most pessimists know and that future dangers are worse than most optimists indicate. Better ideas, new tools, and creative management approaches are popping up all over the world, but the lack of imagination and courage to make serious change is drowning the innovations needed to make the world work for all.

As a global think tank, The Millennium Project gathers insights from a network of more than 4,500 experts who continuously gather and share data via our online Global Futures System (GFS). GFS can be thought of as a global information utility from which different readers can draw different value for improving their understanding and decisions.

The collective intelligence emerges in GFS from synergies among data/information/knowledge, software/hardware, and experts and others with insight that continually learn from feedback to produce just-in-time knowledge for better decisions than any of these elements acting alone.

In addition to succinct but relatively detailed descriptions of the current situation and forecasts, we also formulate recommendations to address the various global challenges. Some of our recommendations are as follows:

- Establish a U.S.–China 10-year environmental security goal to reduce climate change and improve trust.
- Grow meat without growing animals, to reduce water demand and greenhouse-gas emissions.
- Develop seawater agriculture for biofuels, carbon sink, and food without rain.
- Build global collective intelligence systems for input to long-range strategic plans.
- Create tele-nations connecting brains overseas to the development process back home.
- Establish trans-institutions for more effective implementation of strategies.
- Detail and implement a global strategy to counter organized crime.
- Use the State of the Future Index as an alternative to GDP as a measure of progress for the world and nations with 30 variables that include indicators for social equity and well-being.

The World Report Card

The world is in a race between implementing ever-increasing ways to improve the human condition and the seemingly ever-increasing complexity and scale of global problems. So, how is the world doing in this race? What's the score so far?

The Millennium Project's global State of the Future Index (SOFI), produced annually since 2000, measures the 10-year outlook for the future based on historical data on 30 key variables. In the aggregate, these data depict whether the future promises to be better or worse. The SOFI is intended to show the directions and intensity of change and to identify the responsible factors and the relationships among them.

The current SOFI, shown in Figure 1, indicates a slower progress since 2007, although the overall outlook is promising.

Some Key Trends Affecting the State of the Future

- **Computing.** The EU, United States, Japan, and China have announced programs to understand how the brain works and apply that knowledge to make better computers with better computer–user interfaces. Google also is working to create artificial brains that could serve us as personal artificial-intelligence assistants. Another great race is on to make supercomputer power available to the masses with advances in IBM's Watson and with cloud computing by Amazon and others. About 85% of the world's population is expected to be covered by high-speed mobile Internet in 2017.

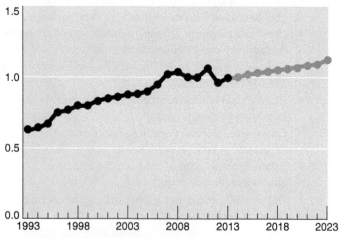

Figure 1   2013 State of the Future Index

Each of the 30 variables making up the index (Box 1) can be examined to show where we are winning, where we are losing, and where there is unclear or little progress.

- **A Web-connected world.** More than 8 billion devices are connected to the Internet of Things, which is expected to grow to 40 billion–80 billion devices by 2020. According to the UN's International Telecommunications Union, nearly 40% of humanity now uses the Internet. This global network is close to becoming the de facto global brain of humanity.

  So what happens when the entire world has access to nearly all the world's knowledge, along with instantaneous access to artificial brains that can solve problems and create new conditions like geniuses, while blurring previous distinctions between virtual realities and physical reality? We have already seen brilliant financial experts—augmented with data and software—making the short-term, selfish, economic decisions that led to the 2008 global financial crisis, continued environmental degradation, and widening income disparities. It is not yet clear that humanity will grow from short-term, me-first thinking to longer-term, we-first, planet-oriented decision making.

> **"It is not yet clear that humanity will grow from short-term, me-first thinking to longer-term, we-first, planet-oriented decision making."**

Humanity may become more responsible and compassionate as the Internet of people and things grows across the planet, making us more aware of humanity as a whole and of our natural and built environments. Yet multi-way interactive media also attracts individuals with common interests into isolated ideological groups, reinforcing social polarization and conflict and forcing some political systems into gridlock.

And although the Internet's growth may make it increasingly difficult for conventional crimes to go undetected, cyberspace has become the medium for new kinds of crimes: According to the cloud-services provider Akamai, there were 628 cyber-attacks over 24 hours on July 24, 2013, the majority of which attacked targets in the United States. Cyber-attacks can be thought of as a new kind of guerrilla warfare. Prevention may involve an endless intellectual arms race of hacking and counter-hacking software, setting cyber traps, exposing sources, and initiating trade sanctions.

- **Civil strife.** The long-range trend toward democracy is strong, but Freedom House reports that world political and civil liberties deteriorated for the eighth consecutive year in 2013, with declines noted in 54 countries and

Box 1
Variables Used in the 2013–14 State of the Future Index

1. GNI per capita, PPP (constant 2005 international $)
2. Economic income inequality (share of top 10%)
3. Unemployment, total (% of world labor force)
4. Poverty headcount ratio at $1.25 a day (PPP) (% of population)
5. Levels of corruption (0 = highly corrupt; 6 = very clean)
6. Foreign direct investment, net inflows (balance of payments, current $, billions)
7. R&D expenditures (% of GDP)
8. Population growth (annual %)
9. Life expectancy at birth (years)
10. Mortality rate, infant (per 1,000 live births)
11. Prevalence of undernourishment
12. Health expenditure per capita (current $)
13. Physicians (per 1,000 people)
14. Improved water source (% of population with access)
15. Renewable internal freshwater resources per capita (thousand cubic meters)
16. Ecological Footprint/Biocapacity ratio
17. Forest area (% of land area)
18. CO_2 emissions from fossil-fuel and cement production (billion tonnes)
19. Energy efficiency [GDP per unit of energy use (constant 2005 PPP $ per kg of oil equivalent)]
20. Electricity production from renewable sources, excluding hydroelectric (% of total)
21. Literacy rate, adult total (% of people ages 15 and above)
22. School enrollment, secondary (% gross)
23. Number of wars (conflicts with more than 1,000 fatalities)
24. Terrorism incidents
25. Number of countries and groups that had or still have intentions to build nuclear weapons
26. Freedom rights (number of countries rated free)
27. Voter turnout (% voting population)
28. Proportion of seats held by women in national parliaments (% of members)
29. Internet users (per 100 people)
30. Prevalence of HIV (% of population age 15–49)

- **Climate change.** The Intergovernmental Panel on Climate Change's *Fifth Assessment Report* found that world greenhouse gas emissions grew by an annual average of 2.2% between 2000 and 2010, up from 1.3% per year between 1970 and 2000. Each decade of the past three was warmer than the previous decade. The past 30 years was likely the warmest period in the Northern Hemisphere in the last 1,400 years.

  Furthermore, even if all CO_2 emissions are stopped today, the IPCC report notes that "most aspects of climate change will persist for many centuries." Hence, the world has to take adaptation far more seriously, in addition to reducing emissions, and creating new methods to reduce the greenhouse gases that are already in the atmosphere.

  Without dramatic changes, UN Environment Program projects a 2°C (3.6°F) rise above preindustrial levels in 20–30 years, accelerating changing climate, ocean acidity, changes in disease patterns, and saltwater intrusions into freshwater areas worldwide. The UN Food and Agriculture Organization reports that 87% of global fish stocks are either fully exploited or overexploited. Oceans absorb about 33% of human-generated CO_2, but their ability to continue doing this is being reduced by changing acidity and the die-offs of coral reefs and other living systems.

- **Energy needs.** The world also needs to create enough electrical production capacity for an additional 3.7 billion people by 2050. There are 1.2 billion people without electricity today (17% of the world), and an additional 2.4 billion people will be added to the world's population between now and 2050.

  Compounding this is the requirement to decommission aging nuclear power plants and to replace or retrofit fossil fuel plants. The cost of nuclear power is increasing, while the cost of renewables is falling—wind power passed nuclear as Spain's leading source of electricity. However, fossil fuels (coal, oil, and natural gas) will continue to supply the vast majority of the world's electricity past 2050 unless there are major social and technological changes. If the long-term trends toward a wealthier and more sophisticated world continue, our energy demands by 2050 could be more than expected. However, the convergences of technologies are accelerating rapidly to make energy efficiencies far greater by 2050 than forecast today.

- **Water stress.** Major progress was made over the past 25 years that provided enough clean water for an additional 2 billion people. But as a result of water pollution, accelerating climate change, falling water tables around the world, and an additional 2.4 billion people in just 36 years, some of the people with safe water today may not have it in the future unless significant changes occur. According to the Organisation for Economic Co-operation

improvements in only 40 countries. At the same time, increasing numbers of educated and mobile-phone/Internet-savvy people are no longer tolerating the abuse of power and may be setting the stage for a long and difficult transition to more global democracy.

and Development (OECD), half the world could be living in areas with severe water stress by 2030.

- **Population growth.** The UN's mid-range forecast is that the world's population, which now totals 7.2 billion people, will number 9.6 billion by 2050. By that date, the number of people over age 65 will equal or surpass the number under 15.

Average life expectancy at birth has increased from 48 years in 1955 to 70.5 years today. Future scientific and medical breakthroughs could give people longer and more productive lives than most would believe possible today. For example, uses of genetic data, software, and nanotechnology will help detect and treat disease at the genetic or molecular level.

> **"It is unreasonable to expect the world to cooperatively create and implement strategies to build a better future without some general agreement about what that desirable future is.**

- **Accelerating technologies.** Science and technology's continued acceleration is fundamentally changing what is possible, and access to this knowledge is becoming universally available. For example, China's Tianhe-2 super-computer is the world's fastest computer, at 33.86 petaflops (quadrillion floating point operations per second)—passing the computational speed of a human brain. Individual gene sequencing is now available for $1,000—and the price could go down much further in coming years—a development that will enable individualized genetic medicine for every patient.

Although advances in synthetic biology, quantum entanglement, Higgs-like particles, and computational science seem remote from improving the human condition, such basic scientific endeavors are necessary to increase the knowledge that scientists can use to develop and improve technologies to benefit humanity. But with little news coverage and educational curricula, the general public seem unaware of the extraordinary changes and consequences that need to be discussed: Is it ethical to clone ourselves, to bring dinosaurs back to life, or to invent thousands of new life forms through synthetic biology? Should basic scientific research be pursued without direct regard for social issues? On the other hand, might social considerations impair progress toward a truthful understanding of reality?

- **Gender equity.** Violence against women is the largest war today, as measured by death and casualties per year. Globally, 35% of women have experienced physical and/

or sexual violence. While the gender gaps for health and educational attainment were closed by 96% and 93% respectively, according to the 2013 Global Gender Gap report by the World Economic Forum, the gap in economic participation has been closed by only 60%, and the gap in political outcomes by only 21%: Women account for only 21.3% of the membership of national legislative bodies worldwide, up from 11.3% in 1997.

- **"Hidden" hunger.** Food markets in much of the developing world exhibit an increasing problem of hidden hunger—that is, the intake of calories is sufficient, but those calories contain little in nutritious value, vitamins, and minerals. Although the share of people in the world who are hungry has fallen from over 30% in 1970 to 15% today, concerns are increasing over the variety and nutritional quality of food. The FAO estimates that some 30% of the world population (2 billion people) suffers from hidden hunger.

- **Vulnerable urban coastal zones.** Human construction is diminishing the land structures that the world's coastal zones rely on to blunt the impacts of hurricanes, tsunamis, and pollution. This is a harmful outcome, not only for flora and fauna, but for us, as well, since more than half the world's people live within 120 miles of a coastline. Without appropriate mitigation, prevention, and management of the natural infrastructure within urban coastal zones, billions of people will be increasingly vulnerable to a range of disasters.

- **"Lone wolf" terrorism.** Individuals acting alone can wield increasing amounts of damage. The number of terrorism incidents increased over the past 20 years, reaching 8,441 in 2012 and more than 5,000 in the first half of 2013.

Of all terrorism, the lone-wolf type is the most insidious, because it is exceedingly difficult to anticipate, given the actions and intent of individuals acting alone. The average opinion of our international panel is that nearly a quarter of terrorist attacks carried out in 2015 might be by a lone wolf, and that the situation might escalate: About half of the participants that we surveyed thought that lone-wolf terrorists might attempt to use weapons of mass destruction by around 2030.

"Global Collective Intelligence Systems" Bring It All Together

It is unreasonable to expect the world to cooperatively create and implement strategies to build a better future without some general agreement about what that desirable future is. Such a future can only be built with awareness of the global situation and of the extraordinary possibilities.

What we need is a global collective intelligence system to track science and technology advances, forecast consequences, and document a range of views on them. The accelerating rates of changes that the world now experiences call for new kinds of decision making with global real-time feedback. The Global Futures System is an early expression of that future direction.

Critical Thinking

1. What critical socioeconomic fault lines are worsening?
2. Pick 10 of the most important variables used in the State of the Future 2013–2014 Index.
3. What are the most important trends in planet-oriented decision-making?

Internet References

The Futures Forum
futuresforum.org
The Millennium Project
Millennium-project.org
The World Futures Society
www.wfs/org/

JEROME C. GLENN is CEO of The Millennium Project and The Global Futures System, www.themp.org. This article is adapted from *2013–14 State of the Future,* co-authored by Glenn with Theodore J. Gordon and Elizabeth Florescu (published by The Millennium Project, millennium-project.org/millennium/201314SOF.html).

Article   Prepared by: Robert Weiner, *University of Massachusetts, Boston*

The Geopolitics of Cyberspace after Snowden

RON DEIBERT

Learning Outcomes

After reading this article, you will be able to:

- Discuss the effects of Snowden's revelations on the freedom of the Internet.

- Explain the major changes that have occurred recently in the use of the Internet.

For several years now, it seems that not a day has gone by without a new revelation about the perils of cyberspace: the networks of Fortune 500 companies breached; cyberespionage campaigns uncovered; shadowy hacker groups infiltrating prominent websites and posting extremist propaganda. But the biggest shock came in June 2013 with the first of an apparently endless stream of riveting disclosures from former US National Security Agency (NSA) contractor Edward Snowden. These alarming revelations have served to refocus the world's attention, aiming the spotlight not at cunning cyber activists or sinister data thieves, but rather at the world's most powerful signals intelligence agencies: the NSA, Britain's Government Communications Headquarters (GCHQ), and their allies.

The public is captivated by these disclosures, partly because of the way in which they have been released, but mostly because cyberspace is so essential to all of us. We are in the midst of what might be the most profound communications evolution in all of human history. Within the span of a few decades, society has become completely dependent on the digital information and communication technologies (ICTs) that infuse our lives. Our homes, our jobs, our social networks—the fundamental pillars of our existence—now demand immediate access to these technologies.

With so much at stake, it should not be surprising that cyberspace has become heavily contested. What was originally designed as a small-scale but robust information-sharing network for advanced university research has exploded into the information infrastructure for the entire planet. Its emergence has unsettled institutions and upset the traditional order of things, while simultaneously contributing to a revolution in economics, a path to extraordinary wealth for Internet entrepreneurs, and new forms of social mobilization. These contrasting outcomes have set off a desperate scramble, as stakeholders with competing interests attempt to shape cyberspace to their advantage. There is a geopolitical battle taking place over the future of cyberspace, similar to those previously fought over land, sea, air, and space.

Three major trends have been increasingly shaping cyberspace: the big data explosion, the growing power and influence of the state, and the demographic shift to the global South. While these trends preceded the Snowden disclosures, his leaks have served to alter them somewhat, by intensifying and in some cases redirecting the focus of the conflicts over the Internet. This essay will identify several focal points where the outcomes of these contests are likely [to] be most critical to the future of cyberspace.

Big Data

Before discussing the implications of cyberspace, we need to first understand its characteristics: What is unique about the ICT environment that surrounds us? There have been many extraordinary inventions that revolutionized communications throughout human history: the alphabet, the printing press, the telegraph, radio, and television all come to mind. But arguably the most far-reaching in its effects is the creation and development of social media, mobile connectivity, and cloud computing—referred to in shorthand as "big data." Although these three technological systems are different in many ways, they share one very important characteristic: a vast and rapidly growing volume of personal information, shared (usually voluntarily) with entities separate from the individuals to whom the information applies. Most of those entities are privately owned companies, often headquartered in political jurisdictions other than the one in which the individual providing the information lives (a critical point that will be further examined below).

We are, in essence, turning our lives inside out. Data that used to be stored in our filing cabinets, on our desktop computers, or even in our minds, are now routinely stored on equipment maintained by private companies spread across the globe. This data we entrust to them includes that which we are conscious of and deliberate about—websites visited, emails sent, texts received, images posted—but a lot of which we are unaware.

For example, a typical mobile phone, even when not in use, emits a pulse every few seconds as a beacon to the nearest WiFi router or cellphone tower. Within that beacon is an extraordinary amount of information about the phone and its owner (known as "metadata"), including make and model, the user's name, and geographic location. And that is just the mobile device itself. Most users have within their devices several dozen applications (more than 50 billion apps have been downloaded from Apple's iTunes store for social networking, fitness, health, games, music, shopping, banking, travel, even tracking sleep patterns), each of which typically gives itself permission to extract data about the user and the device. Some applications take the practice of data extraction several bold steps further, by requesting access to geolocation information, photo albums, contacts, or even the ability to turn on the device's camera and microphone.

We leave behind a trail of digital "exhaust" wherever we go. Data related to our personal lives are compounded by the numerous and growing Internet-connected sensors that permeate our technological environment. The term "Internet of Things" refers to the approximately 15 billion devices (phones, computers, cars, refrigerators, dishwashers, watches, even eyeglasses) that now connect to the Internet and to each other, producing trillions of ever-expanding data points. These data points create an ethereal layer of digital exhaust that circles the globe, forming, in essence, a digital stratosphere.

Given the virtual characteristics of the digital experience, it may be easy to overlook the material properties of communication technologies. But physical geography is an essential component of cyberspace: *Where* technology is located is as important as *what* it is. While our Internet activities may seem a kind of ephemeral and private adventure, they are in fact embedded in a complex infrastructure (material, logistical,

and regulatory) that in many cases crosses several borders. We assume that the data we create, manipulate, and distribute are in our possession. But in actuality, they are transported to us via signals and waves, through cables and wires, from distant servers that may or may not be housed in our own political jurisdiction. It is actual matter we are dealing with when we go online, and that matters—a lot. The data that follow us around, that track our lives and habits, do not disappear; they live in the servers of the companies that own and operate the infrastructure. What is done with this information is a decision for those companies to make. The details are buried in their rarely read terms of service, or, increasingly, in special laws, requirements, or policies laid down by the governments in whose jurisdictions they operate.

The vast majority of Internet users now live in the global South.

Big State

The Internet started out as an isolated experiment largely separate from government. In the early days, most governments had no Internet policy, and those that did took a deliberately laissez-faire approach. Early Internet enthusiasts mistakenly understood this lack of policy engagement as a property unique to the technology. Some even went so far as to predict that the Internet would bring about the end of organized government altogether. Over time, however, state involvement has expanded, resulting in an increasing number of Internet-related laws, regulations, standards, and practices. In hindsight, this was inevitable. Anything that permeates our lives so thoroughly naturally introduces externalities—side effects of industrial or commercial activity—that then require the establishment of government policy. But as history demonstrates, linear progress is always punctuated by specific events—and for cyberspace, that event was 9/11.

We continue to live in the wake of 9/11. The events of that day in 2001 profoundly shaped many aspects of society. But no greater impact can be found than the changes it brought to cyberspace governance and security, specifically with respect to the role and influence of governments. One immediate impact was the acceleration of a change in threat perception that had been building for years.

During the Cold War, and largely throughout the modern period (roughly the eighteenth century onward), the primary threat for most governments was "interstate" based. In this paradigm, the state's foremost concern is a cross-border invasion or attack—the idea that another country's military could use force and violence in order to gain control. After the Cold War, and especially since 9/11, the concern has shifted to a different threat paradigm: that a violent attack could be executed by a small extremist group, or even a single human being who could blow himself or herself up in a crowded mall, hijack an airliner, or hack into critical infrastructure. Threats are now dispersed across all of society, regardless of national borders. As a result, the focus of the state's security gaze has become omnidirectional.

Accompanying this altered threat perception are legal and cultural changes, particularly in reaction to what was widely perceived as the reason for the 9/11 catastrophe in the first place: a "failure to connect the dots." The imperative shifted from the micro to the macro. Now, it is not enough to simply look for a needle in the haystack. As General Keith Alexander (former head of the NSA and the US Cyber Command) said, it is now necessary to collect "the entire haystack." Rapidly, new laws have been introduced that substantially broaden the reach of law enforcement and intelligence agencies, the most notable of them being the Patriot Act in the United States—although many other countries have followed suit.

This imperative to "collect it all" has focused government attention squarely on the private sector, which owns and operates most of cyberspace. States began to apply pressure on companies to act as a proxy for government controls—policing their own networks for content deemed illegal, suspicious, or a threat to national security. Thanks to the Snowden disclosures, we now have a much clearer picture of how this pressure manifests itself. Some companies have been paid fees to collude, such as Cable and Wireless (now owned by Vodafone), which was paid tens of millions of pounds by the GCHQ to install surveillance equipment on its networks. Other companies have been subjected to formal or informal pressures, such as court orders, national security letters, the with-holding of operating licenses, or even appeals to patriotism. Still others became the targets of computer exploitation, such as US-based Google, whose back-end data infrastructure was secretly hacked into by the NSA.

This manner of government pressure on the private sector illustrates the importance of the physical geography of cyberspace. Of course, many of the corporations that own and operate the infrastructure—companies like Facebook, Microsoft, Twitter, Apple, and Google—are headquartered in the United States. They are subject to US national security law and, as a consequence, allow the government to benefit from a distinct home-field advantage in its attempt to "collect it all." And that it does—a staggering volume, as it turns out. One top-secret NSA slide from the Snowden disclosures reveals that by 2011, the United States (with the cooperation of the private sector) was collecting and archiving about 15 billion Internet metadata records *every single day*. Contrary to the expectations of early Internet enthusiasts, the US government's approach to cyberspace—and by extension that of many other governments as well—has been anything but laissez-faire in the post-9/11 era. While cyberspace may have been born largely in the absence of states, as it has matured states have become an inescapable and dominant presence.

Domain Domination

After 9/11, there was also a shift in US military thinking that profoundly affected cyberspace. The definition of cyberspace as a single "domain"— equal to land, sea, air, and space—was formalized in the early 2000s, leading to the imperative to dominate and rule this domain; to develop offensive capabilities to fight and win wars within cyberspace. A Rubicon was crossed with the Stuxnet virus, which sabotaged Iranian nuclear enrichment facilities. Reportedly engineered jointly by the United States and Israel, the Stuxnet attack was the first de facto act of war carried out entirely through cyberspace. As is often the case in international security dynamics, as one country reframes its objectives and builds up its capabilities, other countries follow suit. Dozens of governments now have within their armed forces dedicated "cyber commands" or their equivalents.

The race to build capabilities also has a ripple effect on industry, as the private sector positions itself to reap the rewards of major cyber-related defense contracts. The imperatives of mass surveillance and preparations for cyberwarfare across the globe have reoriented the defense industrial base. It is noteworthy in this regard how the big data explosion and the growing power and influence of the state are together generating a political-economic dynamic. The aims of the Internet economy and those of state security converge around the same functional needs: collecting, monitoring, and analyzing as much data as possible. Not surprisingly, many of the same firms service both segments. For example, companies that market facial recognition systems find their products being employed by Facebook on the one hand and the Central Intelligence Agency on the other.

As private individuals who live, work, and play in the cyber realm, we provide the seeds that are then cultivated, harvested, and delivered to market by a massive machine, fueled by the twin engines of corporate and national security needs. The confluence of these two major trends is creating extraordinary tensions in state-society relations, particularly around privacy. But perhaps the most important implications relate to the fact that the market for the cybersecurity industrial complex knows no boundaries—an ominous reality in light of the shifting demographics of cyberspace.

Southern Shift

While the "what" of cyberspace is critical, the "who" is equally important. There is a major demographic shift happening today that is easily overlooked, especially by users in the West, where the technology originates. The vast majority of Internet users now live in the global South. Of the 6 billion mobile devices in circulation, over 4 billion are located in the developing world. In 2001, 8 of every 100 citizens in developing nations owned a mobile subscription. That number has now jumped to 80. In Indonesia, the number of Internet users increases each month by a stunning 800,000. Nigeria had 200,000 Internet users in 2000; today, it has 68 million.

Remarkably, some of the fastest growing online populations are emerging in countries with weak governmental structures or corrupt, autocratic, or authoritarian regimes. Others are developing in zones of conflict, or in countries that have only recently gone through difficult transitions to democracy. Some of the fastest growth rates are in "failed" states, or in countries riven by ethnic rivalries or challenged by religious differences and sensitivities, such as Nigeria, India, Pakistan, Indonesia, and Thailand. Many of these countries do not have long-standing democratic traditions, and therefore lack proper systems of accountability to guard against abuses of power. In some, corruption is rampant, or the military has disproportionate influence.

Consider the relationship between cyberspace and authoritarian rule. We used to mock authoritarian regimes as slow-footed, technologically challenged dinosaurs that would be inevitably weeded out by the information age. The reality has proved more nuanced and complex. These regimes are proving much more adaptable than expected. National-level Internet controls on content and access to information in these countries are now a growing norm. Indeed, some are beginning to affect the very technology itself, rather than vice versa.

In China (the country with the world's most Internet users), "foreign" social media like Facebook, Google, and Twitter are banned in favor of nationally based, more easily controlled alternatives. For example, We Chat—owned by China-based parent company Tencent—is presently the fifth-largest Internet company in the world after Google, Amazon, Alibaba, and eBay, and as of August 2014 it had 438 million active users (70 million outside China) and a public valuation of over $400 billion. China's popular chat applications and social media are required to police the country's networks with regard to politically sensitive content, and some even have hidden censorship and surveillance functionality "baked" into their software. Interestingly, some of We Chat's users outside China began experiencing the same type of content filtering as users inside China, an issue that Tencent claimed was due to a software bug (which it promptly fixed). But the implication of such extraterritorial applications of national-level controls is certainly worth further scrutiny, particularly as China-based companies begin to expand their service offerings in other countries and regions.

It is important to understand the historical context in which this rapid growth is occurring. Unlike the early adopters of the Internet in the West, citizens in the developing world are plugging in and connecting after the Snowden disclosures, and with the model of the NSA in the public domain. They are coming online with cybersecurity at the top of the international agenda, and fierce international competition emerging throughout cyberspace, from the submarine cables to social media. Political leaders in these countries have at their disposal a vast arsenal of products, services, and tools that provide their regimes with highly sophisticated forms of information control. At the same time, their populations are becoming more savvy about using digital media for political mobilization and protest.

While the digital innovations that we take advantage of daily have their origins in high-tech libertarian and free-market hubs like Silicon Valley, the future of cyberspace innovation will be in the global South. Inevitably, the assumptions, preferences, cultures, and controls that characterize that part of the world will come to define cyberspace as much as those of the early entrepreneurs of the information age did in its first two decades.

Who Rules?

Cyberspace is a complex technological environment that spans numerous industries, governments, and regions. As a consequence, there is no one single forum or international organization for cyberspace. Instead, governance is spread throughout numerous small regimes, standard-setting forums, and technical organizations from the regional to the global. In the early days, Internet governance was largely informal and led by non-state actors, especially engineers. But over time, governments have become heavily involved, leading to more politicized struggles at international meetings.

Although there is no simple division of camps, observers tend to group countries into those that prefer a more open Internet and a tightly restricted role for governments versus those that prefer a more centralized and state-led form of governance, preferably through the auspices of the United Nations. The United States, the United Kingdom, other European nations, and Asian democracies are typically grouped in the former, with China, Russia, Iran, Saudi Arabia, and other nondemocratic countries grouped in the latter. A large number of emerging market economies, led by Brazil, India, and Indonesia, are seen as "swing states" that could go either way.

Prior to the Snowden disclosures, the battle lines between these opposing views were becoming quite acute—especially around the December 2012 World Congress on Information Technology (WCIT), where many feared Internet governance would fall into UN (and thus more state-controlled) hands. But the WCIT process stalled, and those fears never materialized, in part because of successful lobbying by the United States and its allies, and by Internet companies like Google. After the

Snowden disclosures, however, the legitimacy and credibility of the "Internet freedom" camp have been considerably weakened, and there are renewed concerns about the future of cyberspace governance.

> **The original promise of the Internet as a forum for free exchange of information is at risk.**

Meanwhile, less noticed but arguably more effective have been lower-level forms of Internet governance, particularly in regional security forums and standards-setting organizations. For example, Russia, China, and numerous Central Asian states, as well as observer countries like Iran, have been coordinating their Internet security policies through the Shanghai Cooperation Organization (SCO). Recently, the SCO held military exercises designed to counter Internet-enabled opposition of the sort that participated in the "color revolutions" in former Soviet states. Governments that prefer a tightly controlled Internet are engaging in partnerships, sharing best practices, and jointly developing information control platforms through forums like the SCO. While many casual Internet observers ruminate over the prospect of a UN takeover of the Internet that may never materialize, the most important norms around cyberspace controls could be taking hold beneath the spotlight and at the regional level.

Technological Sovereignty

Closely related to the questions surrounding cyberspace governance at the international level are issues of domestic-level Internet controls, and concerns over "technological sovereignty." This area is one where the reactions to the Snowden disclosures have been most palpably felt in the short term, as countries react to what they see as the US "home-field advantage" (though not always in ways that are straightforward). Included among the leaked details of US- and GCHQ-led operations to exploit the global communications infrastructure are numerous accounts of specific actions to compromise state networks, or even the handheld devices of government officials—most notoriously, the hacking of German Chancellor Angela Merkel's personal cellphone and the targeting of Brazilian government officials' classified communications. But the vast scope of US-led exploitation of global cyberspace, from the code to the undersea cables and everything in between, has set off shockwaves of indignation and loud calls to take immediate responses to restore "technological sovereignty."

For example, Brazil has spearheaded a project to lay a new submarine cable linking South America directly to Europe, thus bypassing the United States. Meanwhile, many European politicians have argued that contracts with US-based companies that may be secretly colluding with the NSA should be cancelled and replaced with contracts for domestic industry to implement regional and/or nationally autonomous data-routing policies—arguments that European industry has excitedly supported. It is sometimes difficult to unravel whether such measures are genuinely designed to protect citizens, or are really just another form of national industrial protectionism, or both. Largely obscured beneath the heated rhetoric and underlying self-interest, however, are serious questions about whether any of the measures proposed would have any more than a negligible impact when it comes to actually protecting the confidentiality and integrity of communications. As the Snowden disclosures reveal, the NSA and GCHQ have proved to be remarkably adept at exploiting traffic, no matter where it is based, by a variety of means.

A more troubling concern is that such measures may end up unintentionally legitimizing national cyberspace controls, particularly for developing countries, "swing states," and emerging markets. Pointing to the Snowden disclosures and the fear of NSA-led surveillance can be useful for regimes looking to subject companies and citizens to a variety of information controls, from censorship to surveillance. Whereas policy makers previously might have had concerns about being cast as pariahs or infringers on human rights, they now have a convenient excuse supported by European and other governments' reactions.

Spyware Bazaar

One by-product of the huge growth in military and intelligence spending on cybersecurity has been the fueling of a global market for sophisticated surveillance and other security tools. States that do not have an in-house operation on the level of the NSA can now buy advanced capabilities directly from private contractors. These tools are proving particularly attractive to many regimes that face ongoing insurgencies and other security challenges, as well as persistent popular protests. Since the advertised end uses of these products and services include many legitimate needs, such as network traffic management or the lawful interception of data, it is difficult to prevent abuses, and hard even for the companies themselves to know to what ends their products and services might ultimately be directed. Many therefore employ the term "dual-use" to describe such tools.

> **We leave behind a trail of digital "exhaust" wherever we go.**

Research by the University of Toronto's Citizen Lab from 2012 to 2014 has uncovered numerous cases of human rights activists targeted by advanced digital spyware manufactured by Western companies. Once implanted on a target's device, this spyware can extract files and contacts, send emails and text messages, turn on the microphone and camera, and track the location of the user. If these were isolated incidences, perhaps we could write them off as anomalies. But the Citizen Lab's international scan of the command and control servers of these products—the computers used to send instructions to infected devices—has produced disturbing evidence of a global market that knows no boundaries. Citizen Lab researchers found one product, Finspy, marketed by a UK company, Gamma Group, in a total of 25 countries— some with dubious human rights records, such as Bahrain, Bangladesh, Ethiopia, Qatar, and Turkmenistan. A subsequent Citizen Lab report found that 21 governments are current or former users of a spyware product sold by an Italian company called Hacking Team, including 9 that received the lowest ranking, "authoritarian," in the *Economist's* 2012 Democracy Index.

Meanwhile, a 2014 Privacy International report on surveillance in Central Asia says many of the countries in the region have implemented far-reaching surveillance systems at the base of their telecommunications networks, using advanced US and Israeli equipment, and supported by Russian intelligence training. Products that provide advanced deep packet inspection (the capability to inspect data packets in detail as they flow through networks), content filtering, social network mining, cellphone tracking, and even computer attack targeting are being developed by Western firms and marketed worldwide to regimes seeking to limit democratic participation, isolate and identify opposition, and infiltrate meddlesome adversaries abroad.

Pushing Back

The picture of the cyberspace landscape painted above is admittedly quite bleak, and therefore one-sided. The contests over cyberspace are multidimensional and include many groups and individuals pushing for technologies, laws, and norms that support free speech, privacy, and access to information. Here, too, the Snowden disclosures have had an animating effect, raising awareness of risks and spurring on change. Whereas vague concerns about widespread digital spying were voiced by a minority and sometimes trivialized before Snowden's disclosures, now those fears have been given real substance and credibility, and surveillance is increasingly seen as a practical risk that requires some kind of remediation.

The Snowden disclosures have had a particularly salient impact on the private sector, the Internet engineering community,

and civil society. The revelations have left many US companies in a public relations nightmare, with their trust weakened and lucrative contracts in jeopardy. In response, companies are pushing back. It is now standard for many telecommunications and social media companies to issue transparency reports about government requests to remove information from websites or share user data with authorities. US-based Internet companies even sued the government over gag orders that bar them from disclosing information on the nature and number of requests for user information. Others, including Google, Microsoft, Apple, Facebook, and WhatsApp, have implemented end-to-end encryption.

Internet engineers have reacted strongly to revelations showing that the NSA and its allies have subverted their security standards-setting processes. They are redoubling efforts to secure communications networks wholesale as a way to shield all users from mass surveillance, regardless of who is doing the spying. Among civil society groups that depend on an open cyberspace, the Snowden disclosures have helped trigger a burgeoning social movement around digital-security tool development and training, as well as more advanced research on the nature and impacts of information controls.

Wild Card

The cyberspace environment in which we live and on which we depend has never been more in flux. Tensions are mounting in several key areas, including Internet governance, mass and targeted surveillance, and military rivalry. The original promise of the Internet as a forum for free exchange of information is at risk. We are at a historical fork in the road: Decisions could take us down one path where cyberspace continues to evolve into a global commons, empowering individuals through access to information and freedom of speech and association, or down another path where this ideal meets its eventual demise. Securing cyberspace in ways that encourage freedom, while limiting controls and surveillance, is going to be a serious challenge.

Trends toward militarization and greater state control were already accelerating before the Snowden disclosures, and seem unlikely to abate in the near future. However, the leaks have thrown a wild card into the mix, creating opportunities for alternative approaches emphasizing human rights, corporate social responsibility, norms of mutual restraint, cyberspace arms control, and the rule of law. Whether such measures will be enough to stem the tide of territorialized controls remains to be seen. What is certain, however, is that a debate over the future of cyberspace will be a prominent feature of world politics for many years to come.

Critical Thinking

1. What is the role of the UN in the governance of the Internet?
2. What countries support a free Internet and what countries support state control of the Internet?
3. What countries are considered "swing" states in the use of the Internet?

Internet References

Berkman Center for Internet and Society
https://cyber.law.harvard.edu

Internet Governance Forum
intgovforum.org

Internet Society
internetsociety.org

National Security Agency
https://www.nsa.gov

RON DEIBERT is a professor of political science and director of the Canada Center for Global Security Studies and the Citizen Lab at the University of Toronto. His latest book is *Black Code: Inside the Battle for Cyberspace* (Signal, 2013).

Article

Prepared by: Robert Weiner, *University of Massachusetts, Boston*

The Return of Geopolitics: The Revenge of the Revisionist Powers

WALTER RUSSELL MEAD

Learning Outcomes

After reading this article, you will be able to:

- Understand what the post-Cold War settlement meant.
- Explain what is meant by status quo and revisionism in the international system.

So far, the year 2014 has been a tumultuous one, as geopolitical rivalries have stormed back to center stage. Whether it is Russian forces seizing Crimea, China making aggressive claims in its coastal waters, Japan responding with an increasingly assertive strategy of its own, or Iran trying to use its alliances with Syria and Hezbollah to dominate the Middle East, old-fashioned power plays are back in international relations.

The United States and the EU, at least, find such trends disturbing. Both would rather move past geopolitical questions of territory and military power and focus instead on ones of world order and global governance: trade liberalization, nuclear nonproliferation, human rights, the rule of law, climate change, and so on. Indeed, since the end of the Cold War, the most important objective of U.S. and EU foreign policy has been to shift international relations away from zero-sum issues toward win-win ones. To be dragged back into old-school contests such as that in Ukraine doesn't just divert time and energy away from those important questions; it also changes the character of international politics. As the atmosphere turns dark, the task of promoting and maintaining world order grows more daunting.

But Westerners should never have expected old-fashioned geopolitics to go away. They did so only because they fundamentally misread what the collapse of the Soviet Union meant:

the ideological triumph of liberal capitalist democracy over communism, not the obsolescence of hard power. China, Iran, and Russia never bought into the geopolitical settlement that followed the Cold War, and they are making increasingly forceful attempts to overturn it. That process will not be peaceful, and whether or not the revisionists succeed, their efforts have already shaken the balance of power and changed the dynamics of international politics.

A False Sense of Security

When the Cold War ended, many Americans and Europeans seemed to think that the most vexing geopolitical questions had largely been settled. With the exception of a handful of relatively minor problems, such as the woes of the former Yugoslavia and the Israeli-Palestinian dispute, the biggest issues in world politics, they assumed, would no longer concern boundaries, military bases, national self-determination, or spheres of influence.

One can't blame people for hoping. The West's approach to the realities of the post-Cold War world has made a great deal of sense, and it is hard to see how world peace can ever be achieved without replacing geopolitical competition with the construction of a liberal world order. Still, Westerners often forget that this project rests on the particular geopolitical foundations laid in the early 1990s.

In Europe, the post-Cold War settlement involved the unification of Germany, the dismemberment of the Soviet Union, and the integration of the former Warsaw Pact states and the Baltic republics into NATO and the EU. In the Middle East, it entailed the dominance of Sunni powers that were allied with the United States (Saudi Arabia, its Gulf allies, Egypt, and Turkey) and the double containment of Iran and Iraq. In Asia, it meant the uncontested dominance of the United States, embedded in a series of security relationships with Japan, South Korea, Australia, Indonesia, and other allies.

This settlement reflected the power realities of the day, and it was only as stable as the relationships that held it up. Unfortunately, many observers conflated the temporary geopolitical conditions of the post-Cold War world with the presumably more final outcome of the ideological struggle between liberal democracy and Soviet communism. The political scientist Francis Fukuyama's famous formulation that the end of the Cold War meant "the end of history" was a statement about ideology. But for many people, the collapse of the Soviet Union didn't just mean that humanity's ideological struggle was over for good; they thought geopolitics itself had also come to a permanent end.

At first glance, this conclusion looks like an extrapolation of Fukuyama's argument rather than a distortion of it. After all, the idea of the end of history has rested on the geopolitical consequences of ideological struggles ever since the German philosopher Georg Wilhelm Friedrich Hegel first expressed it at the beginning of the nineteenth century. For Hegel, it was the Battle of Jena, in 1806, that rang the curtain down on the war of ideas. In Hegel's eyes, Napoleon Bonaparte's utter destruction of the Prussian army in that brief campaign represented the triumph of the French Revolution over the best army that prerevolutionary Europe could produce. This spelled an end to history, Hegel argued, because in the future, only states that adopted the principles and techniques of revolutionary France would be able to compete and survive.

Adapted to the post-Cold War world, this argument was taken to mean that in the future, states would have to adopt the principles of liberal capitalism to keep up. Closed, communist societies, such as the Soviet Union, had shown themselves to be too uncreative and unproductive to compete economically and militarily with liberal states. Their political regimes were also shaky, since no social form other than liberal democracy provided enough freedom and dignity for a contemporary society to remain stable.

To fight the West successfully, you would have to become like the West, and if that happened, you would become the kind of wishy-washy, pacifistic milquetoast society that didn't want to fight about anything at all. The only remaining dangers to world peace would come from rogue states such as North Korea, and although such countries might have the will to challenge the West, they would be too crippled by their obsolete political and social structures to rise above the nuisance level (unless they developed nuclear weapons, of course). And thus former communist states, such as Russia, faced a choice. They could jump on the modernization bandwagon and become liberal, open, and pacifistic, or they could

cling bitterly to their guns and their culture as the world passed them by.

At first, it all seemed to work. With history over, the focus shifted from geopolitics to development economics and nonproliferation, and the bulk of foreign policy came to center on questions such as climate change and trade. The conflation of the end of geopolitics and the end of history offered an especially enticing prospect to the United States: the idea that the country could start putting less into the international system and taking out more. It could shrink its defense spending, cut the State Department's appropriations, lower its profile in foreign hotspots—and the world would just go on becoming more prosperous and more free.

This vision appealed to both liberals and conservatives in the United States. The administration of President Bill Clinton, for example, cut both the Defense Department's and the State Department's budgets and was barely able to persuade Congress to keep paying U.S. dues to the UN. At the same time, policymakers assumed that the international system would become stronger and wider-reaching while continuing to be conducive to U.S. interests. Republican neo-isolationists, such as former Representative Ron Paul of Texas, argued that given the absence of serious geopolitical challenges, the United States could dramatically cut both military spending and foreign aid while continuing to benefit from the global economic system.

After 9/11, President George W. Bush based his foreign policy on the belief that Middle Eastern terrorists constituted a uniquely dangerous opponent, and he launched what he said would be a long war against them. In some respects, it appeared that the world was back in the realm of history. But the Bush administration's belief that democracy could be implanted quickly in the Arab Middle East, starting with Iraq, testified to a deep conviction that the overall tide of events was running in America's favor.

President Barack Obama built his foreign policy on the conviction that the "war on terror" was overblown, that history really was over, and that, as in the Clinton years, the United States' most important priorities involved promoting the liberal world order, not playing classical geopolitics. The administration articulated an extremely ambitious agenda in support of that order: blocking Iran's drive for nuclear weapons, solving the Israeli-Palestinian conflict, negotiating a global climate change treaty, striking Pacific and Atlantic trade deals, signing arms control treaties with Russia, repairing U.S. relations with the Muslim world, promoting gay rights, restoring trust with European allies, and ending the war in Afghanistan. At the same time, however, Obama planned to cut defense spending dramatically and reduced U.S. engagement in key world theaters, such as Europe and the Middle East.

An Axis of Weevils?

All these happy convictions are about to be tested. Twenty-five years after the fall of the Berlin Wall, whether one focuses on the rivalry between the EU and Russia over Ukraine, which led Moscow to seize Crimea; the intensifying competition between China and Japan in East Asia; or the subsuming of sectarian conflict into international rivalries and civil wars in the Middle East, the world is looking less post-historical by the day. In very different ways, with very different objectives, China, Iran, and Russia are all pushing back against the political settlement of the Cold War.

The relationships among those three revisionist powers are complex. In the long run, Russia fears the rise of China. Tehran's worldview has little in common with that of either Beijing or Moscow. Iran and Russia are oil-exporting countries and like the price of oil to be high; China is a net consumer and wants prices low. Political instability in the Middle East can work to Iran's and Russia's advantage but poses large risks for China. One should not speak of a strategic alliance among them, and over time, particularly if they succeed in undermining U.S. influence in Eurasia, the tensions among them are more likely to grow than shrink.

What binds these powers together, however, is their agreement that the status quo must be revised. Russia wants to reassemble as much of the Soviet Union as it can. China has no intention of contenting itself with a secondary role in global affairs, nor will it accept the current degree of U.S. influence in Asia and the territorial status quo there. Iran wishes to replace the current order in the Middle East—led by Saudi Arabia and dominated by Sunni Arab states—with one centered on Tehran.

Leaders in all three countries also agree that U.S. power is the chief obstacle to achieving their revisionist goals. Their hostility toward Washington and its order is both offensive and defensive: not only do they hope that the decline of U.S. power will make it easier to reorder their regions, but they also worry that Washington might try to overthrow them should discord within their countries grow. Yet the revisionists want to avoid direct confrontations with the United States, except in rare circumstances when the odds are strongly in their favor (as in Russia's 2008 invasion of Georgia and its occupation and annexation of Crimea this year). Rather than challenge the status quo head on, they seek to chip away at the norms and relationships that sustain it.

Since Obama has been president, each of these powers has pursued a distinct strategy in light of its own strengths and weaknesses. China, which has the greatest capabilities of the three, has paradoxically been the most frustrated. Its efforts to assert itself in its region have only tightened the links between the United States and its Asian allies and intensified nationalism in Japan. As Beijing's capabilities grow, so will its sense of frustration. China's surge in power will be matched by a surge in Japan's resolve, and tensions in Asia will be more likely to spill over into global economics and politics.

Iran, by many measures the weakest of the three states, has had the most successful record. The combination of the United States' invasion of Iraq and then its premature withdrawal has enabled Tehran to cement deep and enduring ties with significant power centers across the Iraqi border, a development that has changed both the sectarian and the political balance of power in the region. In Syria, Iran, with the help of its long-time ally Hezbollah, has been able to reverse the military tide and prop up the government of Bashar al-Assad in the face of strong opposition from the U.S. government. This triumph of realpolitik has added considerably to Iran's power and prestige. Across the region, the Arab Spring has weakened Sunni regimes, further tilting the balance in Iran's favor. So has the growing split among Sunni governments over what to do about the Muslim Brotherhood and its offshoots and adherents.

Russia, meanwhile, has emerged as the middling revisionist: more powerful than Iran but weaker than China, more successful than China at geopolitics but less successful than Iran. Russia has been moderately effective at driving wedges between Germany and the United States, but Russian President Vladimir Putin's preoccupation with rebuilding the Soviet Union has been hobbled by the sharp limits of his country's economic power. To build a real Eurasian bloc, as Putin dreams of doing, Russia would have to underwrite the bills of the former Soviet republics—something it cannot afford to do.

Nevertheless, Putin, despite his weak hand, has been remarkably successful at frustrating Western projects on former Soviet territory. He has stopped NATO expansion dead in its tracks. He has dismembered Georgia, brought Armenia into his orbit, tightened his hold on Crimea, and, with his Ukrainian adventure, dealt the West an unpleasant and humiliating surprise. From the Western point of view, Putin appears to be condemning his country to an ever-darker future of poverty and marginalization. But Putin doesn't believe that history has ended, and from his perspective, he has solidified his power at home and reminded hostile foreign powers that the Russian bear still has sharp claws.

The Powers That Be

The revisionist powers have such varied agendas and capabilities that none can provide the kind of systematic and global opposition that the Soviet Union did. As a result, Americans have been slow to realize that these states have undermined the Eurasian geopolitical order in ways that complicate U.S. and European efforts to construct a post-historical, win-win world.

Still, one can see the effects of this revisionist activity in many places. In East Asia, China's increasingly assertive stance has yet to yield much concrete geopolitical progress, but it has fundamentally altered the political dynamic in the region with the fastest-growing economies on earth. Asian politics today revolve around national rivalries, conflicting territorial claims, naval buildups, and similar historical issues. The nationalist revival in Japan, a direct response to China's agenda, has set up a process in which rising nationalism in one country feeds off the same in the other. China and Japan are escalating their rhetoric, increasing their military budgets, starting bilateral crises with greater frequency, and fixating more and more on zero-sum competition.

Although the EU remains in a post-historical moment, the non-EU republics of the former Soviet Union are living in a very different age. In the last few years, hopes of transforming the former Soviet Union into a post-historical region have faded. The Russian occupation of Ukraine is only the latest in a series of steps that have turned eastern Europe into a zone of sharp geopolitical conflict and made stable and effective democratic governance impossible outside the Baltic states and Poland.

In the Middle East, the situation is even more acute. Dreams that the Arab world was approaching a democratic tipping point—dreams that informed U.S. policy under both the Bush and the Obama administrations—have faded. Rather than building a liberal order in the region, U.S. policymakers are grappling with the unraveling of the state system that dates back to the 1916 Sykes-Picot agreement, which divided up the Middle Eastern provinces of the Ottoman Empire, as governance erodes in Iraq, Lebanon, and Syria. Obama has done his best to separate the geopolitical issue of Iran's surging power across the region from the question of its compliance with the Nuclear Nonproliferation Treaty, but Israeli and Saudi fears about Iran's regional ambitions are making that harder to do. Another obstacle to striking agreements with Iran is Russia, which has used its seat on the UN Security Council and support for Assad to set back U.S. goals in Syria.

Russia sees its influence in the Middle East as an important asset in its competition with the United States. This does not mean that Moscow will reflexively oppose U.S. goals on every occasion, but it does mean that the win-win outcomes that Americans so eagerly seek will sometimes be held hostage to Russian geopolitical interests. In deciding how hard to press Russia over Ukraine, for example, the White House cannot avoid calculating the impact on Russia's stance on the Syrian war or Iran's nuclear program. Russia cannot make itself a richer country or a much larger one, but it has made itself a more important factor in U.S. strategic thinking, and it can use that leverage to extract concessions that matter to it.

If these revisionist powers have gained ground, the status quo powers have been undermined. The deterioration is sharpest in Europe, where the unmitigated disaster of the common currency has divided public opinion and turned the EU's attention in on itself. The EU may have avoided the worst possible consequences of the euro crisis, but both its will and its capacity for effective action beyond its frontiers have been significantly impaired.

The United States has not suffered anything like the economic pain much of Europe has gone through, but with the country facing the foreign policy hangover induced by the Bush-era wars, an increasingly intrusive surveillance state, a slow economic recovery, and an unpopular health-care law, the public mood has soured. On both the left and the right, Americans are questioning the benefits of the current world order and the competence of its architects. Additionally, the public shares the elite consensus that in a post-Cold War world, the United States ought to be able to pay less into the system and get more out. When that doesn't happen, people blame their leaders. In any case, there is little public appetite for large new initiatives at home or abroad, and a cynical public is turning away from a polarized Washington with a mix of boredom and disdain.

Obama came into office planning to cut military spending and reduce the importance of foreign policy in American politics while strengthening the liberal world order. A little more than halfway through his presidency, he finds himself increasingly bogged down in exactly the kinds of geopolitical rivalries he had hoped to transcend. Chinese, Iranian, and Russian revanchism haven't overturned the post-Cold War settlement in Eurasia yet, and may never do so, but they have converted an uncontested status quo into a contested one. U.S. presidents no longer have a free hand as they seek to deepen the liberal system; they are increasingly concerned with shoring up its geopolitical foundations.

The Twilight of History

It was 22 years ago that Fukuyama published *The End of History and the Last Man,* and it is tempting to see the return of geopolitics as a definitive refutation of his thesis. The reality is more complicated. The end of history, as Fukuyama reminded readers, was Hegel's idea, and even though the revolutionary state had triumphed over the old type of regimes for good, Hegel argued, competition and conflict would continue. He predicted that there would be disturbances in the provinces, even as the heartlands of European civilization moved into a post-historical time. Given that Hegel's provinces included China, India, Japan, and Russia, it should hardly be surprising that more than two centuries later, the disturbances haven't ceased. We are living in the twilight of history rather than at its actual end.

A Hegelian view of the historical process today would hold that substantively little has changed since the beginning of the nineteenth century. To be powerful, states must develop the ideas and institutions that allow them to harness the titanic forces of industrial and informational capitalism. There is no alternative; societies unable or unwilling to embrace this route will end up the subjects of history rather than the makers of it. But the road to postmodernity remains rocky. In order to increase its power, China, for example, will clearly have to go through a process of economic and political development that will require the country to master the problems that modern Western societies have confronted. There is no assurance, however, that China's path to stable liberal modernity will be any less tumultuous than, say, the one that Germany trod. The twilight of history is not a quiet time.

The second part of Fukuyama's book has received less attention, perhaps because it is less flattering to the West. As Fukuyama investigated what a post-historical society would look like, he made a disturbing discovery. In a world where the great questions have been solved and geopolitics has been subordinated to economics, humanity will look a lot like the nihilistic "last man" described by the philosopher Friedrich Nietzsche: a narcissistic consumer with no greater aspirations beyond the next trip to the mall.

In other words, these people would closely resemble today's European bureaucrats and Washington lobbyists. They are competent enough at managing their affairs among post-historical people, but understanding the motives and countering the strategies of old-fashioned power politicians is hard for them. Unlike their less productive and less stable rivals, post-historical people are unwilling to make sacrifices, focused on the short term, easily distracted, and lacking in courage.

The realities of personal and political life in post-historical societies are very different from those in such countries as China, Iran, and Russia, where the sun of history still shines. It is not just that those different societies bring different personalities and values to the fore; it is also that their institutions work differently and their publics are shaped by different ideas.

Societies filled with Nietzsche's last men (and women) characteristically misunderstand and underestimate their supposedly primitive opponents in supposedly backward societies—a blind spot that could, at least temporarily, offset their countries' other advantages. The tide of history may be flowing inexorably in the direction of liberal capitalist democracy, and the sun of history may indeed be sinking behind the hills. But even as the shadows lengthen and the first of the stars appears, such figures as Putin still stride the world stage. They will not go gentle into that good night, and they will rage, rage against the dying of the light.

Critical Thinking

1. Why is Francis Fukuyama's idea of the end of history relevant in the contemporary international system?

2. Why should China, Russia, and Iran be considered revisionist powers?

3. What have been the major foreign policy goals of the Obama administration, and why have they shifted?

Internet References

Ministry of Foreign Affairs of the People's Republic of China
www.fmprc.gov.cn/eng/

National Security Strategy 2015
whitehouse.gov

The Ministry of Foreign Affairs of the Russian Federation
en.mid.ru

WALTER RUSSELL MEAD is James Clarke Chace Professor of Foreign Affairs and Humanities at Bard College and Editor-at-Large of The American Interest. Follow him on Twitter @wrmead.

Article   Prepared by: Robert Weiner, *University of Massachusetts, Boston*

The Once and Future Hegemon

SALVATORE BABONES

Learning Outcomes

After reading this article, you will be able to:

- Identify the major elements of the Declinist argument.

- Better understand what is meant by hegemony in global affairs.

I s retreat from global hegemony in America's national interest? No idea has percolated more widely over the past decade—and none is more bogus. The United States is not headed for the skids and there is no reason it should be. The truth is that America can and should seek to remain the world's top dog.

The idea of American hegemony is as old as Benjamin Franklin, but has its practical roots in World War II. The United States emerged from that war as the dominant economic, political and technological power. The only major combatant to avoid serious damage to its infrastructure, its housing stock or its demographic profile, the United States ended the war with the greatest naval order of battle ever seen in the history of the world. It became the postwar home of the United Nations, the International Monetary Fund and the World Bank. And, of course, the United States had the bomb. America was, in every sense of the word, a hegemon.

"Hegemony" is a word used by social scientists to describe leadership within a system of competing states. The Greek historian Thucydides used the term to characterize the position of Athens in the Greek world in the middle of the fifth century BC. Athens had the greatest fleet in the Mediterranean; it was the home of Socrates and Plato, Sophocles and Aeschylus; it crowned its central Acropolis with the solid-marble temple to Athena known to history as the Parthenon. Athens had a powerful rival in Sparta, but no one doubted that Athens was the hegemon of the time until Sparta defeated it in a bitter 27-year war.

America's only global rival in the twentieth century was the Soviet Union. The Soviet Union never produced more than about half of America's total national output. Its nominal allies in Eastern Europe were in fact restive occupied countries, as were many of its constituent republics. Its client states overseas were at best partners of convenience, and at worst expensive drains on its limited resources. The Soviet Union had the power to resist American hegemony, but not to displace it. It had the bomb and an impressive space program, but little else.

When the Soviet Union finally disintegrated in 1991, American hegemony was complete. The United States sat at the top of the international system, facing no serious rivals for global leadership. This "unipolar moment" lasted a mere decade. September 11, 2001, signaled the emergence of a new kind of threat to global stability, and the ensuing rise of China and reemergence of Russia put paid to the era of unchallenged American leadership. Now, America's internal politics have deadlocked and the U.S. government shrinks from playing the role of global policeman. In the second decade of the twenty-first century, American hegemony is widely perceived to be in terminal decline.

> **American hegemony is now as firm as or firmer than it has ever been, and will remain so for a long time to come.**

Or so the story goes. In fact, reports of the passing of U.S. hegemony are greatly exaggerated. America's costly wars in Iraq and Afghanistan were relatively minor affairs considered in long-term perspective. The strategic challenge posed by China has also been exaggerated. Together with its inner circle of unshakable English-speaking allies, the United States possesses near-total control of the world's seas, skies, airwaves and cyberspace, while American universities, think tanks and journals dominate the world of ideas. Put aside all the alarmist punditry. American hegemony is now as firm as or firmer than it has ever been, and will remain so for a long time to come.

The massive federal deficit, negative credit-agency reports, repeated debt-ceiling crises and the 2013 government shutdown all created the impression that the U.S. government is bankrupt, or close to it. The U.S. economy imports half a trillion dollars a year more than it exports. Among the American population, poverty rates are high and ordinary workers' wages have been stagnant (in real terms) for decades. Washington seems to be paralyzed by perpetual gridlock. On top of all this, strategic exhaustion after two costly wars in Afghanistan and Iraq has substantially degraded U.S. military capabilities. Then, at the very moment the military needed to regroup, rebuild and rearm, its budget was hit by sequestration.

If economic power forms the long-term foundation for political and military power, it would seem that America is in terminal decline. But policy analysts tend to have short memories. Cycles of hegemony run in centuries, not decades (or seasons). When the United Kingdom finally defeated Napoleon at Waterloo in 1815, its national resources were completely exhausted. Britain's public-debt-to-GDP ratio was over 250 percent, and early nineteenth-century governments lacked access to the full range of fiscal and financial tools that are available today. Yet the British Century was only just beginning. The *Pax Britannica* and the elevation of Queen Victoria to become empress of India were just around the corner.

By comparison, America's current public-debt-to-GDP ratio of less than 80 percent is relatively benign. Those with even a limited historical memory may remember the day in January 2001 when the then chairman of the Federal Reserve, Alan Greenspan, testified to the Senate Budget Committee that "if current policies remain in place, the total unified surplus will reach $800 billion in fiscal year 2011. . . . The emerging key fiscal policy need is to address the implications of maintaining surpluses." As the poet said, bliss was it in that dawn to be alive! Two tax cuts, two wars and one financial crisis later, America's budget deficit was roughly the size of the projected surplus that so worried Greenspan.

This is not to argue that the U.S. government should ramp up taxes and spending, but it does illustrate the fact that it has enormous potential fiscal resources available to it, should it choose to use them. Deficits come and go. America's fiscal capacity in 2015 is stupendously greater than Great Britain's was in 1815. Financially, there is every reason to think that America's century lies in the future, not in the past.

The same is true of the supposed exhaustion of the U.S. military. On the one hand, thirteen years of continuous warfare have reduced the readiness of many U.S. combat units, particularly in the army. On the other hand, U.S. troops are now far more experienced in actual combat than the forces of any other major military in the world. In any future conflict, the advantage given by this experience would likely outweigh any decline in effectiveness due to deferred maintenance and training. Constant deployment may place an unpleasant and unfair burden on U.S. service personnel and their families, but it does not necessarily diminish the capability of the U.S. military. On the contrary, it may enhance it.

America's limited wars in Afghanistan and Iraq were hardly the final throes of a passing hegemon. They are more akin to Britain's bloody but relatively inconsequential conflicts in Afghanistan and Crimea in the middle of the nineteenth century. Brutal wars like these repeatedly punctured, but never burst, British hegemony. In fact, Britain engaged in costly and sometimes disastrous conflicts throughout the century-long *Pax Britannica*. British hegemony did not come to an end until the country faced Germany head-on in World War I. Even then, Britain ultimately prevailed (with American help). Its empire reached its maximum extent not before World War I but immediately after, in 1922.

Ultimately, it is inevitable that in the long run American power will weaken and American hegemony over the rest of the world will fade. But how long is the long run? There are few factual indications that American decline has begun—or that it will begin anytime soon. Short-term fluctuations should not be extrapolated into long-term trends. Without a doubt, 1991 was a moment of supreme U.S. superiority. But so was 1946, after which came the Soviet bomb, Korea and Vietnam. American hegemony has waxed and waned over the last seventy years, but it has never been eclipsed. And it is unlikely that the eclipse is nigh.

When pundits scope out the imminent threats to U.S. hegemony, the one country on their radar screens is China. While the former Soviet Union never reached above 45 percent of U.S. total national income, the Chinese economy may already have overtaken the American economy, and if not it certainly will soon. If sheer economic size is the foundation of political and military power, China is positioned for future global hegemony. Will it build on this foundation? Can it?

Much depends on the future of China's relationships with its neighbors. China lives in a tough neighborhood. It faces major middle-tier powers on three sides: Russia to the north, South Korea and Japan to the east and Vietnam and India to the south. To the west it faces a series of weak and failing states, but that may be more of a burden than a blessing: China's own western regions are also sites of persistent instability.

It is perhaps realistic to imagine China seeking to expand to the north at the expense of Russia and Mongolia. Ethnic Russians are abandoning Siberia and the Pacific coast in droves, and strategic areas along Russia's border with China have been demographically and economically overwhelmed by Chinese immigration. Twenty-second-century Russia may find it difficult to hold the Far East against China. But that is not a serious threat to U.S. hegemony. If anything, increasing Sino-Russian tensions may reinforce U.S. global hegemony, much as Sino-Soviet tensions did in the 1970s.

There will be no Chinese-sponsored Asian equivalent of NATO or the Warsaw Pact.

To the southeast, China clearly seeks to dominate the South China Sea and beyond. The main barrier to its doing so is the autonomy of Taiwan. Were Taiwan ever to be reintegrated with China, it would be difficult for other regional powers to successfully challenge a united China for control of the basin. In the future, it is entirely possible that China will come to dominate these, its own coastal waters. This would be a minor setback to an America accustomed to dominating all of the world's seas, but it would not constitute a serious strategic threat to the United States.

Across the East China Sea, China faces Japan and South Korea—two of the most prosperous, technologically advanced and militarily best-equipped countries in the world. Historical enmities ensure that China will never expand in that direction. Worse for China, it is quite likely that any increase in China's ability to project power beyond its borders will be matched with similar steps by a wary, remilitarizing Japan.

The countries on China's southern border are so large, populous and poor that it is difficult to imagine China taking much interest in the region beyond simple resource exploitation. Chinese companies may seek profit opportunities in Cambodia, Myanmar and Pakistan, but there is little for China to gain from strategic domination of the region. There will be no Chinese-sponsored Asian equivalent of NATO or the Warsaw Pact.

Farther abroad, much has been made of China's strategic engagement in Africa and Latin America. Investment-starved countries in these regions have been eager to access Chinese capital and in many cases have welcomed Chinese investment, expertise and even immigration. But it is hard to imagine them welcoming Chinese military bases, and equally hard to imagine China asking them for bases. The American presence in Africa is in large part the legacy of centuries of European colonialism. China has no such legacy to build on.

Above all, however, the prospects for future Chinese hegemony depend on the prospects for future Chinese economic growth. Measured in per capita terms, China is still poorer than Mexico. That China will catch up to Mexico seems certain. That China will continue its extraordinary growth trajectory once it has caught up to Mexico is less obvious. In 2011, when the Chinese economy was growing by more than 10 percent a year, I predicted that China was headed for much slower growth. At the time,

the IMF was projecting a long-term growth rate of 9.5 percent. Today, the same IMF projections assume 7 percent growth.

Even at 7 percent annual growth, the Chinese economy would account for more than half of total global output by 2050. The United States in its post-World War II heyday never achieved that level of dominance. But exponential extrapolations are inherently tricky. If China continues to grow at 7 percent while the world economy as a whole grows by 3 percent per year, China will account for 90 percent of global economic output by 2100 and 100 percent by 2110. After that, China's economy will be even larger than the world's economy, which of course is impossible unless China moves a large portion of its production off-planet.

A more reasonable assumption is that China's economic growth will eventually settle down to global average rates. The only question is when. Existing demographic trends make it almost certain that the answer is: soon. The U.S. Census Bureau has projected that China's working-age population would reach its peak in 2014 and then go into long-term decline. In the twenty years from 2014 to 2034, China's working-age population will fall by 87 million, while its elderly population will rise by 149 million. In the language of economic punditry, China will "grow old before it grows rich."

The U.S. population, by contrast, is young and growing. In 2034, the U.S. population is projected to be growing at a rate of 0.6 percent per year (compared to −0.2 percent in China), with substantial immigration of talented, productive people (compared to net emigration from China). The U.S. median age of 39.2 will be significantly younger than the Chinese median age of 44.8. Over the long term these trends may change, but the twenty-year scenario is almost certain, because for the most part it has already happened. Economic trends can turn on a dime, but demographic trends are mostly immutable: tomorrow's child-bearers have already been born.

In the ancient Mediterranean world, Rome rose to regional hegemony a century or two after the passing of the Athenian empire. The hegemonic Roman Republic was a hybrid political entity. It consisted of Rome itself, Roman colonies, Roman protectorates, cities conquered by Rome and cities allied to Rome. For four hundred years before 91 BC, the Italian cities allied to Rome were effectively part of the Roman state despite their formal political independence. They participated in Rome's wars under Roman command. They did not pay taxes or tribute to Rome, but they were fully incorporated into a political system centered on Rome. When Hannibal crossed the Alps in 218 BC, most of the Italian cities did not rise up against Rome as he expected. They stood with Rome because they were effectively part of Rome.

In a similar way, the effective borders of the American polity extend well beyond the Atlantic and Pacific coasts. If the

Edward Snowden leaks have revealed nothing else, they have shown the depth of intelligence cooperation between the United States and its English-speaking allies Australia, Canada, New Zealand and the United Kingdom. These are the so-called Five Eyes countries. These English-speaking allies work so closely with the United States on security issues that they resemble ancient Rome's Italian allies. Despite their formal political independence, they do not make major strategic decisions without considering America's interests as well as their own.

Curiously, America's English-speaking allies resemble the United States in their demographic structures as well. While East Asia's birthrates have fallen well below replacement levels and parts of continental Europe face outright depopulation, the English-speaking countries have stable birthrates and substantial immigration. The most talented people in the world don't always move to the United States, but more often than not they move to English-speaking countries. It doesn't hurt that English is the global lingua franca as well as the language of the Internet.

One surprising result of these trends is that the once-unfathomable demographic gap between China and the English-speaking world is narrowing. According to U.S. Census Bureau projections, in 2050 the U.S. population will be 399 million and rising by 0.5 percent per year while the Chinese population will be 1.304 billion and falling by 0.5 percent per year. Throw in America's four English-speaking allies, and the combined five-country population will be 546 million—nearly 42 percent of China's population—with a growth rate of 0.4 percent per year. No longer will China have the overwhelming demographic advantage that has historically let it punch above its economic weight.

Is it reasonable to treat America's English-speaking allies as integrated components of the U.S. power structure? Of course, they are not formally integrated into the U.S. state. But the real, effective borders of countries are much fuzzier than the legal lines drawn on maps. The United States exercises different levels of influence over its sovereign territory, extraterritorial possessions, the English-speaking allies, NATO allies, other treaty allies, nontreaty allies, client states, spheres of influence, exclusionary zones and even enemy territories. All of these categories are fluid in their memberships and meanings, but taken together they constitute more than just a network of relationships. They constitute a cooperative system of shared sovereignty, something akin to the power structure of the Roman Republic.

America's allies constitute more than just a network of relationships. They are a cooperative system of shared sovereignty, something akin to the power structure of the Roman Republic.

No other country in the world possesses, has ever possessed, or is likely to possess in this century such a world-straddling vehicle for the enforcement of its will. More to the point, the U.S.-dominated system shows no signs of falling apart. Even the revelation that America and its English-speaking allies have been spying on the leaders of their NATO peers has not led to calls for the dissolution of NATO. The American system may not last forever, but its remaining life may be measured in centuries rather than decades. Cycles of hegemony turn very slowly because systems of hegemony are very robust. The American power network is much bigger, much stronger and much more resilient than the formal American state as such.

A recurring meme is the idea that the whole world should be able to vote in U.S. presidential elections because the whole world has a stake in the outcome. This argument is not meant to be taken seriously. It is made to prove a point: that the United States is uniquely and pervasively important in the world. At least since the Suez crisis of 1956, it has been clear to everyone that the other countries of the world, whether alone or in concert, are unable to project power beyond their shores without American support. Mere American acquiescence is not enough. In global statecraft, the United States is the indispensable state.

One widely held definition of a state is that a state is a body that successfully claims a monopoly on the legitimate use of force within a territory. The German sociologist Max Weber first proposed this definition in 1919, in the chaotic aftermath of World War I. Interestingly, he included the qualifier "successfully" in his definition. To constitute a real state, a government cannot merely claim the sole right to use force; it must make this claim stick. It must be successful in convincing its people, civil-society groups and, most importantly, other states to accept its claim.

In the twenty-first century, the United States effectively claims a monopoly on the legitimate use of force worldwide. Whether or not it makes this claim in so many words, it makes it through its policies and actions, and America's monopoly on the legitimate use of force is generally accepted by most of the governments (if not the peoples) of the world. That is not to say that all American uses of force are accepted as legitimate, but that all uses of force that are accepted as legitimate are either American or actively supported by the United States. The world condemns Russian intervention in Ukraine but accepts Saudi intervention in Yemen, and of course it looks to the United States to solve conflicts in places like Libya, Syria and Iraq. The United States has not conquered the world, but most of the world's governments (with the exceptions of countries such as Russia, Iran and China) and major intergovernmental organizations accept America's lead. Very often they ask for it.

This American domination of global affairs extends well beyond hegemony. In the nineteenth century, the United Kingdom was a global hegemon. Britannia ruled the waves, and from its domination of the oceans it derived extraordinary influence over global affairs. But China, France, Germany, Russia and later Japan continually challenged the legitimacy of British domination and tested it at every turn. Major powers certainly believed that they could engage independently in global statecraft and acted on that belief. France did not seek British permission to conquer its colonies; Germany did not seek British permission to conquer France.

No one ever likes an empire, but despite Ronald Reagan's memorable phrase, the word "empire" is not inseparably linked to the word "evil."

Twenty-first-century America dominates the world to an extent completely unmatched by nineteenth-century Britain. There is no conflict anywhere in the world in which the United States is not in some way involved. More to the point, participants in conflicts everywhere in the world, no matter how remote, expect the United States to be involved. Revisionists ranging from pro-Russian separatists in eastern Ukraine to Bolivian peasant farmers who want to chew coca leaves see the United States as the power against which they are rebelling. The United States is much more than the world's policeman. It is the world's lawgiver.

The world state of so many fictional utopias and dystopias is here, and it is not a nameless postmodern entity called global governance. It is America. Another word for a world state that dominates all others is an "empire," a word that Americans of all political persuasions abhor. For FDR liberals it challenges cherished principles of internationalism and fair play. For Jeffersonian conservatives it reeks of foreign adventurism. For today's neoliberals it undermines faith in the primacy of market competition over political manipulation. And for neoconservatives it implies an unwelcome responsibility for the welfare of the world beyond America's shores.

In fact, it is difficult to avoid the conclusion that the United States has become an imperial world state—a world-empire—that sets the ground rules for smooth running of the global economy, imposes its will largely without constraint and without consideration of the reasonable desires of other countries, and severely punishes those few states and nonstate actors that resist its dictates.

No one ever likes an empire, but despite Ronald Reagan's memorable phrase, the word "empire" is not inseparably linked to the word "evil." When it comes to understanding empire, history is probably a better guide than science fiction. Consider the Roman Empire. For several centuries after the ascension of Augustus, life under Rome was generally freer, safer and more prosperous than it had been under the previously independent states. Perhaps it was not better for the enslaved or for the Druids, and certainly not for the Jews, but for most people of the ancient Mediterranean, imperial Rome brought vast improvements most of the time.

Ancient analogies notwithstanding, no one would seriously suggest that the United States should attempt to directly rule the rest of the world, and there is no indication that the rest of the world would let it. But the United States could manage its empire more effectively, which is something that the rest of the world would welcome. A winning strategy for low-cost, effective management of empire would be for America to work with and through the system of global governance that America itself has set up, rather than systematically seeking to blunt its own instruments of power.

For example, the United States was instrumental in setting up the International Criminal Court, yet Washington will not place itself under the jurisdiction of the ICC and will not allow its citizens to be subject to the jurisdiction of the ICC. Similarly, though the United States is willing to use UN Security Council resolutions to censure its enemies, it is not willing to accept negotiated limits on its own freedom of action. From a purely military-political standpoint, the United States is sufficiently powerful to go it alone. But from a broader realist standpoint that takes account of the full costs and unintended consequences of military action, that is a suboptimal strategy. Had the United States listened to dissenting opinions on the Security Council before the invasion of Iraq, it would have saved hundreds of billions of dollars and hundreds of thousands of lives. The United States might similarly have done well to have heeded Russian reservations over Libya, as it ultimately did in responding to the use of chemical weapons in Syria.

A more responsible (and consequently more effective) United States would subject itself to the international laws and agreements that it expects others to follow. It would genuinely seek to reduce its nuclear arsenal in line with its commitments under the Nuclear Non-Proliferation Treaty. It would use slow but sure police procedures to catch terrorists, instead of quick but messy drone strikes. It would disavow all forms of torture. All of these policies would save American treasure while increasing American power. They would also increase America's ability to say "no" to its allies when they demand expensive U.S. commitments to protect their interests abroad.

Such measures would not ensure global peace, nor would they necessarily endear the United States to everyone across the world. But they would reduce global tensions and make it easier for America to act in its national interests where those interests are truly at stake. Both the United States and the world as a whole would be better off if Washington did not waste time, money and diplomatic capital on asserting every petty sovereign right it is capable of enforcing. A more strategic United States would preside over a more peaceful and prosperous world.

In pondering its future course, Washington might consider this tale from the ancient world: When Cyrus the Great conquered the neighboring kingdom of Lydia, he allowed his army to loot and pillage Lydia's capital city, Sardis. The deposed Lydian king Croesus became his captive and slave. After Cyrus taunted Croesus by asking him how it felt to see his capital city being plundered, Croesus responded: "It's not my city that your troops are plundering; it's your city." Cyrus ordered an immediate end to the destruction.

Critical Thinking

1. Compare the differences and similarities between British and U.S. hegemony.

2. Does the United States have an empire?
3. Why isn't China a strategic threat to the U.S.?

Internet References

Center for a New American Security
www.cnas.org/
The Atlantic Council
www.atlanticcouncil.org
The Hudson Institute
www.hudson.org/

SALVATORE BABONES is an associate professor of sociology and social policy at the University of Sydney and an associate fellow at the Institute for Policy Studies.

Article   Prepared by: Robert Weiner, *University of Massachusetts, Boston*

The Unraveling: How to Respond to a Disordered World

RICHARD N. HAASS

Learning Outcomes

After reading this article, you will be able to:

- Explain what is meant by disorder in the global system.
- Gain insights into what the post-Cold War international order consists of.

I n his classic *The Anarchical Society,* the scholar Hedley Bull argued that there was a perennial tension in the world between forces of order and forces of disorder, with the details of the balance between them defining each era's particular character. Sources of order include actors committed to existing international rules and arrangements and to a process for modifying them; sources of disorder include actors who reject those rules and arrangements in principle and feel free to ignore or undermine them. The balance can also be affected by global trends, to varying degrees beyond the control of governments, that create the context for actors' choices. These days, the balance between order and disorder is shifting toward the latter. Some of the reasons are structural, but some are the result of bad choices made by important players—and at least some of those can and should be corrected.

The chief cauldron of contemporary disorder is the Middle East. For all the comparisons that have been made to World War I or the Cold War, what is taking place in the region today most resembles the Thirty Years' War, three decades of conflict that ravaged much of Europe in the first half of the seventeenth century. As with Europe back then, in coming years, the Middle East is likely to be filled with mostly weak states unable to police large swaths of their territories, militias and terrorist groups acting with increasing sway, and both civil war and interstate strife. Sectarian and communal identities will be more powerful than national ones. Fueled by vast supplies of natural resources, powerful local actors will continue to meddle in neighboring countries' internal affairs, and major outside actors will remain unable or unwilling to stabilize the region.

There is also renewed instability on the periphery of Europe. Under President Vladimir Putin, Russia appears to have given up on the proposition of significant integration into the current European and global orders and chosen instead to fashion an alternative future based on special ties with immediate neighbors and clients. The crisis in Ukraine may be the most pronounced, but not the last, manifestation of what could well be a project of Russian or, rather, Soviet restoration.

In Asia, the problem is less current instability than the growing potential for it. There, most states are neither weak nor crumbling, but strong and getting stronger. The mix of several countries with robust identities, dynamic economies, rising military budgets, bitter historical memories, and unresolved territorial disputes yields a recipe for classic geopolitical maneuvering and possibly armed conflict. Adding to the challenges in this stretch of the world are a brittle North Korea and a turbulent Pakistan, both with nuclear weapons (and one with some of the world's most dangerous terrorists). Either could be the source of a local or global crisis, resulting from reckless action or state collapse.

Some contemporary challenges to order are global, a reflection of dangerous aspects of globalization that include cross-border flows of terrorists, viruses (both physical and virtual), and greenhouse gas emissions. With few institutional mechanisms available for stanching or managing them, such flows hold the potential to disrupt and degrade the quality of the system as a whole. And the rise of populism amid economic stagnation and increasing inequality makes improving global governance even more challenging.

The principles informing international order are also in contention. Some consensus exists about the unacceptability of acquiring territory by force, and it was such agreement that undergirded the broad coalition supporting the reversal of Saddam Hussein's attempt to absorb Kuwait into Iraq in 1990. But the consensus had frayed enough over the succeeding generation to allow Russia to escape similar universal condemnation after its taking of Crimea last spring, and it is anyone's guess how much of the world would respond to an attempt by China to muscle in on contested airspace, seas, or territory. International agreement on sovereignty breaks down even more when it comes to the question of the right of outsiders to intervene when a government attacks its own citizens or otherwise fails to meet its sovereign obligations. A decade after UN approval, the concept of "the responsibility to protect" no longer enjoys broad support, and there is no shared agreement on what constitutes legitimate involvement in the affairs of other countries.

To be sure, there are forces of order at work as well. There has been no great-power war for many decades, and there is no significant prospect of one in the near future. China and the United States cooperate on some occasions and compete on others, but even in the latter case, the competition is bounded. Interdependence is real, and both countries have a great deal invested (literally and figuratively) in the other, making any major and prolonged rupture in the relationship a worrisome possibility for both.

Russia, too, is constrained by interdependence, although less so than China given its energy-concentrated economy and more modest levels of external trade and investment. That means sanctions have a chance of influencing its behavior over time. Putin's foreign policy may be revanchist, but Russia's hard- and soft-power resources are both anything that appeals to anyone other than ethnic Russians, and as a result, the geopolitical troubles it can cause will remain on Europe's periphery, without touching the continent's core. Indeed, the critical elements of Europe's transformation over the past 70 years—the democratization of Germany, Franco-German reconciliation, economic integration—are so robust that they can reasonably be taken for granted.

Europe's parochialism and military weakness may make the region a poor partner for the United States in global affairs, but the continent itself is no longer a security problem, which is a huge advance on the past.

It would also be wrong to look at the Asia-Pacific and assume the worst. The region has been experiencing unprecedented economic growth for decades and has managed it peacefully. Here, too, economic interdependence acts as a brake on conflict. And there is still time for diplomacy and creative policymaking to create institutional shock absorbers that can help reduce the risk of confrontation stemming from surging nationalism and spiraling distrust.

The global economy, meanwhile, has stabilized in the aftermath of the financial crisis, and new regulations have been put in place to reduce the odds and scale of future crises. U.S. and European growth rates are still below historical norms, but what is holding the United States and Europe back is not the residue of the crisis so much as various policies that restrict robust growth.

North America could once again become the world's economic engine, given its stable, prosperous, and open economy; its 470 million people; and its emerging energy self-sufficiency. Latin America is, for the most part, at peace. Mexico is a far more stable and successful country than it was a decade ago, as is Colombia. Questions hovering over the futures of such countries as Brazil, Chile, Cuba, and Venezuela do not alter the fundamental narrative of a region heading in the right direction. And Africa, too, has a growing number of countries in which better governance and economic performance are becoming the norm rather than the exception.

Traditional analytic approaches have little to offer in making sense of these seemingly contradictory trends. One conventional route, for example, would be to frame the international dynamic as one of rising and falling powers, pitting China's advance against the United States' decline. But this exaggerates the United States' weaknesses and underestimates China's. For all its problems, the United States is well positioned to thrive in the twenty-first century, whereas China faces a multitude of challenges, including slowing growth, rampant corruption, an aging population, environmental degradation, and wary neighbors. And no other country is even close to having the necessary mix of capacity and commitment to be a challenger to the United States for global preeminence. U.S. President Barack Obama was recently quoted as brushing off concerns that things are falling apart, noting that "the world has always been messy" and that what is going on today "is not something that is comparable to the challenges we faced during the Cold War." Such sanguinity is misplaced, however, as today's world is messier, thanks to the emergence of a greater number of meaningful actors and the lack of overlapping interests or mechanisms to constrain the capacity or moderate the behavior of the most radical ones.

Indeed, with U.S. hegemony waning but no successor waiting to pick up the baton, the likeliest future is one in which the current international system gives way to a disorderly one with a larger number of power centers acting with increasing autonomy, paying less heed to U.S. interests and preferences. This will cause new problems even as it makes existing ones more difficult to solve. In short, the post-Cold War order is unraveling, and while not perfect, it will be missed.

The Causes of the Problem

Just why have things begun to unravel? For various reasons, some structural, others volitional. In the Middle East, for example, order has been undermined by a tradition of top-heavy, often corrupt, and illegitimate governments; minimal civil society; the curse of abundant energy resources (which often retard economic and political reform); poor educational systems; and various religion-related problems, such as sectarian division, fights between moderates and radicals, and the lack of a clear and widely accepted line between religious and secular spheres. But outside actions have added to the problems, from poorly drawn national borders to recent interventions.

With more than a decade of hindsight, the decision of the United States to oust Saddam and remake Iraq looks even more mistaken than it did at the time. It is not just that the articulated reason for the war—ridding Saddam of weapons of mass destruction—was shown to be faulty. What looms even larger in retrospect is the fact that removing Saddam and empowering Iraq's Shiite majority shifted the country from balancing Iranian strategic ambitions to serving them, in the process exacerbating frictions between Sunni and Shiite Muslims within the country and the region at large.

Nor did regime change have better results in two other countries where it was achieved. In Egypt, the American call for President Hosni Mubarak to leave office contributed to the polarization of the society. Subsequent events demonstrated that Egypt was not yet ready for a democratic transition, and U.S. withdrawal of support from a longtime friend and ally raised questions elsewhere (most notably in other Arab capitals) about the dependability of Washington's commitments. In Libya, meanwhile, the removal of Muammar al-Qaddafi by a combined U.S. and European effort helped create a failed state, one increasingly dominated by militias and terrorists. The uncertain necessity of the intervention itself was compounded by the lack of effective follow-up, and the entire exercise—coming as it did a few years after Qaddafi had been induced to give up his unconventional weapons programs—probably increased the perceived value of nuclear weapons and reduced the likelihood of getting other states to follow Qaddafi's example.

In Syria, the United States expressed support for the ouster of President Bashar al-Assad and then did little to bring it about. Obama went on to make a bad situation worse by articulating a set of redlines involving Syrian use of chemical munitions and then failing to act even when those lines were clearly crossed. This demoralized what opposition there was, forfeited a rare opportunity to weaken the government and change the momentum of the civil war, and helped usher in a context in which the Islamic State of Iraq and al-Sham (ISIS), which has declared itself the Islamic State, could flourish. The gap between rhetoric and action also further contributed to perceptions of American unreliability.

In Asia, too, the chief criticism that can be levied against U.S. policy is one of omission. As structural trends have increased the risks of traditional interstate conflict, Washington has failed to move in a determined fashion to stabilize the situation—not raising the U.S. military's presence in the region significantly in order to reassure allies and ward off challengers, doing little to build domestic support for a regional trade pact, and pursuing insufficiently active or sustained consultations to shape the thinking and actions of local leaders.

With regard to Russia, both internal and external factors have contributed to the deterioration of the situation. Putin himself chose to consolidate his political and economic power and adopt a foreign policy that increasingly characterizes Russia as an opponent of an international order defined and led by the United States. But U.S. and Western policy have not always encouraged more constructive choices on his part. Disregarding Winston Churchill's famous dictum about how to treat a beaten enemy, the West displayed little magnanimity in the aftermath of its victory in the Cold War. NATO enlargement was seen by many Russians as a humiliation, a betrayal, or both. More could have been made of the Partnership for Peace, a program designed to foster better relations between Russia and the alliance. Alternatively, Russia could have been asked to join NATO, an outcome that would have made little military difference, as NATO has become less of an alliance in the classic sense than a standing pool of potential contributors to "coalitions of the willing." Arms control, one of the few domains in which Russia could lay claim to still being a great power, was shunted to the side as unilateralism and minimalist treaties became the norm. Russian policy might have evolved the way it has anyway, even if the United States and the West overall had been more generous and welcoming, but Western policy increased the odds of such an outcome.

As for global governance, international accords are often hard to come by for many reasons. The sheer number of states makes consensus difficult or impossible. So, too, do divergent national interests. As a result, attempts to construct new global arrangements to foster trade and frustrate climate change have foundered. Sometimes countries just disagree on what is to be done and what they are prepared to sacrifice to achieve a goal, or they are reluctant to support an initiative for fear of setting a precedent that could be used against them later. There is thus decidedly less of an "international community" than the frequent use of the phrase would suggest.

Once again, however, in recent years, developments in and actions by the United States have contributed to the problem. The post-Cold War order was premised on U.S. primacy, which was a function of not just U.S. power but also U.S. influence, reflecting a willingness on the part of others to accept the

United States' lead. This influence has suffered from what is generally perceived as a series of failures or errors, including lax economic regulation that contributed to the financial crisis, overly aggressive national security policies that trampled international norms, and domestic administrative incompetence and political dysfunction.

Order has unraveled, in short, thanks to a confluence of three trends. Power in the world has diffused across a greater number and range of actors. Respect for the American economic and political model has diminished. And specific U.S. policy choices, especially in the Middle East, have raised doubts about American judgment and the reliability of the United States' threats and promises. The net result is that while the United States' absolute strength remains considerable, American influence has diminished.

What Is to Be Done?

Left unattended, the current world turbulence is unlikely to fade away or resolve itself. Bad could become worse all too easily should the United States prove unwilling or unable to make wiser and more constructive choices. Nor is there a single solution to the problem, as the nature of the challenges varies from region to region and issue to issue. In fact, there is no solution of any sort to a situation that can at best be managed, not resolved. But there are steps that can and should be taken. In the Middle East, the United States could do worse than to adopt the Hippocratic oath and try above all to do no further harm. The gap between U.S. ambitions and U.S. actions needs to be narrowed, and it will normally make more sense to reduce the former than increase the latter. The unfortunate reality is that democratic transformations of other societies are often beyond the means of outsiders to achieve. Not all societies are equally well positioned to become democratic at any given moment. in place; an adverse political culture can pose obstacles. Truly liberal democracies may make for better countries international get citizens, to that point but is helping more difficult than often recognized-and the attempts often riskier, as immature can be hijacked by demagoguery or nationalism. Promoting order among states—shaping their foreign policies more than their internal politics—is an ambitious enough goal for U.S. policy to pursue.

But if attempts at regime change should be jettisoned, so, too, should calendar-based commitments. U.S. interests in Iraq were not well served by the inability to arrange for the ongoing presence of a residual U.S. force there, one that might have dampened the feuding of Iraqi factions and provided much-needed training for Iraqi security forces. The same holds for Afghanistan, where all U.S. forces are due to exit by the end of 2016. Such decisions should be linked to interests and

conditions rather than timelines. Doing too little can be just as costly and risky as doing too much.

Other things outsiders could usefully do in the region include promoting and supporting civil society, helping refugees and displaced people, countering terrorism and militancy, and working to stem the proliferation of weapons of mass destruction (such as by trying to place a meaningful ceiling on the Iranian nuclear program). Degrading ISIS will require regular applications of U.S. airpower against targets inside both Iraq and Syria, along with coordinated efforts with countries such as Saudi Arabia and Turkey to stem the flow of recruits and dollars. There are several potential partners on the ground in Iraq, but fewer in Syria—where action against ISIS must be undertaken in the midst of a civil war. Unfortunately, the struggle against ISIS and similar groups is likely to be difficult, expensive, and long.

In Asia, the prescription is considerably simpler: implement existing policy assiduously. The Obama administration's "pivot," or "rebalance," to Asia was supposed to involve regular high-level diplomatic engagement to address and calm the region's all-too-numerous disputes; an increased U.S. air and naval presence in the region; and the building of domestic and international support for a regional trade pact. All these actions can and should be higher administration priorities, as should a special attempt to explore the conditions under which China might be prepared to reconsider its commitment to a divided Korean Peninsula.

With Russia and Ukraine, what is required is a mixture of efforts designed to shore up Ukraine economically and militarily, strengthen NATO, and sanction Russia. At the same time, Russia should also be offered a diplomatic exit, one that would include assurances that Ukraine would not become a member of NATO anytime soon or enter into exclusive ties with the EU. Reducing European energy dependence on Russia should also be a priority—something that will necessarily take a long time but should be started now. In dealing with Russia and other powers, meanwhile, Washington should generally eschew attempts at linkage, trying to condition cooperation in one area on cooperation in another. Cooperation of any sort anywhere is too difficult to achieve these days to jeopardize it by overreaching.

At the global level, the goal of U.S. policy should still be integration, trying to bring others into arrangements to manage global challenges such as climate change, terrorism, proliferation, trade, public health, and maintaining a secure and open commons. Where these arrangements can be global, so much the better, but where they cannot, they should be regional or selective, involving those actors with significant interests and capacity and that share some degree of policy consensus.

The United States also needs to put its domestic house in order, both to increase Americans' living standards and to generate the resources needed to sustain an active global role. A stagnant and unequal society will be unlikely to trust its government or favor robust efforts abroad. This need not mean gutting defense budgets, however; to the contrary, there is a strong case to be made that U.S. defense spending needs to be increased somewhat. The good news is that the United States can afford both guns and butter, so long as resources are allocated appropriately and efficiently. Another reason to get things right at home is to reduce American vulnerability. U.S. energy security has improved dramatically in recent years, thanks to the oil and gas revolutions, but the same cannot be said about other problems, such as the country's aging public infrastructure, its inadequate immigration policy, and its long-term public finances.

. . . American political dysfunction is increasing rather than decreasing, thanks to weakened parties, powerful interest groups, political finance rules, and demographic changes. Those who suggest that the country is only a budget deal away from comity are as mistaken as those who suggest that the country is only one crisis away from restored national unity. The world can see this, and see as well that a majority of the American public has grown skeptical of global involvement, let alone leadership. Such an attitude should hardly be surprising given the persistence of economic difficulties and the poor track record of recent U.S. interventions abroad. But it is up to the president to persuade a war-weary American society that the world still matters—for better and for worse—and that an active foreign policy can and should be pursued without undermining domestic well-being.

In fact, sensible foreign and domestic policies are mutually reinforcing: a stable world is good for the home front, and a successful home front provides the resources needed for American global leadership. Selling this case will be difficult, but one way to make it easier is to advance a foreign policy that tries to reorder the world rather than remake it. But even if this is done, it will not be enough to prevent the further erosion of order, which results as much from a wider distribution of power and a decentralization of decisionmaking as it does from how the United States is perceived and acts. The question is not whether the world will continue to unravel but how fast and how far.

Critical Thinking

1. What mistakes has the Obama administration made in the Middle East, if any?
2. What is meant by U.S. political dysfunction, and why is it important in U.S. foreign policy?
3. What can the wars in the Middle East be compared to and why?

Internet References

Council on Foreign Relations
   www.cfr.org/
Foreign Policy Research Institute
   www.fpri.org/
U.S. State Department
   www.state.gov/

RICHARD N. HAASS is President of the Council on Foreign Relations. Follow him on Twitter @RichardHaass.

Article   Prepared by: Robert Weiner, *University of Massachusetts, Boston*

A Kinder, Gentler Immigration Policy: Forget Comprehensive Reform—Let the States Compete

Jagdish Bhagwati and Francisco Rivera-Batiz

Learning Outcomes

After reading this article, you will be able to:

- Explain what lies behind the influx of refugees from Central America.

- Discuss why the Obama administration is having a difficult time in reforming immigration policy.

E ver since Congress passed the Immigration Reform and Control Act, in 1986, attempts at a similar comprehensive reform of U.S. immigration policies have failed. Yet today, as the Republican Party smarts from its poor performance among Hispanic voters in 2012 and such influential Republicans as former Florida Governor Jeb Bush have come out in favor of a new approach, the day for comprehensive immigration reform may seem close at hand. President Barack Obama was so confident about its prospects that he asked for it in his State of the Union address in February 2013. Now, the U.S. Senate looks poised to offer illegal immigrants a pathway to citizenship.

But a top-down legislative approach to immigration could nonetheless easily die in Congress, just as the last serious one did, in 2007. Indeed, the president's domestic problems with health care and foreign problems with Syria have already cast a shadow over the prospects for reform.

Even if a bill did manage to pass, the sad fact is that it would work no better than the 1986 law did. That act was based on the assumption that punishments, such as sanctions on employers and heightened border security, and incentives, such as an increase in the number of legal immigrants allowed to enter the country and amnesty for illegal immigrants already there, could eliminate illegal immigration altogether. That assumption proved illusory: the offer of amnesty may have temporarily reduced the stock of illegal immigrants, but it was not enough to eliminate it. Nor did employer sanctions and border enforcement reduce the flow of new illegal immigrants.

The challenges to eliminating illegal immigration are, if anything, greater today than they were in 1986. For one thing, in order to make today's proposals politically feasible, their authors decided to offer illegal immigrants not immediate unconditional amnesty but a protracted process of legalization. Confronted with this approach, a large share of the estimated 11 million illegal immigrants now living in the United States would likely choose to remain illegal rather than gamble on the distant promise of naturalization.

Nor would reform dissuade new illegal immigrants from joining those already in the country. Extrapolating from the recent drop in apprehensions near the Rio Grande, some analysts have argued that since the flow of illegal immigrants has already slowed to a trickle, the issue has lost its urgency. This notion is misguided. One cannot focus just on the area around the Rio Grande, since only half of all illegal immigrants residing in the United States entered the country by unlawfully crossing the U.S.-Mexican border, according to a 2006 study by the Pew Research Center. Moreover, whatever drop-off has occurred is mostly the result of the recent economic slowdown in the United States and will not prove permanent. As long as wages in the United States greatly outstrip those in poor countries, the United States will remain a mecca for potential immigrants, legal and illegal.

Not only would immigration reform fail to achieve its goal of eliminating illegal immigrants, it would also lead to increasingly draconian treatment of them. In order to appease anti-immigrant groups, the Senate's immigration reform bill provides for stricter enforcement of the U.S.-Mexican border, along with $40 billion in funding. But past experience suggests that such regulations are an exercise in futility: they do little to slow the influx of illegal immigrants while greatly increasing the risk to their lives as they try to cross the border over more dangerous terrain, aided by unscrupulous smugglers who may abandon them mid-journey.

Given these realities, the United States should stop attempting to eliminate illegal immigrants—since that will never work—and focus instead on policies that treat them with humanity. Doing so would mean adopting a variety of measures to diminish the public's hostility to illegal immigrants. Principal among them would be a shift from a top-down approach to a bottom-up one: letting states compete for illegal immigrants. States with laws that were unfriendly to illegal immigrants would lose them and their badly needed labor to states that were more welcoming. The result would be a competition that would do far more to improve the treatment of illegal immigrants than anything coming from Washington.

Impossible Difficulties

Americans can be schizophrenic when it comes to illegal immigration, suffering from a sort of right-brain, left-brain problem. The right brain sympathizes with illegal immigrants, since they are immigrants, after all, and the United States was founded on immigration. But the left brain fixates on their illegality, which offends Americans' respect for the rule of law. Negotiating a viable compromise between those who wish to throw illegal immigrants out and those who wish to embrace them has always proved exceptionally difficult. As the historian Mae Ngai has shown, U.S. immigration policy in the 1920s and 1930s was as conflicted as it is today, with proponents of deportations pitted against proimmigrant humanitarian groups.

Further complicating matters is Americans' sense of fairness. Liberals have called on Congress to offer illegal immigrants a path to citizenship, but unlike most other countries, the United States has an enormous backlog of potential immigrants who have dutifully lined up for entry—an issue that Spain, for example, did not face when it granted its illegal immigrants amnesty in 2005. Many Americans consider it unfair to let immigrants who have broken the law join the same line that those who followed the rules are in. The proponents of amnesty have, in consequence, cluttered up their proposed policy with various restrictions and requirements that make it far less attractive than a forthright granting of full citizenship.

Like past reform proposals, the current one offers illegal immigrants a long road to legality. But the longer the process, the greater the risk that a new Congress will reverse the old. Many illegal immigrants may prefer not to accept that risk and instead stay illegal. Furthermore, as the immigration scholars Mark Rosenzweig, Guillermina Jasso, Douglas Massey, and James Smith have shown, around 30 percent of U.S. immigrants achieve legal status despite having violating immigration laws in the past. Taking these factors into account, it is reasonable to predict that of the estimated 11 million illegal immigrants, only half would take an offer of amnesty, perhaps less.

Just like the chimera of legalizing away the stock of illegal immigrants, the notion that the flow of new illegal immigrants can be shut off is also deeply impractical. For instance, attempts at expanding legal immigration in the hope that it will reduce the incentive for illegal immigration would require, at minimum, vastly expanded legal admissions. Yet even though trade unions have given up their long-standing opposition to legalizing illegal immigrants—which they figure will boost their membership—they oppose significantly expanding legal admissions. Unions have long blamed immigration for the stagnation of workers' wages, just as they have blamed outsourcing and trade liberalization. In fact, the AFL-CIO recently suggested that it should be involved in determining how many legal guest workers the United States will admit in the future. When President George W. Bush proposed a more expansive guest worker program, unions helped kill the measure, and they would fiercely fight any efforts to liberalize legal immigration this time, too.

It is also dubious that draconian enforcement measures, at the border or internally, would actually intimidate would-be illegal immigrants, no matter what mix of punishments and inducements Congress legislates. Unlike in 1986, almost every U.S. immigrant is now more secure: their ethnic compatriots will, as they already do, go to bat for better treatment, raising their voices against such measures.

But the biggest hurdle that immigration reform faces is that as long as immigration restrictions exist, people will continue to enter the United States illegally. The government can send as many Eliot Nesses to Chicago to nab as many Al Capones as it wants, but the bootlegged liquor will keep flowing across the Canadian border as long as Prohibition remains in place.

Immigration Inhumanity

Short of dismantling all border restrictions, then, no policy could magically eliminate illegal immigration. Yet not only would a reform bill be ineffective, it could also be harmful. If a comprehensive reform bill were passed, there is a serious danger that policymakers, operating on the flawed assumption that there should then be no reason for illegal immigrants to exist, might enact even harsher measures against them.

In fact, merely attempting to secure support for a reform bill is certain to harm illegal immigrants. Their experiences under President Bill Clinton and Obama have not been reassuring. Although Democrats have generally been more sympathetic to illegal immigrants than have Republicans, both Clinton and Obama, in their attempts to secure bipartisan consensus on immigration reform, implemented ruthless measures against illegal immigrants.

In the wake of the Immigration Reform and Control Act, the U.S. government ramped up enforcement at the border, which reached new heights during the Clinton administration. Ditches were built and fences constructed. To seal off common routes of entry into the United States, the government mounted military-style actions with names that seemed straight out of a war room: Operation Blockade in El Paso in 1993; Operation Gatekeeper in San Diego in 1994; and many more. The border security budget skyrocketed, rising from $326 million in 1992 to $1.1 billion by the time President George W. Bush took office in 2001. The number of U.S. Border Patrol agents stationed at the southwestern border nearly tripled. In the end, these measures did little to stem the inflow. Demographer Jeffrey Passel of the Pew Research Center has estimated that the average net annual influx of illegal immigrants crossing the Rio Grande rose from 324,000 in the first half of the 1990s to 654,800 in the second half of the decade.

What stricter enforcement did do was force illegal immigrants to bypass safer crossing points and travel through the desert instead. Desperate immigrants made no secret of their desire to keep trying to sneak across the border despite heightened enforcement, often attempting again and again until they got through. But crossing the desert meant that they had to pay smugglers, known as coyotes, who left carloads of illegal immigrants for dead when they feared apprehension by U.S. Border Patrol personnel. At best, the Clinton administration's policies had a marginal impact on illegal border traffic and led to a major decline in the welfare of those trying to enter the country illegally. They also failed to achieve their larger objective of getting legislation through Congress; the "keep them out" and "throw them out" lobbies were too strongly opposed to any compromise.

Obama has ramped up border enforcement, too, but he has also deported record numbers of illegal immigrants already living in the United States. In 2011, he expanded the Secure Communities initiative, a joint effort between state and local governments—the federal authorities have even ordered uncooperative states, such as New York, to fall in line—that uses integrated databases to track down illegal immigrants. According to official statistics, the number of deportations (excluding apprehensions at the border itself) has risen under Obama, to 395,000 in 2009. In 2001, under George W. Bush, deportations numbered only 189,000.

The focus on border enforcement is misguided. In part, it owes to the false equation of lax border control with the influx of terrorists. There is little evidence of that link: even the 9/11 hijackers entered the United States legally. Moreover, correcting for the effect of the recession on attempted crossings, it is clear that the impact of Obama's policies has been far from dramatic in deterring illegal immigration. But the distress caused to illegal immigrants has been great. As a 2011 report from Human Rights Watch detailed, tens of thousands of immigrants are shuffled from jail to jail awaiting deportation. Once again, the country has gained little and lost much.

Race to the Top

With top-down immigration reform unworkable and inhumane, Americans need to shift their focus to treating their inevitable neighbors with humanity. That objective cannot be pursued through Washington. It must come from elsewhere: competition among states. States that harass illegal immigrants, such as Alabama, Arizona, Georgia, Indiana, and South Carolina, will drive illegal immigrants to more welcoming states, such as Maryland, New York, Utah, and Washington. As the former lose badly needed cheap labor to the latter, the political equilibrium will shift toward those who favor policies that help retain and attract illegal immigrants.

Of course, states cannot intrude on the parts of immigration enforcement over which the federal government has exclusive authority, such as border control and civil rights. But there are a number of steps states can take to make life easier for illegal immigrants, such as issuing them driver's licenses and making accessible to them everything from health care to university scholarships.

Illegal immigrants are already voting with their feet, leaving or bypassing states that treat them harshly and flocking to those with more benign policies. In 2011, hours after a federal judge in Alabama upheld most of the state's strict immigration law, illegal immigrants began fleeing. Frightened families, *The New York Times* reported, "left behind mobile homes, sold fully furnished for a 1,000 dollars or even less." The article continued: "2, 5, 10 years of living here, and then gone in a matter of days, to Tennessee, Illinois, Oregon, Florida, Arkansas, Mexico—who knows? Anywhere but Alabama."

Ample statistical evidence demonstrates this pattern. From 1990 to 2010, when tough border-enforcement policies (which naturally focused on the border states) were in vogue, Arizona, California, New Mexico, and Texas saw their collective share of illegal immigrants decline by 17 percent. In California alone, the percentage of all illegal immigrants residing there fell from 43 percent to 23 percent. Similarly, economists Sarah Bohn,

Magnus Lofstrom, and Steven Raphael have calculated that Arizona's 2007 Legal Arizona Workers Act, which banned businesses from hiring illegal immigrants, led to a notable decline in the proportion of the state's foreign-born Hispanic population.

The resulting blow to economic activity has often been drastic; employers in agriculture and construction, for example, regularly complain about the absence of workers. Fortunately, however, as business interests begin to agitate in favor of easing up on illegal immigrants, state capitals will start taking note. Already, many groups in the unwelcoming states have begun to question their states' draconian immigration enforcement laws and argue for more modest measures. After Alabama passed its immigration law, for example, business leaders complained to lawmakers about the resulting labor shortages. After the Legal Arizona Workers Act went into effect, in 2008, the state's contractors' trade association even joined civil rights groups in seeking the law's repeal. That same year, the U.S. Chamber of Commerce filed a lawsuit challenging the constitutionality of an Oklahoma law that required employers to verify the work status of their employees.

As this dynamic plays out, states will begin to compete for illegal immigrants, who will then face less harassment and be able to better integrate into their communities. Democrats and Republicans who care about human rights should welcome this change. More important, so should Republicans who prize states' rights. A race to the top in the treatment of illegal immigrants is a viable path to reform that would greatly advance human rights in the United States.

There are other ways to improve the lives of illegal immigrants that also do not involve Washington. Consider the problem of Mexicans who risk their lives traveling through the desert while attempting to cross the border and who occasionally damage the property of Texan ranchers. With no method to recoup their losses, the affected ranchers found it tempting to join forces with the Minutemen vigilantes who used to patrol the border. To reduce ranchers' hostility toward illegal immigrants, the Mexican government should set up a fund that compensates ranchers who can establish credible claims of damage. Since the stories of such damages tend to outstrip the reality, the fund need not be particularly large to go a long way in defusing the hostility.

Another way to improve the plight of illegal immigrants would be for Mexico to help pay for the education and medical expenses of those illegal immigrants coming from Mexico that are otherwise borne by the U.S. government. Although a number of studies show that illegal immigrants represent a net contribution to U.S. government coffers, the common perception that American taxpayers must bear these costs and that Mexico should share some of the burden of its own citizens breeds resentment. Were the Mexican government to make such a contribution, it would serve as a gesture of goodwill that could help reduce the hostility toward illegal immigrants.

"Give me your tired, your poor, your huddled masses," reads the poem by Emma Lazarus that adorns the Statue of Liberty, which once welcomed the millions of immigrants arriving at Ellis Island. It is well past time to revive that sense of humanity, and the diverse recommendations outlined here can help the United States do just that. Whether or not they come with Washington's permission, immigrants to the United States nonetheless deserve the compassion Lazarus promised.

Critical Thinking

1. Do you think that the suggestion for other states to accept illegal immigrants is feasible?
2. What can be done at the international level to deal with illegal immigration?
3. What is the policy of the Obama Administration toward illegal immigration?

Create Central

www.mhhe.com/createcentral

Internet References

Department of Homeland Security
http://www.dhs.gov

International Organization on Migration
http://www.iom/nt/cms/en/sites/iom/home

JAGDISH BHAGWATI is Senior Fellow for International Economics at the Council on Foreign Relations and University Professor of Economics, Law, and International Affairs at Columbia University. **FRANCISCO RIVERA-BATIZ** is Professor Emeritus of Economics and Education at Teachers College at Columbia University.

Article   Prepared by: Robert Weiner, *University of Massachusetts, Boston*

The Information Revolution and Power

JOSEPH S. NYE JR.

Learning Outcomes

After reading this article, you will be able to:

- Understand the relationship between the information revolution and soft power.

- Understand what two power shifts are occurring in the 21st century.

O ne of the notable trends of the past century that will likely continue to strongly influence global politics in this century is the current information revolution. And with this information revolution comes an increase in the role of soft power—the ability to obtain preferred outcomes by attraction and persuasion rather than coercion and payment.

Information revolutions are not new—one can think back to the dramatic effects of Gutenberg's printing press in the 16th century. But today's information revolution is changing the nature of power and increasing its diffusion. Sometimes called "the third industrial revolution," the current transformation is based on rapid technological advances in computers and communications that in turn have led to extraordinary declines in the costs of creating, processing, transmitting, and searching for information.

One could date the ongoing information revolution from Intel cofounder Gordon Moore's observation in the 1960s that the number of transistors fitting on an integrated circuit doubles approximately every 2 years. As a result of Moore's Law, computing power has grown enormously, and by the beginning of the 21st century doubling this power cost one-thousandth of what it did in the early 1970s.

Meanwhile, computer-networked communications have spread worldwide. In 1993, there were about 50 websites in the world; by 2000, the number had surpassed 5 million, and a decade later had exceeded 500 million. Today, about a third of the global population is online; by 2020 that share is projected to grow to 60 percent, or 5 billion people, many connected with multiple devices.

The key characteristic of this information revolution is not the *speed* of communications among the wealthy and the powerful; for a century and a half, instantaneous communication by telegraph has been possible between Europe and North America. The crucial change, rather, is the radical and ongoing reduction in the *cost* of transmitting information. If the price of an automobile had declined as rapidly as the price of computing power, one could buy a car today for $10 to 15.

When the price of a technology shrinks so rapidly, it becomes readily accessible and the barriers to entry are reduced. For all practical purposes, transmission costs have become negligible; hence the amount of information that can be transmitted worldwide is effectively infinite.

Winning Stories

In the middle of the 20th century, people feared that the computers and communications of the information revolution would create the central governmental control dramatized in George Orwell's dystopian novel *1984*. Instead, as computing power has decreased in cost and computers have shrunk to the size of smartphones and other portable devices, their decentralizing effects have outweighed their centralizing effects, as WikiLeaks and Edward Snowden have demonstrated.

Power over information is much more widely distributed today than even a few decades ago. Information can often provide a key power resource, and more people have access to more information than ever before. This has led to a diffusion of power away from governments to nonstate actors, ranging from large corporations to nonprofits to informal ad hoc groups.

This does not mean the end of the nation-state. Governments will remain the most powerful actors on the global stage. However, the stage will become more crowded, and many nonstate actors will compete effectively for influence. They will do so mostly in the realm of soft power.

The increasingly important cyber domain provides a good example. A powerful navy is important in controlling sea-lanes; it does not provide much help on the internet. The historian A.J.P. Taylor wrote that in 19th century Europe, the mark of a great power was the ability to prevail in war. Yet, as the American defense analyst John Arquilla has noted, in today's global information age, victory may sometimes depend not on whose army wins, but on whose story wins.

Sources of Power

I first coined the term "soft power" in my 1990 book *Bound to Lead,* which challenged the then-conventional view of the decline of US power. After looking at American military and economic power resources, I felt that something was still missing—the ability to affect others by attraction and persuasion rather than just coercion and payment. I thought of soft power as an analytic concept to fill a deficiency in the way analysts thought about power.

The term was eventually used by European leaders to describe some of their power resources, as well as by other governments, such as Japan and Australia. But I was surprised when President Hu Jintao told the Chinese Communist Party's 17th Party Congress in 2007 that his country needed to increase its soft power.

This is a smart strategy, because as China's hard military and economic power grows, it may frighten its neighbors into balancing coalitions. If China can accompany its rise with an increase in its soft power, it can weaken the incentives for these coalitions. Consequently, the Chinese government has invested billions of dollars in this task, and Chinese journals and papers are filled with hundreds of articles about soft power. But what, precisely, is it?

Power is the ability to affect others to obtain the outcomes you want. You can affect their behavior in three main ways: threats of coercion (sticks), inducements or payments (carrots), and attraction that makes others want what you want. A country may obtain the outcomes it desires in world politics because other countries want to follow it—admiring its values, emulating its example, and aspiring to its level of prosperity and openness.

More people have access to more information than ever before.

In this sense, it is important to set the agenda and attract others in world politics, and not only to force them to change through the threat or use of military or economic weapons. This soft power—getting others to want the outcomes that you want—co-opts countries rather than coerces them.

Soft power rests on the ability to shape the preferences of others. It is not the possession of any one country, nor only of countries. For example, companies invest heavily in their brands, and nongovernmental activists often attack their brands to press them to change their practices. In international politics, a nation's soft power rests primarily on three resources: its culture (in places where it is attractive to others), its political values (when it lives up to them at home and abroad), and its foreign policies (when they are seen as legitimate and having moral authority).

Propaganda Ploys

China is doing well in terms of culture, but is having difficulty with values and policies. The world's most populous country has always had an attractive traditional culture; now it has created hundreds of Confucius Institutes around the world to teach its language and culture. Beijing is also increasing its international radio and television broadcasting. Moreover, China's economic success has attracted others. This attraction was reinforced by China's successful response to the 2008 global financial crisis—maintaining growth while much of the West fell into recession—and by its economic aid and investment in poor countries. In the past decade, it became common to refer to these efforts as "China's charm offensive."

Yet, as the University of Denver's Jing Sun observed in the September 2013 issue of *Current History,* China has not reaped a good return on its investment. This is not because soft power is becoming less important in world politics. It is a result of limitations in China's strategy—a strategy that overly stresses culture while neglecting civil society and the damage done by nationalistic policies.

In 2009, Beijing announced plans to spend huge sums to develop global media giants to compete with Bloomberg, Time Warner, and Viacom, using soft power rather than military might to win friends abroad. As George Washington University's David Shambaugh has documented, China has invested billions in external publicity work, including a 24-hour Xinhua cable news channel.

China's soft power, however, still has a long way to go. A recent BBC poll shows that opinions of China's influence are positive in much of Africa and Latin America, but predominantly negative in the United States and everywhere in Europe, as well as in India, Japan, and South Korea. Similarly, a poll taken in Asia after the 2008 Beijing Olympics found that Beijing's charm offensive had not been effective.

China does not yet have global cultural industries on the scale of Hollywood, and its universities are not yet the equal of America's. But more important, it lacks the many nongovernmental organizations that generate much of America's soft power. Chinese officials seem to think that soft power is generated primarily by government policies and public diplomacy, whereas much of America's soft power is generated by its civil society rather than its government.

Great powers try to use culture and narrative to create soft power that promotes their advantage, but it is not an easy sell when it is inconsistent with their domestic realities. For example, while the 2008 Olympic Games were a great success, Beijing's crackdowns shortly thereafter in Tibet, in Xinjiang, and on human rights activists undercut its soft power gains. The Shanghai Expo in 2010 likewise was judged a success, but it was followed by the jailing of Nobel Peace laureate Liu Xiaobo and the artist Ai Weiwei. In the world of communications theory, this is called "stepping on your own message."

And for all the efforts to turn Xinhua and China Central Television into competitors of CNN and the BBC, there is not much of an international audience for brittle propaganda. As *The Economist* reported, "the party has not bought into Mr. Nye's view that soft power springs largely from individuals, the private sector, and civil society. So the government has taken to promoting ancient cultural icons whom it thinks might have global appeal."

Given a political system that relies on one-party control, it is difficult to tolerate dissent and diversity. Moreover, the Chinese Communist Party has based its legitimacy on high rates of economic growth and appeals to nationalism. The nationalism reduces the universal appeal of "the Chinese Dream" promoted by President Xi Jinping, and encourages policies in the South China Sea and elsewhere that antagonize its neighbors. For example, when Chinese ships drove Philippine fishing boats from the Scarborough Shoal in 2012, China gained control of the remote area, and from a domestic nationalist point of view, the action was a success. However, it came at the cost of reduced Chinese soft power in Manila.

Russian President Vladimir Putin has recently called for an effort to increase his country's soft power, but he might consider lessons from China the next time he locks up dissidents or bullies neighbors such as Georgia or Ukraine. A successful soft power strategy must attend to all three resources: culture, political values, and foreign policies that are seen as legitimate in the eyes of others. Investment in government propaganda is not a successful strategy for increasing a country's soft power.

Positive Sums

The development of soft power need not be a zero-sum game. All countries can gain from finding attraction in each other. Just as the national interests of China and the United States are partly congruent and partly conflicting, their soft powers are reinforcing each other in some issue areas and contradicting each other in others.

This is not something unique to soft power. In general, power relationships can be zero- or positive-sum depending on the objectives of the actors. For example, if two countries both desire stability, a balance of military power in which neither side fears attack by the other can be a positive-sum relationship. Likewise, if China and the United States both become more attractive in each other's eyes, the prospects of damaging conflicts will be reduced. If the rise of China's soft power reduces the likelihood of conflict, it can be part of a positive-sum relationship.

In the long term, there will always be elements of both competition and cooperation in the US-China relationship, but the two countries have more to gain from the cooperative element, and this can be strengthened by the rise in both countries' soft power. Prudent policies would aim to make that a trend in coming decades.

The 21st century is experiencing two great power shifts: a "horizontal" transition among countries from west to east, as Asia recovers its historic proportion of the world economy, and a "vertical" diffusion of power away from states to nongovernmental actors. This diffusion is fueled by the current information revolution, and it is creating an international politics that will involve many more actors than in the several centuries since the Treaty of Westphalia enshrined the norm of sovereignty.

But power diffusion also affects relations among states. It strengthens transnational actors and puts new transnational issues on the agenda, such as terrorism, global financial stability, cyber-conflict, pandemics, and climate change. No government can solve these problems acting on its own. In seeking to organize coalitions and networks to deal with such challenges, governments will need to exercise the powers not only of coercion and payment, but also of attraction and persuasion.

Critical Thinking

1. How is the information revolution contributing to the diffusion of power in the international system?

2. Why is information a key power source?

3. Are states still important in view of the information revolution?

Create Central

www.mhhe.com/createcentral

Internet References

Department of Homeland Security
http://www.dhs.gov

International Corporation for Names and Numbers
http://www.icann.org

International Organization on Migration
http://www.iom/nt/cms/en/sites/iom/home

International Telecommunications Union
http://www.itu.int

National Security Agency
http://www.nsa.gov

World Wide Web Consortium
www.w3.org

JOSEPH S. NYE JR., a Current History contributing editor, is a professor of political science at Harvard University and the author most recently of *Presidential Leadership and the Creation of the American Era* (Princeton University Press, 2013).

Unit 2

Prepared by: Robert Weiner, *University of Massachusetts, Boston*

UNIT

Population, Natural Resources, and Climate Change

After World War II, the global population reached an estimated two billion people. It had taken 250 years to triple the population to that level. In the six decades following World War II, the population tripled again to six billion. By 2050, or about 100 years after World War II, some analysts forecast that the population will go up to 10 to 12 billion. While demographers develop various scenarios forecasting population growth, it is important to remember that there are circumstances that could lead not to growth but to significant decline in global population. The spread of AIDS and other infectious diseases like the Ebola virus are cases in point. The lead article in this unit provides an overview of general demographic trends, with a special focus on issues related to aging. Making predictions about the future of the world's population is a complicated task, for there are a variety of forces at work and considerable variation from region to region. The dangers of oversimplification must be overcome if governments and international organizations are going to respond with meaningful policies. Perhaps one could say that this is not a global population challenge, but many population challenges that vary from country to country and region to region.

The increase in population has also put pressure on countries to gain access to vital resources such as oil and natural gas. A recent trend has been for multinational corporations to prospect for oil in more advanced economies such as New Zealand. Multinational corporations find advanced economies more stable and less corrupt than developing countries. Rather than scarcity, advances in energy technology have also resulted in a significant increase in oil and natural gas production in the United States. This is done through the technique of fracking natural gas and oil from shale rocks. Fracking, however, results in the release of toxic chemicals which affect the water supplies of surrounding communities. Analysts predict that the United States has the capacity to become the leading producer of natural gas and oil in the world and that it will surpass Saudi Arabia as the world's leading exporter of energy. This has important geopolitical implications because an increase in the export of U.S. natural gas to Europe could reduce European dependence on Russian natural gas imports. Increased U.S. production of natural gas and oil, and other developments such as reduced demand in Europe, created a glut of oil in the world market in 2015. The result was a significant reduction in the benchmark price of Brent crude oil, contributing to a decline in the Russian economy. The oversupply of oil in the world market has also created a crisis for the major oil cartel, The Organization of Petroleum Exporting Countries (OPEC), which has faced pressure from some of its members to reduce the production of oil in order to keep prices up.

The world's population also faces an existential threat from global warming, according to credible scientists. The main question is whether the international community is willing to reach the consensus that will effectively regulate the emission of greenhouse gases into the atmosphere. In 2014–2015, the United States and China, the two largest emitters of greenhouse gases reached an agreement to reduce the amount released into the atmosphere. China has also promised, in a statement made by the President of China in a state visit to the United States in September 2015, to provide the developing countries with about $3 billion in aid to help reduce the emission of greenhouse gases. Some analysts, however, are rather pessimistic about the prospects for environmental diplomacy that will replace the Kyoto Protocol of 1997 with a new climate treaty at a conference scheduled to meet in Paris in 2015. David Schorr, in an article in this unit, for example, argues that "idealized multilateralism" does not work in a world based on realist principles. Climate warming is here now, and whether the international community can come up with a new effective treaty remains to be seen.

Article   Prepared by: Robert Weiner, *University of Massachusetts, Boston*

The New Population Bomb: The Four Megatrends That Will Change the World

JACK A. GOLDSTONE

Learning Outcomes

After reading this article, you will be able to:

- Identify the four demographic trends.
- Discuss how international politics is changing due to these trends.
- Summarize the Afghanistan case study.

Forty-two years ago, the biologist Paul Ehrlich warned in The Population Bomb that mass starvation would strike in the 1970s and 1980s, with the world's population growth outpacing the production of food and other critical resources. Thanks to innovations and efforts such as the "green revolution" in farming and the widespread adoption of family planning, Ehrlich's worst fears did not come to pass. In fact, since the 1970s, global economic output has increased and fertility has fallen dramatically, especially in developing countries.

The United Nations Population Division now projects that global population growth will nearly halt by 2050. By that date, the world's population will have stabilized at 9.15 billion people, according to the "medium growth" variant of the UN's authoritative population database World Population Prospects: The 2008 Revision. (Today's global population is 6.83 billion.) Barring a cataclysmic climate crisis or a complete failure to recover from the current economic malaise, global economic output is expected to increase by two to three percent per year, meaning that global income will increase far more than population over the next four decades.

But twenty-first-century international security will depend less on how many people inhabit the world than on how the global population is composed and distributed: where populations are declining and where they are growing, which countries are relatively older and which are more youthful, and how demographics will influence population movements across regions.

These elements are not well recognized or widely understood. A recent article in *The Economist*, for example, cheered the decline in global fertility without noting other vital demographic developments. Indeed, the same UN data cited by *The Economist* reveal four historic shifts that will fundamentally alter the world's population over the next four decades: the relative demographic weight of the world's developed countries will drop by nearly 25 percent, shifting economic power to the developing nations; the developed countries' labor forces will substantially age and decline, constraining economic growth in the developed world and raising the demand for immigrant workers; most of the world's expected population growth will increasingly be concentrated in today's poorest, youngest, and most heavily Muslim countries, which have a dangerous lack of quality education, capital, and employment opportunities; and, for the first time in history, most of the world's population will become urbanized, with the largest urban centers being in the world's poorest countries, where policing, sanitation, and health care are often scarce. Taken together, these trends will pose challenges every bit as alarming as those noted by Ehrlich. Coping with them will require nothing less than a major reconsideration of the world's basic global governance structures.

Europe's Reversal of Fortunes

At the beginning of the eighteenth century, approximately 20 percent of the world's inhabitants lived in Europe (including Russia). Then, with the Industrial Revolution, Europe's population boomed, and streams of European emigrants set off for the Americas. By the eve of World War I, Europe's population had more than quadrupled. In 1913, Europe had more people than China, and the proportion of the world's population living in Europe and the former European colonies of North America had risen to over 33 percent. But this trend reversed after World War I, as basic health care and sanitation began to spread to poorer countries. In Asia, Africa, and Latin America, people began to live longer, and birthrates remained high or fell only slowly. By 2003, the combined populations of Europe, the United States, and Canada accounted for just 17 percent of the global population. In 2050, this figure is expected to be just 12 percent—far less than it was in 1700. (These projections, moreover, might even understate the reality because they reflect the "medium growth" projection of the UN forecasts, which assumes that the fertility rates of developing countries will decline while those of developed countries will increase. In fact, many developed countries show no evidence of increasing fertility rates.) The West's relative decline is even more dramatic if one also considers changes in income. The Industrial Revolution made Europeans not only more numerous than they had been but also considerably richer per capita than others worldwide. According to the economic historian Angus Maddison, Europe, the United States, and Canada together produced about 32 percent of the world's GDP at the beginning of the nineteenth century. By 1950, that proportion had increased to a remarkable 68 percent of the world's total output (adjusted to reflect purchasing power parity).

This trend, too, is headed for a sharp reversal. The proportion of global GDP produced by Europe, the United States, and Canada fell from 68 percent in 1950 to 47 percent in 2003 and will decline even more steeply in the future. If the growth rate of per capita income (again, adjusted for purchasing power parity) between 2003 and 2050 remains as it was between 1973 and 2003—averaging 1.68 percent annually in Europe, the United States, and Canada and 2.47 percent annually in the rest of the world—then the combined GDP of Europe, the United States, and Canada will roughly double by 2050, whereas the GDP of the rest of the world will grow by a factor of five. The portion of global GDP produced by Europe, the United States, and Canada in 2050 will then be less than 30 percent—smaller than it was in 1820.

These figures also imply that an overwhelming proportion of the world's GDP growth between 2003 and 2050—nearly 80 percent—will occur outside of Europe, the United States, and Canada. By the middle of this century, the global middle class—those capable of purchasing durable consumer products, such as cars, appliances, and electronics—will increasingly be found in what is now considered the developing world. The World Bank has predicted that by 2030 the number of middle-class people in the developing world will be 1.2 billion—a rise of 200 percent since 2005. This means that the developing world's middle class alone will be larger than the total populations of Europe, Japan, and the United States combined. From now on, therefore, the main driver of global economic expansion will be the economic growth of newly industrialized countries, such as Brazil, China, India, Indonesia, Mexico, and Turkey.

Aging Pains

Part of the reason developed countries will be less economically dynamic in the coming decades is that their populations will become substantially older. The European countries, Canada, the United States, Japan, South Korea, and even China are aging at unprecedented rates. Today, the proportion of people aged 60 or older in China and South Korea is 12–15 percent. It is 15–22 percent in the European Union, Canada, and the United States and 30 percent in Japan. With baby boomers aging and life expectancy increasing, these numbers will increase dramatically. In 2050, approximately 30 percent of Americans, Canadians, Chinese, and Europeans will be over 60, as will more than 40 percent of Japanese and South Koreans.

Over the next decades, therefore, these countries will have increasingly large proportions of retirees and increasingly small proportions of workers. As workers born during the baby boom of 1945–65 are retiring, they are not being replaced by a new cohort of citizens of prime working age (15–59 years old).

Industrialized countries are experiencing a drop in their working-age populations that is even more severe than the overall slowdown in their population growth. South Korea represents the most extreme example. Even as its total population is projected to decline by almost 9 percent by 2050 (from 48.3 million to 44.1 million), the population of working-age South Koreans is expected to drop by 36 percent (from 32.9 million to 21.1 million), and the number of South Koreans aged 60 and older will increase by almost 150 percent (from 7.3 million to 18 million). By 2050, in other words, the entire working-age population will barely exceed the 60-and-older population. Although South Korea's case is extreme, it represents an increasingly common fate for developed countries. Europe is expected to lose 24 percent of its prime working-age population (about 120 million workers) by 2050, and its 60-and-older population is expected to increase by 47 percent. In the United States, where higher fertility and more immigration are expected than in Europe, the working-age population will grow by 15 percent over the next four decades—a steep decline from its growth of 62 percent between 1950 and 2010. And by 2050, the United States' 60-and-older population is expected to double.

All this will have a dramatic impact on economic growth, health care, and military strength in the developed world. The forces that fueled economic growth in industrialized countries during the second half of the twentieth century—increased productivity due to better education, the movement of women into the labor force, and innovations in technology—will all likely weaken in the coming decades. College enrollment boomed after World War II, a trend that is not likely to recur in the twenty-first century; the extensive movement of women into the labor force also was a one-time social change; and the technological change of the time resulted from innovators who created new products and leading-edge consumers who were willing to try them out—two groups that are thinning out as the industrialized world's population ages.

Overall economic growth will also be hampered by a decline in the number of new consumers and new households. When developed countries' labor forces were growing by 0.5–1.0 percent per year, as they did until 2005, even annual increases in real output per worker of just 1.7 percent meant that annual economic growth totaled 2.2–2.7 percent per year. But with the labor forces of many developed countries (such as Germany, Hungary, Japan, Russia, and the Baltic states) now shrinking by 0.2 percent per year and those of other countries (including Austria, the Czech Republic, Denmark, Greece, and Italy) growing by less than 0.2 percent per year, the same 1.7 percent increase in real output per worker yields only 1.5–1.9 percent annual overall growth. Moreover, developed countries will be lucky to keep productivity growth at even that level; in many developed countries, productivity is more likely to decline as the population ages.

A further strain on industrialized economies will be rising medical costs: as populations age, they will demand more health care for longer periods of time. Public pension schemes for aging populations are already being reformed in various industrialized countries—often prompting heated debate. In theory, at least, pensions might be kept solvent by increasing the retirement age, raising taxes modestly, and phasing out benefits for the wealthy. Regardless, the number of 80- and 90-year-olds—who are unlikely to work and highly likely to require nursing-home and other expensive care—will rise dramatically. And even if 60- and 70-year-olds remain active and employed, they will require procedures and medications—hip replacements, kidney transplants, blood-pressure treatments—to sustain their health in old age.

All this means that just as aging developed countries will have proportionally fewer workers, innovators, and consumerist young households, a large portion of those countries' remaining economic growth will have to be diverted to pay for the medical bills and pensions of their growing elderly populations. Basic services, meanwhile, will be increasingly costly because fewer young workers will be available for strenuous and labor-intensive jobs. Unfortunately, policymakers seldom reckon with these potentially disruptive effects of otherwise welcome developments, such as higher life expectancy.

Youth and Islam in the Developing World

Even as the industrialized countries of Europe, North America, and Northeast Asia will experience unprecedented aging this century, fast-growing countries in Africa, Latin America, the Middle East, and Southeast Asia will have exceptionally youthful populations. Today, roughly nine out of ten children under the age of 15 live in developing countries. And these are the countries that will continue to have the world's highest birthrates. Indeed, over 70 percent of the world's population growth between now and 2050 will occur in 24 countries, all of which are classified by the World Bank as low income or lower-middle income, with an average per capita income of under $3,855 in 2008.

Many developing countries have few ways of providing employment to their young, fast-growing populations. Would-be laborers, therefore, will be increasingly attracted to the labor markets of the aging developed countries of Europe, North America, and Northeast Asia. Youthful immigrants from nearby regions with high unemployment—Central America, North Africa, and Southeast Asia, for example—will be drawn to those vital entry-level and manual-labor jobs that sustain advanced economies: janitors, nursing-home aides, bus drivers, plumbers, security guards, farm workers, and the like. Current levels of immigration from developing to developed countries are paltry compared to those that the forces of supply and demand might soon create across the world.

These forces will act strongly on the Muslim world, where many economically weak countries will continue to experience dramatic population growth in the decades ahead. In 1950, Bangladesh, Egypt, Indonesia, Nigeria, Pakistan, and Turkey had a combined population of 242 million. By 2009, those six countries were the world's most populous Muslim-majority countries and had a combined population of 886 million. Their populations are continuing to grow and indeed are expected to increase by 475 million between now and 2050—during which time, by comparison, the six most populous developed countries are projected to gain only 44 million inhabitants. Worldwide, of the 48 fastest-growing countries today—those with annual population growth of two percent or more—28 are majority Muslim or have Muslim minorities of 33 percent or more.

It is therefore imperative to improve relations between Muslim and Western societies. This will be difficult given that many Muslims live in poor communities vulnerable to radical appeals and many see the West as antagonistic and militaristic. In the 2009 Pew Global Attitudes Project survey, for example, whereas 69 percent of those Indonesians and Nigerians surveyed reported viewing the United States favorably, just 18 percent of those polled in Egypt, Jordan, Pakistan, and Turkey (all U.S.

allies) did. And in 2006, when the Pew survey last asked detailed questions about Muslim-Western relations, more than half of the respondents in Muslim countries characterized those relations as bad and blamed the West for this state of affairs.

But improving relations is all the more important because of the growing demographic weight of poor Muslim countries and the attendant increase in Muslim immigration, especially to Europe from North Africa and the Middle East. (To be sure, forecasts that Muslims will soon dominate Europe are outlandish: Muslims compose just three to ten percent of the population in the major European countries today, and this proportion will at most double by midcentury.) Strategists worldwide must consider that the world's young are becoming concentrated in those countries least prepared to educate and employ them, including some Muslim states. Any resulting poverty, social tension, or ideological radicalization could have disruptive effects in many corners of the world. But this need not be the case; the healthy immigration of workers to the developed world and the movement of capital to the developing world, among other things, could lead to better results.

Urban Sprawl

Exacerbating twenty-first-century risks will be the fact that the world is urbanizing to an unprecedented degree. The year 2010 will likely be the first time in history that a majority of the world's people live in cities rather than in the countryside. Whereas less than 30 percent of the world's population was urban in 1950, according to UN projections, more than 70 percent will be by 2050.

Lower-income countries in Asia and Africa are urbanizing especially rapidly, as agriculture becomes less labor intensive and as employment opportunities shift to the industrial and service sectors. Already, most of the world's urban agglomerations—Mumbai (population 20.1 million), Mexico City (19.5 million), New Delhi (17 million), Shanghai (15.8 million), Calcutta (15.6 million), Karachi (13.1 million), Cairo (12.5 million), Manila (11.7 million), Lagos (10.6 million), Jakarta (9.7 million)—are found in low-income countries. Many of these countries have multiple cities with over one million residents each: Pakistan has eight, Mexico 12, and China more than 100. The UN projects that the urbanized proportion of sub-Saharan Africa will nearly double between 2005 and 2050, from 35 percent (300 million people) to over 67 percent (1 billion). China, which is roughly 40 percent urbanized today, is expected to be 73 percent urbanized by 2050; India, which is less than 30 percent urbanized today, is expected to be 55 percent urbanized by 2050. Overall, the world's urban population is expected to grow by 3 billion people by 2050.

This urbanization may prove destabilizing. Developing countries that urbanize in the twenty-first century will have far lower per capita incomes than did many industrial countries when they first urbanized. The United States, for example, did not reach 65 percent urbanization until 1950, when per capita income was nearly $13,000 (in 2005 dollars). By contrast, Nigeria, Pakistan, and the Philippines, which are approaching similar levels of urbanization, currently have per capita incomes of just $1,800–$4,000 (in 2005 dollars).

According to the research of Richard Cincotta and other political demographers, countries with younger populations are especially prone to civil unrest and are less able to create or sustain democratic institutions. And the more heavily urbanized, the more such countries are likely to experience Dickensian poverty and anarchic violence. In good times, a thriving economy might keep urban residents employed and governments flush with sufficient resources to meet their needs. More often, however, sprawling and impoverished cities are vulnerable to crime lords, gangs, and petty rebellions. Thus, the rapid urbanization of the developing world in the decades ahead might bring, in exaggerated form, problems similar to those that urbanization brought to nineteenth-century Europe. Back then, cyclical employment, inadequate policing, and limited sanitation and education often spawned widespread labor strife, periodic violence, and sometimes—as in the 1820s, the 1830s, and 1848—even revolutions.

International terrorism might also originate in fast-urbanizing developing countries (even more than it already does). With their neighborhood networks, access to the Internet and digital communications technology, and concentration of valuable targets, sprawling cities offer excellent opportunities for recruiting, maintaining, and hiding terrorist networks.

Defusing the Bomb

Averting this century's potential dangers will require sweeping measures. Three major global efforts defused the population bomb of Ehrlich's day: a commitment by governments and nongovernmental organizations to control reproduction rates; agricultural advances, such as the green revolution and the spread of new technology; and a vast increase in international trade, which globalized markets and thus allowed developing countries to export foodstuffs in exchange for seeds, fertilizers, and machinery, which in turn helped them boost production. But today's population bomb is the product less of absolute growth in the world's population than of changes in its age and distribution. Policymakers must therefore adapt today's global governance institutions to the new realities of the aging of the industrialized world, the concentration of the world's economic and population growth in developing countries, and the increase in international immigration.

During the Cold War, Western strategists divided the world into a "First World," of democratic industrialized countries;

a "Second World," of communist industrialized countries; and a "Third World," of developing countries. These strategists focused chiefly on deterring or managing conflict between the First and the Second Worlds and on launching proxy wars and diplomatic initiatives to attract Third World countries into the First World's camp. Since the end of the Cold War, strategists have largely abandoned this three-group division and have tended to believe either that the United States, as the sole superpower, would maintain a Pax Americana or that the world would become multipolar, with the United States, Europe, and China playing major roles.

Unfortunately, because they ignore current global demographic trends, these views will be obsolete within a few decades. A better approach would be to consider a different three-world order, with a new First World of the aging industrialized nations of North America, Europe, and Asia's Pacific Rim (including Japan, Singapore, South Korea, and Taiwan, as well as China after 2030, by which point the one-child policy will have produced significant aging); a Second World comprising fast-growing and economically dynamic countries with a healthy mix of young and old inhabitants (such as Brazil, Iran, Mexico, Thailand, Turkey, and Vietnam, as well as China until 2030); and a Third World of fast-growing, very young, and increasingly urbanized countries with poorer economies and often weak governments. To cope with the instability that will likely arise from the new Third World's urbanization, economic strife, lawlessness, and potential terrorist activity, the aging industrialized nations of the new First World must build effective alliances with the growing powers of the new Second World and together reach out to Third World nations. Second World powers will be pivotal in the twenty-first century not just because they will drive economic growth and consume technologies and other products engineered in the First World; they will also be central to international security and cooperation. The realities of religion, culture, and geographic proximity mean that any peaceful and productive engagement by the First World of Third World countries will have to include the open cooperation of Second World countries.

Strategists, therefore, must fundamentally reconsider the structure of various current global institutions. The G-8, for example, will likely become obsolete as a body for making global economic policy. The G-20 is already becoming increasingly important, and this is less a short-term consequence of the ongoing global financial crisis than the beginning of the necessary recognition that Brazil, China, India, Indonesia, Mexico, Turkey, and others are becoming global economic powers. International institutions will not retain their legitimacy if they exclude the world's fastest-growing and most economically dynamic countries. It is essential, therefore, despite European concerns about the potential effects on immigration, to take steps such as admitting Turkey into the European Union. This would add youth and economic dynamism to the EU—and would prove that Muslims are welcome to join Europeans as equals in shaping a free and prosperous future. On the other hand, excluding Turkey from the EU could lead to hostility not only on the part of Turkish citizens, who are expected to number 100 million by 2050, but also on the part of Muslim populations worldwide.

NATO must also adapt. The alliance today is composed almost entirely of countries with aging, shrinking populations and relatively slow-growing economies. It is oriented toward the Northern Hemisphere and holds on to a Cold War structure that cannot adequately respond to contemporary threats. The young and increasingly populous countries of Africa, the Middle East, Central Asia, and South Asia could mobilize insurgents much more easily than NATO could mobilize the troops it would need if it were called on to stabilize those countries. Long-standing NATO members should, therefore—although it would require atypical creativity and flexibility—consider the logistical and demographic advantages of inviting into the alliance countries such as Brazil and Morocco, rather than countries such as Albania. That this seems far-fetched does not minimize the imperative that First World countries begin including large and strategic Second and Third World powers in formal international alliances.

The case of Afghanistan—a country whose population is growing fast and where NATO is currently engaged—illustrates the importance of building effective global institutions. Today, there are 28 million Afghans; by 2025, there will be 45 million; and by 2050, there will be close to 75 million. As nearly 20 million additional Afghans are born over the next 15 years, NATO will have an opportunity to help Afghanistan become reasonably stable, self-governing, and prosperous. If NATO's efforts fail and the Afghans judge that NATO intervention harmed their interests, tens of millions of young Afghans will become more hostile to the West. But if they come to think that NATO's involvement benefited their society, the West will have tens of millions of new friends. The example might then motivate the approximately one billion other young Muslims growing up in low-income countries over the next four decades to look more kindly on relations between their countries and the countries of the industrialized West.

Creative Reforms at Home

The aging industrialized countries can also take various steps at home to promote stability in light of the coming demographic trends. First, they should encourage families to have more children. France and Sweden have had success providing child care, generous leave time, and financial allowances to families with young children. Yet there is no consensus among policymakers—and certainly not among demographers—about what policies best encourage fertility.

More important than unproven tactics for increasing family size is immigration. Correctly managed, population movement can benefit developed and developing countries alike. Given the dangers of young, underemployed, and unstable populations in developing countries, immigration to developed countries can provide economic opportunities for the ambitious and serve as a safety valve for all. Countries that embrace immigrants, such as the United States, gain economically by having willing laborers and greater entrepreneurial spirit. And countries with high levels of emigration (but not so much that they experience so-called brain drains) also benefit because emigrants often send remittances home or return to their native countries with valuable education and work experience.

One somewhat daring approach to immigration would be to encourage a reverse flow of older immigrants from developed to developing countries. If older residents of developed countries took their retirements along the southern coast of the Mediterranean or in Latin America or Africa, it would greatly reduce the strain on their home countries' public entitlement systems. The developing countries involved, meanwhile, would benefit because caring for the elderly and providing retirement and leisure services is highly labor intensive. Relocating a portion of these activities to developing countries would provide employment and valuable training to the young, growing populations of the Second and Third Worlds.

This would require developing residential and medical facilities of First World quality in Second and Third World countries. Yet even this difficult task would be preferable to the status quo, by which low wages and poor facilities lead to a steady drain of medical and nursing talent from developing to developed countries. Many residents of developed countries who desire cheaper medical procedures already practice medical tourism today, with India, Singapore, and Thailand being the most common destinations. (For example, the international consulting firm Deloitte estimated that 750,000 Americans traveled abroad for care in 2008.)

Never since 1800 has a majority of the world's economic growth occurred outside of Europe, the United States, and Canada. Never have so many people in those regions been over 60 years old. And never have low-income countries' populations been so young and so urbanized. But such will be the world's demography in the twenty-first century. The strategic and economic policies of the twentieth century are obsolete, and it is time to find new ones.

Reference

Goldstone, Jack A. "The new population bomb: the four megatrends that will change the world." *Foreign Affairs* 89.1 (2010): 31. *General OneFile*. Web. 23 Jan. 2010. http://0-find.galegroup .com.www.consuls.org/gps/start.do?proId=IPS& userGroupName=a30wc.

Critical Thinking

1. Using the websites below, develop a comparison of demographic trends between two different regions of the world.
2. Compare the projected demographic makeup of the United States in 2050 with China, Mexico, Pakistan, and Russia.

Create Central

www.mhhe.com/createcentral

Internet References

INED (French Institute for Demographic Studies)
www.ined.fr/en/everything_about_population
PRB World Population
www.prb.org/Publications/Datasheets/2011/world-population-data-sheet/world-map.aspx#/map/population
Worldmapper
www.worldmapper.org

Article   Prepared by: Robert Weiner, *University of Massachusetts, Boston*

Climate Change Politics on the Road to Paris

Lorraine Elliott

Learning Outcomes

After reading this article, you will be able to:

- Discuss the actions that have been taken by developed and developing countries to deal with climate change.

- Understand the issues involved in climate justice.

2015 will be a crucial year for global climate change politics. The Fifth Assessment Report of the Intergovernmental Panel on Climate Change (IPCC), released in 2014, confirmed that warming of the climate system is unequivocal and that the role of human influence in that trend is clear. By the end of the year in Paris, at the 21st Conference of the Parties (COP) to the United Nations Framework Convention on Climate Change (UNFCCC), 195 governments plus the European Union are expected to adopt a new, legally binding agreement to replace the 1997 Kyoto Protocol. There will be at least three inter-sessional negotiating sessions between now and then.

Assuming that the parties meet their Paris objective (though almost nothing can be certain in climate diplomacy), this agreement on what UN Secretary-General Ban Ki-moon has called the defining issue of our age would not have to come into legal force until 2020. That will be almost 30 years after governments adopted the UNFCCC, in which they promised to prevent dangerous anthropogenic interference with the climate system and to protect that system for the benefit of present and future generations. If anything characterizes climate diplomacy, it is that haste is made very, very slowly.

The trajectory of COPs, MOPs (Meetings of Parties), inter-sessional consultations, ad hoc working groups, and subsidiary bodies on the way to Paris began seven years ago with the adoption of the Bali Roadmap and Bali Action Plan at the 13th COP

in 2007. The excessive ambition of that meeting—to have a new climate agreement in place within two years—was exposed at Copenhagen in 2009. Although the Copenhagen COP did succeed in getting in-principal financial commitments to a Green Climate Fund, it is best known for its lack of consensus on any new agreement. The final political statement—rather misleadingly called the Copenhagen Accord—could only be "noted" rather than adopted. It was negotiated behind closed doors by a self-selected group of heads of state and government. Countries excluded from the small group objected, some quite vociferously, to this "plurilateral" model that ran counter to the UN norm of inclusive multilateralism.

Halting Steps

To be fair, the 2015 Paris agreement (or whatever it might be called) can build on some collective action successes since 2009. Over this period, literally hundreds of other climate change meetings have been held and arrangements adopted at local, municipal, national, sub-regional, and regional levels, either adjacent to or independently of the UNFCCC process. Within the UNFCCC, parties have adopted the Cancun Agreements, the Durban Platform for Enhanced Action, and the Doha Climate Gateway, along with the Warsaw International Mechanism for Loss and Damage and the Warsaw Framework for REDD+ (Reducing Emissions from Deforestation and Forest Degradation plus strategies for forest conservation and sustainable management). The latest variant is the Lima Call for Climate Action, which was adopted in December after tense and extended debate at the 20th COP in Peru.

A small number of developed country governments submitted binding individual emissions reduction targets for a second budget period under the Kyoto Protocol (2012 to December 2020), although together they account for only a small proportion of global greenhouse emissions (about 14 percent). A larger number of governments, including some representing major developing-country economies, have made voluntary commitments. Governments have also committed to providing fast-track finance, and to long-term targets of $100 billion a year by 2020 to support mitigation and adaptation actions, particularly in those countries whose economies and people are most vulnerable to both the sudden-event and slow-onset impacts of climate change.

Some progress has been made on a whole range of technical issues: monitoring, verification, and reporting; the Technology Mechanism designed to facilitate the development and transfer of technology for mitigation and adaptation; Nationally Appropriate Mitigation Actions for developing countries; and action to address loss and damage. One of the most important outcomes was the commitment at Cancun in 2010 that climate change actions under the Paris agreement should be of a magnitude that will limit increases in average global temperature to no more than 2 degrees Celsius above pre-industrial levels. And in late 2014, following months of discussions, the US and Chinese governments made a surprise joint announcement in which they committed to ambitious goals on emissions reductions and caps.

But generally, this road to Paris has followed the pattern of the past 30 or so years of climate diplomacy: moments of high acclaim and last-minute diplomatic breakthroughs, followed by more talks about talks, delegation walkouts, constant struggles over technical and procedural details, and regular (some might say monotonous) declarations that climate change is one of the greatest challenges of our time. The so-called Kyoto era has been no different. The COP in Lima codified some of the elements of a new agreement, amid claims that the world was now on a firmer footing in the run-up to Paris. But many of those elements had already been agreed informally, and many points of difference remain.

Notwithstanding the US-China cooperation pact, negotiations this year will continue in an environment of fragile trust between developing and developed countries, especially since some of the latter have withdrawn from the Kyoto Protocol or downsized their voluntary commitments. Industrialized countries, among them the leading global emitters on per capita and historical bases (and often in terms of gross emissions as well), demand commitments from developing countries, particularly the so-called emerging economies whose emissions are increasing. But this is not a simple matter. While China is often labeled the world's largest gross emitter, it still lags well behind many developed countries and some developing nations in terms of its per capita emissions and historical contributions to greenhouse gas concentrations. China and other developing countries, through the Group of 77, therefore continue to argue that any new agreement must be true to the principles of common but differentiated responsibilities and respective capabilities embodied in the UNFCCC and in the Kyoto Protocol. Under those agreements, industrialized countries committed to taking the lead on climate change mitigation on the basis of their historical and current contributions to emissions and concentrations (flows and stocks) and because of the financial and technological resources that they command.

A focus on threat rather than vulnerability emphasizes symptoms instead of causes.

Commitment Gap

What, then, are the big issues for 2015? Progress on one of the biggest—the level of greenhouse gas mitigation commitments to meet the 2-degree threshold—has been too slow. Indeed, it has been manifestly inadequate. The IPCC has called for a reduction in carbon emissions to almost zero by the end of the current century. A review by the United Nations Environment Program of the mitigation gaps arising from informal Copenhagen commitments suggests that the 2 degrees Celsius target will not be reached, and anticipates an increase as high as 5 degrees if targets and emissions are not tightly controlled.

In a change from the top-down approach of the Kyoto Protocol, the Paris process requires governments to submit indicative Intended Nationally Determined Contributions (INDCs) for the new agreement by the end of March this year. The term "contribution" is contentious. Developed countries interpret it to mean mitigation targets only. Developing countries have argued that it should entail bundles of commitments including adaptation support and technological and financial assistance for their transitions to low-carbon economies. Until the INDCs are submitted, no calculations can be done to determine if they will be sufficient collectively. That is unlikely, which means that more negotiation will be required amid robust defense of national interests.

Governments have long accepted that adaptation is an integral component of global actions to manage the impacts of climate change. It is especially crucial for those countries and people who are already most vulnerable to the consequences of climate change—including food insecurity, water deprivation, loss of livelihoods, increasing health burdens, and the threat of displacement—but who have the fewest resources to respond or adapt to those impacts. Although there is now an Adaption Framework (set at Cancun in 2010) and an Adaptation Committee (Durban, 2011), significant gaps remain in the understanding and implementation of adaptation strategies.

There are also limits to adaptation, particularly but not exclusively in low-lying island states and in the world's most arid regions. As a result of both disaster and slow-onset impacts of climate change, people and communities will experience irreversible economic shocks, infrastructure damage, and loss of productive lands. In the most extreme cases, countries will become uninhabitable. There has been a tendency to securitize these issues in order to mobilize action—to focus on the potential for social disruption and political violence in the face of internal displacement and competition for scarce resources, or the need to protect borders against forced mobility and migration. The danger is that a focus on threat rather than vulnerability emphasizes symptoms instead of causes and overlooks the human security dimensions of climate change.

Climate Justice

Funding for mitigation and adaptation is expected to be part of the Paris package. But there is still little certainty on predictable and adequate sources of finance to meet the $100 billion a year target for 2020 and beyond, despite pledges made by a number of governments at the UN secretary-general's 2014 Climate Summit and since. Much more needs to be done to help those affected by climate change adapt to its impacts and deal with loss and damage. In part, these are issues about strengthening capacity and governance within recipient countries and communities, linked to global discussions on development assistance and new sustainable development goals. But when those who are most affected by climate change are those who have contributed least to its causes, these become matters that also go to the heart of global equity and climate justice.

The climate-change challenges ahead require sustained political momentum, effective institutional and ethical leadership, and real action on a long-term global transformation to a low-carbon economy. If these are not forthcoming—and they have been in somewhat short supply over the past three decades of climate change negotiations—it is hard to be optimistic that there will be a meaningful, effective, equitable, and binding successor to the Kyoto Protocol adopted at the Paris COP in December 2015.

Critical Thinking

1. How do national interests affect efforts to reach a new agreement on climate change?
2. What are the major differences between developed and developing countries on climate change?
3. What are the limits of adaptation strategies?

Internet References

Climate Summit
   http:www.un.org/climatechange/summit
UN Framework Convention on Climate Change
   http://unfccc.int/2860.php
University of Massachusetts/Boston Center for Governance and Sustainability
   http://www.umb.edu/cgs

LORRAINE ELLIOTT is a professor of international relations and a public policy fellow at Australian National University.

Article   Prepared by: Robert Weiner, *University of Massachusetts, Boston*

Welcome to the Revolution: Why Shale Is the Next Shale

Edward L. Morse

Learning Outcomes

After reading this article, you will be able to:

- Understand what the shale revolution is all about.

- Gain an insight into the relationship between the shale revolution and the global oil and natural gas situation.

Despite its doubters and haters, the shale revolution in oil and gas production is here to stay. In the second half of this decade, moreover, it is likely to spread globally more quickly than most think. And all of that is, on balance, a good thing for the world.

The recent surge of U.S. oil and natural gas production has been nothing short of astonishing. For the past 3 years, the United States has been the world's fastest-growing hydrocarbon producer, and the trend is not likely to stop anytime soon. U.S. natural gas production has risen by 25 percent since 2010, and the only reason it has temporarily stalled is that investments are required to facilitate further growth. Having already outstripped Russia as the world's largest gas producer, by the end of the decade, the United States will become one of the world's largest gas exporters, fundamentally changing pricing and trade patterns in global energy markets. U.S. oil production, meanwhile, has grown by 60 percent since 2008, climbing by three million barrels a day to more than eight million barrels a day. Within a couple of years, it will exceed its old record level of almost 10 million barrels a day as the United States overtakes Russia and Saudi Arabia and becomes the world's largest oil producer. And U.S. production of natural gas liquids, such as propane and butane, has already grown by one million barrels per day and should grow by another million soon.

What is unfolding in reaction is nothing less than a paradigm shift in thinking about hydrocarbons. A decade ago, there was a near-global consensus that U.S. (and, for that matter, non-OPEC) production was in inexorable decline. Today, most serious analysts are confident that it will continue to grow. The growth is occurring, to boot, at a time when U.S. oil consumption is falling. (Forget peak oil production; given a combination of efficiency gains, environmental concerns, and substitution by natural gas, what is foreseeable is peak oil demand.) And to cap things off, the costs of finding and producing oil and gas in shale and tight rock formations are steadily going down and will drop even more in the years to come.

The evidence from what has been happening is now overwhelming. Efficiency gains in the shale sector have been large and accelerating and are now hovering at around 25 percent per year, meaning that increases in capital expenditures are triggering even more potential production growth. It is clear that vast amounts of hydrocarbons have migrated from their original source rock and become trapped in shale and tight rock, and the extent of these rock formations, like the extent of the original source rock, is enormous—containing resources far in excess of total global conventional proven oil reserves, which are 1.5 trillion barrels. And there are already signs that the technology involved in extracting these resources is transferable outside the United States, so that its international spread is inevitable.

In short, it now looks as though the first few decades of the 21st century will see an extension of the trend that has persisted for the past few millennia: the availability of plentiful energy at ever-lower cost and with ever-greater efficiency, enabling major advances in global economic growth.

Why the Past Is Prologue

The shale revolution has been very much a "made in America" phenomenon. In no other country can landowners also own mineral rights. In only a few other countries (such as Australia,

Canada, and the United Kingdom) is there a tradition of an energy sector featuring many independent entrepreneurial companies, as opposed to a few major companies or national champions. And in still fewer countries are there capital markets able and willing to support financially risky exploration and production.

This powerful combination of indigenous factors will continue to drive U.S. efforts. A further 30 percent increase in U.S. natural gas production is plausible before 2020, and from then on, it should be possible to maintain a constant or even higher level of production for decades to come. As for oil, given the research and development now under way, it is likely that U.S. production could rise to 12 million barrels per day or more in a few years and be sustained there for a long time. (And that figure does not include additional potential output from deepwater drilling, which is also seeing a renaissance in investment.)

Two factors, meanwhile, should bring prices down for a long time to come. The first is declining production costs, a consequence of efficiency gains from the application of new and growing technologies. And the second is the spread of shale gas and tight oil production globally. Together, these suggest a sustainable price of around $5.50 per 1,000 cubic feet for natural gas in the United States and a trading range of $70–90 per barrel for oil globally by the end of this decade.

These trends will provide a significant boost to the U.S. economy. Households could save close to $30 billion annually in electricity costs by 2020, compared to the U.S. Energy Information Administration's current forecast. Gasoline costs could fall from an average of 5 to 3 percent of real disposable personal income. The price of gasoline could drop by 30 percent, increasing annual disposable income by $750, on average, per driving household. The oil and gas boom could add about 2.8 percent in cumulative GDP growth by 2020 and bolster employment by some three million jobs.

Beyond the United States, the spread of shale gas and tight oil exploitation should have geopolitically profound implications. There is no longer any doubt about the sheer abundance of this new accessible resource base, and that recognition is leading many governments to accelerate the delineation and development of commercially available resources. Countries' motivations are diverse and clear. For Saudi Arabia, which is already developing its first power plant using indigenous shale gas, the exploitation of its shale resources can free up more oil for exports, increasing revenues for the country as a whole. For Russia, with an estimated 75 billion barrels of recoverable tight oil (50 percent more than the United States), production growth spells more government revenue. And for a host of other countries, the motivations range from reducing dependence on imports to increasing export earnings to enabling domestic economic development.

Risky Business?

Skeptics point to three problems that could lead the fruits of the revolution to be left to wither on the vine: environmental regulation, declining rates of production, and drilling economics. But none is likely to be catastrophic.

Hydraulic fracturing, or "fracking"—the process of injecting sand, water, and chemicals into shale rocks to crack them open and release the hydrocarbons trapped inside—poses potential environmental risks, such as the draining or polluting of underground aquifers, the spurring of seismic activity, and the spilling of waste products during their aboveground transport. All these risks can be mitigated, and they are in fact being addressed in the industry's evolving set of best practices. But that message needs to be delivered more clearly, and best practices need to be implemented across the board, in order to head off local bans or restrictive regulation that would slow the revolution's spread or minimize its impact.

As for declining rates of production, fracking creates a surge in production at the beginning of a well's operation and a rapid drop later on, and critics argue that this means that the revolution's purported gains will be illusory. But there are two good reasons to think that high production will continue for decades rather than years. First, the accumulation of fracked wells with a long tail of production is building up a durable base of flows that will continue over time, and second, the economics of drilling work in favor of drilling at a high and sustained rate of production.

Finally, some criticize the economics of fracking, but these concerns have been exaggerated. It is true that through 2013, the upstream sector of the U.S. oil and gas industry has been massively cash-flow negative. In 2012, for example, the industry spent about $60 billion more than it earned, and some analysts believe that such trends will continue. But the costs were driven by the need to acquire land for exploration and to pursue unproductive drilling in order to hold the acreage. Now that the land-grab days are almost over, the industry's cash flow should be increasingly positive.

It is also true that traditional finding and development costs indicate that natural gas prices need to be above $4 per 1,000 cubic feet and oil prices above $70 per barrel for the economics of drilling to work—which suggests that abundant production might drive prices down below what is profitable. But as demand grows for natural gas—for industry, residential and commercial space heating, the export market, power generation, and transportation—prices should rise to a level that can sustain increased drilling: the $5–6 range, which is about where prices were this past winter. Efficiency gains stemming from new technology, meanwhile, are driving down break-even drilling costs. In the oil sector, most drilling now brings an adequate return on investment at prices below $50 per barrel, and within a few years, that level could be under $40 per barrel.

Think Globally

Since shale resources are found around the globe, many countries are trying to duplicate the United States' success in the sector, and it is likely that some, and perhaps many, will succeed. U.S. recoverable shale resources constitute only about 15 percent of the global total, and so if the true extent and duration of even the U.S. windfall are not yet measurable, the same applies even more so for the rest of the world. Many countries are already taking early steps to develop their shale resources, and in several, the results look promising. It is highly likely that Australia, China, Mexico, Russia, Saudi Arabia, and the United Kingdom will see meaningful production before the end of this decade. As a result, global trade in energy will be dramatically disrupted.

A few years ago, hydrocarbon exports from the United States were negligible. But by the start of 2013, oil, natural gas, and petrochemicals had become the single largest category of U.S. exports, surpassing agricultural products, transportation equipment, and capital goods. The shift in the U.S. trade balance for petroleum products has been stunning. In 2008, the United States was a net importer of petroleum products, taking in about two million barrels per day; by the end of 2013, it was a net exporter, with an outflow of more than two million barrels per day. By the end of 2014, the United States should overtake Russia as the largest exporter of diesel, jet fuel, and other energy products, and by 2015, it should overtake Saudi Arabia as the largest exporter of petrochemical feedstocks. The U.S. trade balance for oil, which in 2011 was −$354 billion, should flip to +$5 billion by 2020.

By then, the United States will be a net exporter of natural gas, on a scale potentially rivaling both Qatar and Russia, and the consequences will be enormous. The U.S. gas trade balance should shift from −$8 billion in 2013 to +$14 billion by 2020. U.S. pipeline exports to Mexico and eastern Canada are likely to grow by 400 percent, to eight billion cubic feet per day, by 2018, and perhaps to 10 billion by 2020. U.S. exports of liquefied natural gas (lng) look likely to reach nine billion cubic feet per day by 2020.

Sheer volume is important, but not as much as two other factors: the pricing basis and the amount of natural gas that can be sold in a spot market. Most LNG trade links the price of natural gas to the price of oil. But the shale gas revolution has delinked these two prices in the United States, where the traditional 7:1 ratio between oil and gas prices has exploded to more than 20:1. That makes lng exports from the United States competitive with LNG exports from Qatar or Russia, eroding the oil link in LNG pricing. What's more, traditional LNG contracts are tied to specific destinations and prohibit trading. U.S. LNG (and likely also new LNG from Australia and Canada) will not come with anticompetitive trade restrictions, and so a spot market should emerge quickly. And U.S. LNG exports to Europe should erode the Russian state oil company Gazprom's pricing hold on the continent, just as they should bring down prices of natural gas around the world.

In the geopolitics of energy, there are always winners and losers. OPEC will be among the latter, as the United States moves from having had a net hydrocarbon trade deficit of some nine million barrels per day in 2007, to having one of under six million barrels today, to enjoying a net positive position by 2020. Lost market share and lower prices could pose a devastating challenge to oil producers dependent on exports for government revenue. Growing populations and declining per capita incomes are already playing a central role in triggering domestic upheaval in Iraq, Libya, Nigeria, and Venezuela, and in that regard, the years ahead do not look promising for those countries.

At the same time, the U.S. economy might actually start approaching energy independence. And the shale revolution should also lead to the prevalence of market forces in international energy pricing, putting an end to OPEC's 40-year dominance, during which producers were able to band together to raise prices well above production costs, with negative consequences for the world economy. When it comes to oil and natural gas, we now know that though much is taken, much abides—and the shale revolution is only just getting started.

Critical Thinking

1. Will the shale revolution allow the United States to become energy independent?

2. What are some of the environmental drawbacks associated with the mining of shale?

3. What will be the effect of a glut of oil on the energy market as a result of the shale revolution?

Create Central

www.mhhe.com/createcentral

Internet References

Fracking's Future
http://harvardmagazine.com/2013/01/frackings.future

Hydraulic Fracturing
http://www.dangersoffracking.com

National Renewable Energy Laboratory
http://www.nrel.gov

International Energy Agency
http://www.iea.org

Organization of Petroleum Exporting Countries
http://www.opec_web/en

EDWARD L. MORSE is Global Head of Commodities Research at Citi.

Article   Prepared by: Robert Weiner, *University of Massachusetts, Boston*

Think Again: Climate Treaties

DAVID SHORR

Learning Outcomes

After reading this article, you will be able to:

- Explain what is meant by climate change.

- Understand what causes climate change.

Time is running short for the international community to tackle climate change.

Pressure to act comes from rising temperatures and sea levels, superstorms, brutal droughts, and diminishing food crops. It also comes from fears that these problems are going to get worse. Modern economies have already boosted the concentration of carbon dioxide (CO_2) in the atmosphere by 40 percent since the Industrial Revolution. If the world stays on its current course, CO_2 levels could double by century's end, potentially raising global temperatures several more degrees. (The last time the planet's CO_2 levels were so high was 15 million years ago, when temperatures were 5 to 10 °F higher than they are today.)

Another source of pressure, however, is self-imposed. Under the auspices of the United Nations, the next global climate treaty—to be negotiated among some 200 countries, with the central goal of cutting greenhouse gas emissions—should be enacted in 2015, to replace the now-outmoded 1997 Kyoto Protocol. (Once passed by state parties, the new treaty would actually go into effect in 2020.)

The race against both nature and the diplomatic clock is stressful. But in the rush to do something, the international community—most notably, and ironically, those individuals and organizations most fervent about combating global warming—is often doing the wrong thing. It has become fixated on the notion of consensus codified in international law.

The U.N. process for climate diplomacy has been in place for more than two decades, punctuated since 1995 by annual meetings at which countries assess global progress in protecting the environment and negotiate treaties and other agreements to keep the ball rolling. Kyoto was finalized at the third such conference. A milestone, it established targets for country-based emissions cuts. Its signal failure, however, was leaving the world's three largest emitters of greenhouse gases unconstrained, two of them by design. Kyoto gave developing countries, including China and India, a blanket exemption from cutting emissions. Meanwhile, the United States bristled at its obligations—particularly in light of the free pass given to China and India—and refused to ratify the treaty.

Still, Kyoto was lauded by many because it was a legally binding accord, a high bar to clear in international diplomacy. The agreement's provisions were compulsory for countries that ratified it; violating them would invite a stigma—a reputation for weaseling out of promises deemed essential to saving the planet.

Today, the principle of "if you sign it, you stick to it" continues to guide a lot of conventional thinking about climate diplomacy, particularly among the political left and international NGOs, which have been driving forces of U.N. climate negotiations, and among leaders of developing countries that are not yet major polluters but are profoundly affected by global warming. For instance, in the lead-up to the last annual U.N. climate conference—held in Warsaw, Poland, in November 2013, Oxfam International's executive director, Winnie Byanyima, said the world should not accept a successor agreement to Kyoto that has anything less than the force of international law: "Of course not . . . . If it's not legally binding, then what is it?" Ultimately, Byanyima and other civil society leaders walked out of the conference to protest what they viewed as a failure to take steps toward a new, ironclad treaty.

The frustration in Warsaw showed an ongoing failure among many staunch advocates of climate diplomacy to learn the key lesson of Kyoto: Legal force is the wrong litmus test for judging an international framework. Idealized multilateralism has become a trap. It only leads to countries agreeing to the lowest common denominator—or balking altogether.

Evidence shows that a drive for the tightest possible treaty obligations has the perverse effect of provoking resistance. In a seminal 2011 study of climate diplomacy, David Victor of the University of California, San Diego, concluded, "The very attributes that made targets and timetables so attractive to environmentalists—that they set clear, binding goals without much attention to cost—made the Kyoto treaty brittle because countries that discovered they could not honor their commitments had few options but to exit."

This argument may sound like one made by many political conservatives, who opposed Kyoto and have long been wary of treaties in general. But the point is not that international efforts are useless. It is that global agreements are most useful when they include a healthy measure of realism in the demands that they make of countries. Instead of insisting on a binding agreement, diplomats must identify what governments and other actors, like the private sector, are willing to do to combat global warming and develop mechanisms to choreograph, incentivize, and monitor them as they do it. Otherwise, U.N. talks will remain a dialogue of the deaf, as the Earth keeps cooking.

To explain multilateralism's recent failures, from the Kyoto Protocol to the Warsaw conference, its most fervent advocates often take aim at the same purported stumbling block: the spinelessness of politicians. Fainthearted presidents and prime ministers shy away from commitments to protect the planet because it is more politically expedient to focus on economic growth, no matter the environmental consequences.

Thanks to this conventional wisdom, "political will" has become a loaded term. If a leader doesn't sign on to a tough, legally binding treaty, he or she must be morally bankrupt. Mary Robinson, a former president of Ireland who now runs a foundation dedicated to climate change issues, has called the "legal character" of climate agreements "an expression of or an extension of political will." Meanwhile, Kumi Naidoo, executive director of Greenpeace International, has written that he hopes governments will "find the political will to act beyond short-sighted electoral cycles and the corrupting influence of some business elites."

The fallacy of the political will argument, however, is that it assumes everyone already agrees on the steps necessary to address climate change and that the only remaining task is follow-through. It is true that the weight of scientific evidence tells us humanity can only spew so many more gigatons of CO_2 into the air before subjecting the planet and its inhabitants to dire consequences. But the only guidance this gives policymakers is that they must transition to low-carbon economies, stat. It does not tell them how they should do this or how they can do it most efficiently, with the least cost incurred. As a result, advocates

of strict climate treaties hammer home the imperative for environmental action without providing for discussion about how countries can actually transform their economies in practice.

Consider environmental author and activist Bill McKibben's comments in early 2013 praising Germany for using more renewable energy: "There were days last summer when Germany generated more than half the power it used from solar panels within its borders. What does that tell you about the relative role of technological prowess and political will in solving this?"

Unfortunately, it tells us very little. It doesn't tell us what it would take to stretch the reliance on solar energy beyond some sunny German days or the subsidy levels required to make solar power a more widely used energy source. It also tells us nothing about how we could translate Germany's accomplishments to countries with very different political and economic circumstances. And it doesn't explain what would induce those diverse countries to accept a multilateral arrangement boosting the global use of renewable energy. All McKibben's factoid tells us is that the myth of political will is quite powerful.

Certainly, economic imperatives should not override environmental ones. Yet the standard for climate diplomacy should not be broad appeals for boldness that ask policymakers to deny trade-offs rather than wrestle with them—particularly in the countries that the world needs most in the fight against global warming. Last fall, after the Warsaw meeting, many experts and pundits were quick to place blame for the gathering's tumult. "The India Problem: Why is it thwarting every international climate agreement?" a headline on Slate demanded. Other observers scorned India and China for saying they would not make "commitments" to greenhouse gas cuts in the 2015 climate agreement. (The meeting's attendees ultimately settled on the word "contributions.")

These complaints, however, are increasingly out of date.

It's true that, throughout most of the 2000s, China and India clung to the exemption that the Kyoto Protocol had granted them, arguing that the industrialized world had caused global warming and that developing countries shouldn't be deprived of their own chance to prosper. This has induced great anxiety because, since 2005, China's annual share of CO_2 emissions has grown from around 16 percent to more than 25 percent, while India has emerged as the world's third-largest carbon emitter. In short, without China and India, progress on climate change will be virtually impossible.

By 2010, however, Beijing and New Delhi had begun to change their stance. A desire to save face diplomatically, combined with increasing pollution at home and domestic need for energy efficiency, have made China and India more willing to cut emissions than ever before.

Chinese leaders in particular are eager to recast their country as an environmental paragon, rather than a pariah. Some analysts attribute this shift to China's aspirations to global prominence. Playing off the popular idea of the "Chinese century," Robert Stavins, director of the Harvard Project on Climate Agreements, has said, "If it's your century, you don't obstruct—you lead." Recently, China has taken significant steps forward with green energy, mimicking many of the regulations and mandates that have helped the United States achieve environmental progress. Wind, solar, and hydroelectric power now provide one-quarter of China's electricity-generating capacity. More energy is being added to China's grid each year from clean sources than from fossil fuels. And in a show of its willingness to step up to the diplomatic plate, China signed an accord with the United States in 2013 that scales down emissions of hydrofluorocarbons, which are so-called super-greenhouse gases.

Yet these changes have not substantially bent the curve of China's total emissions. According to Chris Nielsen and Mun Ho of Harvard University's China Project, this is largely because the country's rapid economic growth makes the tools that have slowed emissions in other economies less effective in China: "[T]he unprecedented pace of China's economic transformation makes improving China's air quality a moving target." Ultimately, Nielsen and Ho argue, the only way for China to rein in emissions will be to attach a price to carbon, through either a tax or a cap-and-trade system. As if on cue, China is now setting up municipal and provincial markets in which polluters can trade emissions credits, with the goal of creating a national market by 2016.

The point here is that the leaders of countries with rapidly developing economies cannot predict environmental payoffs with any real confidence. Tools that work well for others may not for them. That's why China and India are hesitant to sign legally binding treaties, which would put them on the hook to hit targets that could prove much harder to reach than anticipated. They don't want to undertake costly reforms that might not have the predicted benefits, and they do not want to risk the hefty criticism that failure to abide by a treaty would surely bring.

Chinese and Indian leaders realize they'll be judged by their contributions to a cleaner environment, and they embrace the challenge. (Recently in India, more than 20 major industry players launched an initiative to cut emissions.) And they are apt to be less guarded on the international stage if a new climate agreement functions as a measuring stick, not a bludgeon—much like the 2009 Copenhagen accord has done.

In December 2008, the U.N.'s annual cunate conference, hosted in Copenhagen, produced an agreement that is still roundly condemned by environmentalists, the leaders of developing countries, and political liberals alike. Unlike the Kyoto Protocol, the agreement let countries voluntarily set their own targets for emissions cuts over 10 years. "The city of Copenhagen is a crime scene tonight," the executive director of Greenpeace U.K. declared when the deal was reached. Lumumba Di-Aping, the chief negotiator for a group of developing countries known as the G-77, which had wanted major polluters like the United States to take greater responsibility for global warming, said the agreement had "the lowest level of ambition you can imagine."

In reality, however, the conference wasn't a fiasco. It offered the basis for a promising, more flexible regime for climate action that could be a model for the 2015 agreement.

The Copenhagen agreement had a number of advantages. It didn't have to be ratified by governments, which can delay implementation by years. Moreover, in an important new benchmark for climate negotiations, the agreement set the goal of preventing a global average temperature rise of more than 2 °C, with all countries' emission cuts to be gauged against that objective. This provision went to the heart of climate diplomacy's collective-action problem: Apportioning responsibility for cutting emissions among countries is always tricky, but the 2° target creates a shared definition of success.

Most importantly, however, the shift to voluntary pledges showed the first glimmers of lessons learned from the most common mistakes of climate negotiations. In the U.N. process, countries usually operate by consensus: They must all agree on each other's respective climate goals, a surefire recipe for dysfunction. (In 2010, the chair of annual climate talks refused to let a single delegation—Bolivia—block consensus, which counts as a daring move at U.N. conferences.)

Under Copenhagen, by contrast, countries can pledge to do their share while remaining within their comfort zones as dictated by circumstances back home. For instance, faced with economic imperatives to continue delivering high growth, China and India pledged at Copenhagen to reach targets pegged relative to carbon intensity (emissions per unit of economic output) rather than absolute levels of greenhouse gases. This was as far as they were willing to go—but it was further than they'd ever gone before.

Admittedly, the Copenhagen conference wasn't perfect. The deal was struck on the conference's tail end, after U.S. President Barack Obama barged in on a meeting already under way among the leaders of China, India, Brazil, and South Africa. Many of the other delegates registered outrage that the five leaders had negotiated a deal in private by having the conference merely "take note" of the accord.

But the following year's U.N. conference fleshed out the Copenhagen framework, and it has since gained enough legitimacy that 114 countries have agreed to the accord and another

27 have expressed their intention to agree. Taken together, this includes the world's 17 largest emitters, responsible for 80 percent of carbon-based pollution.

The Copenhagen accord will expire as the Kyoto successor agreement takes effect in 2020. But it shouldn't be viewed as just a stopgap. In giving governments more flexibility, Copenhagen offers the chance to build more confidence—and ambition—where historically there has only been uncertainty and rancor. Any future climate agreement should do the same.

"Countries Will Never Keep Mere Promises to Cut Emissions."

Never say never

The most obvious criticism of Copenhagen's system, of course, is that, while it is nice for countries to set voluntary goals, they will never meet them unless they are legally compelled to do so. That is why, just after the Copenhagen deal was reached, then-British Prime Minister Gordon Brown hastily said, "I know what we really need is a legally binding treaty as quickly as possible."

To date, there has been progress on meeting targets set under Copenhagen. The United States and the European Union, for instance, are all within reach of meeting their 10-year goals, perhaps even ahead of schedule. Meanwhile, China's pledge to cut carbon intensity, based on 2005 levels, has become the framework for the country's new emissions-trading markets.

But the most important reason to have confidence in the Copenhagen deal lies in its provisions for measurement, reporting, and verification. If done right, these so-called MRV mechanisms will alert the world as to how countries are (or are not) reducing greenhouse gases, while also pushing states to keep pace toward pledged cuts.

MRVs rely on peer pressure. Countries report to and monitor one another, tracking and urging progress. This kind of system has already proved effective in a variety of international policy areas. For instance, the Mutual Assessment Process of the G-20 and International Monetary Fund brings together the major economic powers to discuss whether their respective policies are helping to maximize global economic growth or are instead widening imbalances between export and consumer-based economies. The process is fairly new, but already, it is widely credited with prodding China long reluctant to discuss these issues in multilateral forums (sound familiar?)—to let its currency appreciate and to make boosting domestic consumption a main plank of its 5-year (2011–2015) plan.

MRVs have also proved valuable in narrower climate regimes, such as the European Union's cap-and-trade mechanism. As a 2012 Environmental Defense Fund report explained, "[B]ecause EU governments based the system's initial caps and emissions allowance allocation on estimates of regulated entities' emissions . . . governments issued too many emissions allowances (over-allocation). Now, however, caps are established on the basis of measured and verified past emissions and best-practices benchmarks, so over-allocation is less of a problem." In other words, MRVs have helped the European Union tighten market standards, correcting an earlier miscalculation and actually heightening the system's ambition.

The Copenhagen agreement enhanced the utility of global, climate-related MRVs by requiring greater transparency from developing countries. Under Kyoto, these countries were only required to provide a summary of their emissions for 2 years: a choice of either 1990 or 1994, and 2000. Copenhagen, by contrast, committed developing countries to report on their emissions biennially—the first reports are due in December—narrowing the gap with the requirement for annual reports that Kyoto imposed on developed countries.

Copenhagen's MRVs are not yet as strong as they could be. For instance, they should require annual reports from all countries, no matter their stages of development. These reports should also include a breakdown of information according to economic subsectors and different greenhouse gases, along with supporting details about data-collection methods. In addition, the process of reviewing reports needs to be fleshed out, taking cues from other strong MRVs that already exist, and wealthier countries should help underwrite the cost to developing countries of preparing comprehensive reports.

The good news is that, given the ongoing nature of U.N. climate diplomacy, it's still possible to strengthen Copenhagen's MRVs. Important new principles and guidelines for peer review have been established in negotiations since 2009, and those involved in climate diplomacy should now buckle down to finish the job. Robust MRVs would guarantee that the world makes the most of the next few years and draws on that experience to chart a new phase of climate action anchored in a 2015 agreement.

"Forget treaties. Solutions will come from the bottom up."

Don't get carried away

Some critics of the U.N. process, hailing from conservative political ranks, the private sector, and other areas, have lost all patience and think that a top-down process, particularly one negotiated in an international forum, is the wrong way to go. They point out that, while national leaders negotiated the Copenhagen deal, actual progress toward its goals is being cobbled together by actors at lower levels—in cities, states, markets, and industries. They are choosing which energy will generate electricity, honing farming practices, improving industrial efficiency, and the like.

Indeed, some policymakers and climate analysts point to the influence of local authorities as a game-changer for climate action. After all, Chinese cities and provinces have begun building emissions-trading markets, and California has passed a law establishing one of the most robust such markets in the world. Meanwhile, leaders of the world's megacities have banded together to cut emissions in what's known as the C40 group, established in 2005. As C40 chair and Rio de Janeiro Mayor Eduardo Paes has put it, "C40's networks and efforts on measurement and reporting are accelerating city-led action at a transformative scale around the world."

Given this sort of local progress, it is certainly worth asking whether diplomats and national policymakers should just get out of the way. Maybe a thoroughly bottom-up approach would be better for the planet than an international climate regime, no matter how flexible. David Hodgkinson, a law professor and executive director of the nonprofit EcoCarbon, which focuses on market solutions for reducing emissions, has argued that such an approach has "more substance" and "probably holds out more hope than a top-down UN deal."

Ultimately, however, this view is misguided. There is no substitute for high-level diplomacy in getting everyone to do their utmost and in keeping track of their efforts. In particular, as Copenhagen reminded the world, the value of the agenda setting, peer pressure, and leverage unique to international diplomacy shouldn't be overlooked. Moreover, we've seen in other policy spheres how the international community can first establish fundamental principles, which then sharpen over time with the aid of global coordinating bodies and more localized initiatives. For instance, the nonbinding 1948 Universal Declaration of Human Rights established a framework for a host of subsequent international treaties, U.N. agencies, regional charters and courts, national policies, and, more recently, corporate responsibility efforts.

Practically speaking, it would also be shortsighted to rely on an assortment of subnational actors to tackle a global problem like climate change. Determining how the work of these actors intersects, what it adds up to, and who monitors that sum are critical matters best managed from the top-down. As the goal of preventing a global average temperature rise of 2 °C reminds us, it is the aggregate of countries' reduced emissions that will be the ultimate test of success.

Even so, the status quo of climate talks, focused on badgering countries to join another legally binding treaty, represents diplomatic overreach. This hasn't worked in the past, and it won't in the future. The international community should give up the quest to sign a legally binding treaty in 2015. Stop fretting about political will and acknowledge the various pressures different countries face. Focus on fully implementing Copenhagen's pledge-and-review system and use that as a model for the successor to Kyoto. Then, allow that new pact to be what steers action and innovation.

Interest in this approach is slowly mounting, including in the U.S. government. Todd Stern, the State Department's special envoy for climate change, said in a 2013 speech, "An agreement that is animated by the progressive development of norms and expectations rather than by the hard edge of law, compliance, and penalty has a much better chance of working." Still, there's a long way to go before the all-or-nothing attitude that has dominated climate diplomacy for so long disappears for good.

In the meantime, the environmental clock keeps ticking.

Critical Thinking

1. Do you think that climate treaties can deal with the problem? Why or why not?

2. Why is it so difficult to persuade developing countries to reduce emissions of greenhouse gases into the atmosphere?

3. Is climate warming a danger now or in the future?

Create Central

www.mhhe.com/createcentral

Internet References

Arctic Council
    http://www.arctic-council.org/index.php/en
Center for Governance and Sustainability
    http://www.umb.edu/cgs
Climate Summit 2014
    http://www.un.org/climatechange/summit
UN Framework Convention on Climate Change
    http://unfccc.int/2860.php

DAVID SHORR has been analyzing multilateral affairs for over 25 years. He has worked with a range of international organizations and participates in Think20, a global meeting of leading think-tank representatives.

David Schorr, "Think Again: Climate Treaties," *Foreign Policy*, March/April 2014, pp. 38–39, 41–43. Copyright © 2014 by Foreign Policy. All rights reserved. Used with permission.

Unit 3

UNIT

Prepared by: Robert Weiner, *University of Massachusetts, Boston*

The Global Political Economy

A defining characteristic of the 20th century was an intense struggle between proponents of two economic ideologies. At the heart of the conflict was the question of what role government should play in the management of a country's economy. For some, the dominant capitalist economic system appeared to be organized primarily for the benefit of a few wealthy people. From their perspective, the masses were trapped in poverty providing cheap labor to further enrich the privileged elite. These critics argued that the capitalist system could be changed only by giving control of the political system to the state and having the state own the means of production. In striking contrast to this perspective, others argued that the best way to create wealth and eliminate poverty was through the profit motive. The profit motive encouraged entrepreneurs to create products and businesses at the cutting edge of new technologies. An example of this is "The Internet of Things," which can be seen as a new industrial revolution in which devices are connected to the Internet. An open and competitive marketplace, from this point of view, minimized government interference and was the best system for making decisions about production, wages, and the distribution of goods and services.

Conflict at times characterized the contest between capitalism and communism/socialism. The Russian and Chinese revolutions ended the old social order and created radical changes in the political and economic systems in these two important countries. The political structures that were created to support new systems of agricultural and industrial production, along with the centralized planning of virtually all aspects of economic activity eliminated most private ownership of property. These two revolutions were, in short, unparalleled experiments in social engineering.

The economic collapse of the Soviet Union and the dramatic market reforms in China have recast the debate about how to best structure contemporary economic systems. Some believe that with the end of communism and the resulting participation of hundreds of millions of consumers in the global market, an unprecedented era has begun.

Many have noted that this process of globalization is being accelerated by an evolution in communications and computer technology such as the "Internet of Things" and the use of mobile devices. Proponents of this view argue that a new global economy will ultimately eliminate national economic systems. Others are less optimistic about the process of globalization. They argue that the creation of a single economic system where there are no boundaries to impede the flow of capital and goods and services does not mean a closing of the gap between the world's rich and poor. Rather, they argue that multinational corporations and global financial institutions will have fewer legal constraints on their behavior, and this will lead to not only increased risks of periodic financial crises (such as 2008) but also greater expectations of workers and accelerated destruction of the environment. Other analysts of globalization argue that economic development is resulting in the emergence of a global middle class that is closing the economic gap between nations, while the income gap within states is increasing.

The use of the term political economy in this unit recognizes that economic and political systems are not separate. All economic systems have some type of marketplace where goods and services are bought and sold. Governments, whether national or international, regulate these transactions to some degree: that is, government sets the rules that regulate the marketplace. One of the most important concepts in assessing the contemporary political economy is development. Developed economies, such as the members of the European Union, are characterized by a profile that includes lower infant mortality rates, longer life expectancies, lower disease rates, higher rates of literacy, and healthier sanitation systems. As the process of globalization proceeds, the question is whether control of the economy and financial system of the world by the West is hindering the development of third world countries. Also, rather than a political revolution taking place on a global scale as envisioned by Marx, economic development around the globe has resulted in the emergence of a global middle class that could find itself in conflict with global elites who control the world's economic and financial system.

There are two articles in this unit that deal with the European Union. The European movement, which emerged after World War II, had the goal of creating a working peace system on the continent. It was based on the idea, beginning with the European Coal and Steel Community, that economic and technical cooperation between France and Germany would spill over into political cooperation. Since then, what is now called the European Union has expanded from its original six members to include 28 states, some drawn from the former communist world in Eastern Europe. However, by 2015, questions had been raised about the future of the European project. The Eurozone continued to be plagued by the financial difficulties of some of its members which were particularly acute in the southern periphery of the economic organization. In 2015, virtually bankrupt members like Greece had to be rescued by a third bailout by an infusion of financial aid from the European Central Bank, the European Commission, and the International Monetary Fund. The process of European economic and political integration that had been proceeding in fits and starts over the past 60 years faced a further challenge when the United Kingdom threatened to withdraw from the organization altogether. The United Kingdom had already had an ambivalent relationship with the European movement and did not join the organization until 1973. There existed a strong current of Euro skepticism in Britain, aptly illustrated by the emergence of the UK Independence Party, which, running on a platform of withdrawal from the European Union, reportedly enjoyed over 20% support among the British public.

One article in this unit is more optimistic about the prospects of the European Union, while another is somewhat more skeptical. The British general elections were held in May 2015. To the surprise of analysts and pollsters, the incumbent Conservative Party, led by Prime Minister David Cameron, did much better than anticipated. The Conservatives won a majority in the House of Commons, although not a very large one. However, the margin of victory was sufficient for the Conservatives to govern by themselves, without the support of the Liberal Democrats, ending the coalition which had existed for the preceding five years. In the general elections, the UK Independence Party garnered about 12% of the vote rather than 20%. Nonetheless, the way was still left open for a referendum, probably in 2017, as to whether the United Kingdom would stay in the European Union or withdraw from it.

Although globalization has contributed to the emergence of new economic centers in the world economic system, such as the BRICS (Brazil, Russia, India, China, and South Africa), poverty still affects the bottom 1 billion members of world society. The drive of developing countries in what is still called the third world, to achieve the goal of economic modernization, in a number of cases still has a considerable way to go. The collapse of the Western colonial empires, especially since the end of World War II, found a number of colonies enjoying nominal political independence, but still continuing to exist in a condition of economic dependency on the former metropolitan or imperial powers and the advanced economies of the industrialized core of world society. In a number of cases, the developing countries were relegated to the periphery of the world economic system as exporters of commodities and raw materials to the industrialized sector. Developing countries were caught in an economic bind in which the prices that they received for their commodities and raw materials did not keep pace with the prices they had to pay for the importation of manufactured and semi-manufactured goods from the economically advanced states. Multinational corporations were also able to take advantage of the process of globalization to exploit the resources and labor of the developing countries.

By the 1960s and 1970s, the developing countries constituted a majority of the membership of the United Nations. The third world viewed the United Nations, which had been created primarily as a political organization in 1945 (some 70 years ago) as an institution that could be used to mobilize the international community to promote economic justice by redistributing wealth from the industrialized core of world society to countries on the periphery and semi-periphery of the system. In 1964, the developing countries founded the Group of 77 to function as a "poor man's" lobbying group within the framework of the United Nations Conference on Trade and Development (UNCTAD). The Group of 77, which consisted of states drawn from Africa, Asia, Latin America, and the Middle East, was designed to be used as a bargaining tool to extract economic concessions from the richer states in the world economic system. The Group of 77 focused on such issues as better prices for commodities and raw materials exported by the developing countries, a reduction in the prices of manufactured and semi-manufactured goods, access to the technology of the West to be transferred to them on easy terms, a code of conduct for multinational corporations, and the forgiveness or easier terms for the repayment of loans from Western banks and international financial institutions. With the passage of time, in 2014, as UNCTAD observed its 50th anniversary, it was clear that the G-77 was not a cohesive body. For example, some of its members had moved into the ranks of the Newly Industrialized Countries (NICS), the petroleum exporting countries, or the "Asian Tigers."

In the 1970s, the developing countries demanded the creation of a New International Economic Order (NIEO). The purpose of the NIEO was to transfer wealth from the developed economies to the developing states in the Third World. Among other things, the drive for the new International Economic Order was based on a philosophy of economic justice, that the former imperial powers and the developed world owed reparations to their former colonies for centuries of economic exploitation. Advocates of the NIEO argued that the economic development of the former metropolitan powers would not have been possible without the exploitation of their former colonies. The mobilization of the developing countries to build a NIEO was based also on the success of OPEC (Organization of Petroleum Exporting Countries) in raising the price of a barrel of oil fourfold in 1973. Some of the developing countries advocated the creation of similar commodity-based cartels to extract wealth from the industrialized states. The Group of 77 used its majority of votes in the United Nations General assembly to push through omnibus resolutions demanding the creation of the NIEO. This could be viewed as an effort to rewrite the Charter of the United Nations to focus on the Economic Rights and Duties of States, such as the recognition of a state's sovereign right of ownership of its natural resources. However, the drive for a New

International Economic Order failed, due to the unwillingness of the industrialized countries to recognize an obligation to provide the finances necessary to implement the NIEO.

The 1980s were dubbed a "lost decade" as a number of developing countries found themselves unable to repay the loans that they had received from private banks, which were awash with petrodollars that had been recycled to them by the petroleum exporting states. Eventually, through a combination of default (in the case of Mexico), debt forgiveness, and repayment of loans on far less than the value of the loan itself, the external debt crisis was contained.

The next major phase in the international political economy shifted to the concept of sustainable development. Sustainable development means that economic development should be implemented based on the protection of the resources of the environment for future generations. Developing countries, however, had a tendency to balk at the benchmarks that were set by the developed countries as a hindrance to their efforts to move ahead in economic development.

In spite of the global economic meltdown that occurred in 2008, one of the surprising economic success stories in the Third World has been the economic progress that has been made in sub-Saharan Africa. The economic renaissance of Africa has been stimulated by such factors as a commodity boom, increased foreign direct investment, and the widespread use of new communications technology, such as cell phones for microfinancial purposes such as banking. However, the African states still need to further develop the institutional structure necessary to promote economic development. The G-77 has also called for a reform of the Bretton Woods institutions' voting system on a more equitable basis to reflect the realities of the globalized economy of the 21st century.

Article Prepared by: Robert Weiner, *University of Massachusetts, Boston*

Think Again: European Decline

Sure, it may seem as if Europe is down and out. But things are far, far better than they look.

MARK LEONARD AND HANS KUNDNANI

Learning Outcomes

After reading this article, you will be able to:

- Briefly describe the historical role of Europe in international affairs.

- Identify reasons that predictions of Europe's economic decline are likely premature.

- Discuss the political and economic challenges the European Union faces.

"Europe Is History."

No

These days, many speak of Europe as if it has already faded into irrelevance. In the words of American pundit Fareed Zakaria, "it may well turn out that the most consequential trend of the next decade will be the economic decline of Europe." According to Singaporean scholar Kishore Mahbubani, Europe "does not get how irrelevant it is becoming to the rest of the world." Not a day went by on the 2012 U.S. campaign trail, it seemed, without Republican challenger Mitt Romney warning that President Barack Obama was—gasp—turning the United States into a "European social welfare state."

With its anemic growth, ongoing eurocrisis, and the complexity of its decision-making, Europe is admittedly a fat target right now. And the stunning rise of countries like Brazil and China in recent years has led many to believe that the Old World is destined for the proverbial ash heap. But the declinists would do well to remember a few stubborn facts. Not only does the European Union remain the largest single economy in

the world, but it also has the world's second-highest defense budget after the United States, with more than 66,000 troops deployed around the world and some 57,000 diplomats (India has roughly 600). The EU's GDP per capita in purchasing-power terms is still nearly four times that of China, three times Brazil's, and nearly nine times India's. If this is decline, it sure beats living in a rising power.

Power, of course, depends not just on these resources but on the ability to convert them to produce outcomes. Here too Europe delivers: Indeed, no other power apart from the United States has had such an impact on the world in the last 20 years. Since the end of the Cold War, the EU has peacefully expanded to include 15 new member states and has transformed much of its neighborhood by reducing ethnic conflicts, exporting the rule of law, and developing economies from the Baltic to the Balkans. Compare that with China, whose rise is creating fear and provoking resistance across Asia. At a global level, many of the rules and institutions that keep markets open and regulate world trade, limit carbon emissions, and prosecute human rights abusers were created by the European Union. Who was behind the World Trade Organization and the International Criminal Court? Not the United States or China. It's Europe that has led the way toward a future run by committees and statesmen, not soldiers and strongmen.

Yes, the EU now faces an existential crisis. Even as it struggles, however, it is still contributing more than other powers to solving both regional conflicts and global problems. When the Arab revolutions erupted in 2011, the supposedly bankrupt EU pledged more money to support democracy in Egypt and Tunisia than the United States did. When Libya's Muammar al-Qaddafi was about to carry out a massacre in Benghazi in March 2011, it was France and Britain that led from the front. This year, France acted to prevent a takeover of southern Mali by jihadists and drug smugglers. Europeans may not have done enough to stop the conflict in Syria, but they have done as much as anyone else in this tragic story.

In one sense, it is true that Europe is in inexorable decline. For four centuries, Europe was the dominant force in international relations. It was home to the Renaissance and the Enlightenment. It industrialized first and colonized much of the world. As a result, until the 20th century, all the world's great powers were European. It was inevitable—and desirable—that other players would gradually narrow the gap in wealth and power over time. Since World War II, that catch-up process has accelerated. But Europeans benefit from this: Through their economic interdependence with rising powers, including those in Asia, Europeans have continued to increase their GDP and improve their quality of life. In other words, like the United States—and unlike, for example, Russia on the continent's eastern frontier—Europe is in relative though not absolute decline.

The EU is an entirely unprecedented phenomenon in world affairs: a project of political, economic, and above all legal integration among 27 countries with a long history of fighting each other. What has emerged is neither an intergovernmental organization nor a superstate, but a new model that pools resources and sovereignty with a continent-sized market and common legislation and budgets to address transnational threats from organized crime to climate change. Most importantly, the EU has revolutionized the way its members think about security, replacing the old traditions of balance-of-power politics and noninterference in internal affairs with a new model under which security for all is guaranteed by working together. This experiment is now at a pivotal moment, and it faces serious, complex challenges—some related to its unique character and some that other major powers, particularly Japan and the United States, also face. But the EU's problems are not quite the stuff of doomsday scenarios.

"The Eurozone Is an Economic Basket Case."

Only Part of It

Many describe the eurozone, the 17 countries that share the euro as a common currency, as an economic disaster. As a whole, however, it has lower debt and a more competitive economy than many other parts of the world. For example, the International Monetary Fund projects that the eurozone's combined 2013 government deficit as a share of GDP will be 2.6 percent—roughly a third of that of the United States. Gross government debt as a percentage of GDP is around the same as in the United States and much lower than that in Japan.

Nor is Europe as a whole uncompetitive. In fact, according to the latest edition of the World Economic Forum's Global Competitiveness Index, three eurozone countries (Finland, the Netherlands, and Germany) and another two EU member states (Britain and Sweden) are among the world's 10 most competitive economies. China ranks 29th. The eurozone accounts for 15.6 percent of the world's exports, well above 8.3 percent for the United States and 4.6 percent for Japan. And unlike the United States, its current trade account is roughly in balance with the rest of the world.

These figures show that, in spite of the tragically counterproductive policies imposed on Europe's debtor countries and despite whatever happens to the euro, the European economy is fundamentally sound. European companies are among the most successful exporters anywhere. Airbus competes with Boeing; Volkswagen is the world's third largest automaker and is forecast to extend its lead in sales over Toyota and General Motors in the next five years; and European luxury brands (many from crisis-wracked Italy) are coveted all over the world. Europe has a highly skilled workforce, with universities second only to America's, well-developed systems of vocational training, empowered women in the workforce, and excellent infrastructure. Europe's economic model is not unsustainable simply because its GDP growth has slowed of late.

The real difference between the eurozone and the United States or Japan is that it has internal imbalances but is not a country, and that it has a common currency but no common treasury. Financial markets therefore look at the worst data for individual countries—say, Greece or Italy—rather than aggregate figures. Due to uncertainty about whether the eurozone's creditor countries will stand by its debtors, spreads—that is, the difference in bond yields between countries with different credit ratings—have increased since the crisis began. Creditor countries such as Germany have the resources to bail out the debtors, but by insisting on austerity measures, they are trapping debtor countries like Spain in a debt-deflation spiral. Nobody knows whether the eurozone will be able to overcome these challenges, but the pundits who confidently predicted a "Grexit" or a complete breakup of the single currency have been proved wrong thus far. Above all, the eurocrisis is a political problem rather than an economic one.

"Europeans Are From Venus."

Hardly

In 2002, American author Robert Kagan famously wrote, "Americans are from Mars and Europeans are from Venus." More recently, Robert Gates, then U.S. defense secretary, warned in 2010 of the "demilitarization" of Europe. But not only are European militaries among the world's strongest, these assessments also overlook one of the great achievements of human civilization: A continent that gave us the most destructive conflicts in history has now basically agreed to give up war on its own turf. Besides, within Europe there are huge differences in attitudes toward the uses and abuses of hard power. Hawkish countries such as Poland and Britain are closer to the United States than they are to dovish Germany, and many continue to foresee a world where a strong military is an indispensable component of security. And unlike rising powers such as China that proclaim the principle of noninterference, Europeans are still prepared to use force to intervene abroad. Ask the people of the Malian city of Gao, which had been occupied for nearly a year by hard-line Islamists until French troops ejected them, whether they see Europeans as timid pacifists.

At the same time, Americans have changed much in the decade since Kagan said they are from Mars. As the United States draws down from the wars in Afghanistan and Iraq and focuses on "nation-building at home," it looks increasingly Venusian. In fact, attitudes toward military intervention are converging on both sides of the Atlantic. According to the most recent edition of Transatlantic Trends, a regular survey by the German Marshall Fund, only 49 percent of Americans think that the intervention in Libya was the right thing to do, compared with 48 percent of Europeans. Almost as many Americans (68 percent) as Europeans (75 percent) now want to withdraw troops from Afghanistan.

Many American critics of Europe point to the continent's low levels of military spending. But it only looks low next to the United States—by far the world's biggest spender. In fact, Europeans collectively accounted for about 20 percent of the world's military spending in 2011, compared with 8 percent for China, 4 percent for Russia, and less than 3 percent for India, according to the Stockholm International Peace Research Institute. It is true that, against the background of the crisis, many EU member states are now making dramatic cuts in military spending, including, most worryingly, France. Britain and Germany, however, have so far made only modest cuts, and Poland and Sweden are actually increasing military spending. Moreover, the crisis is accelerating much-needed pooling and sharing of capabilities, such as air policing and satellite navigation. As for those Martians in Washington, the U.S. Congress is cutting military spending by $487 billion over the next 10 years and by $43 billion this year alone—and the supposedly warlike American people seem content with butter's triumph over guns.

No, But It Has a Legitimacy Problem

Skeptics have claimed for years that Europe has a "democratic deficit" because the European Commission, which runs the EU, is unelected or because the European Parliament, which approves and amends legislation, has insufficient powers. But European Commission members are appointed by directly elected national governments, and European Parliament members are elected directly by voters. In general, EU-level decisions are made jointly by democratically elected national governments and the European Parliament. Compared with other states or even an ideal democracy, the EU has more checks and balances and requires bigger majorities to pass legislation. If Obama thinks it's tough assembling 60 votes to get a bill through the Senate, he should try putting together a two-thirds majority of Europe's governments and then getting it ratified by the European Parliament. The European Union is plenty democratic.

The eurozone does, however, have a more fundamental legitimacy problem due to the way it was constructed. Although decisions are made by democratically elected leaders, the EU is a fundamentally technocratic project based on the "Monnet method," named for French diplomat Jean Monnet, one of the founding fathers of an integrated Europe. Monnet rejected grand plans and instead sought to "build Europe" step by step through "concrete achievements." This incremental strategy— first a coal and steel community, then a single market, and finally a single currency—took ever more areas out of the political sphere. But the more successful this project became, the more it restricted the powers of national governments and the more it fueled a populist backlash.

To solve the current crisis, member states and EU institutions are now taking new areas of economic policymaking out of the political sphere. Led by Germany, eurozone countries have signed up to a "fiscal compact" that commits them to austerity indefinitely. There is a real danger that this approach will lead to democracy without real choices: Citizens will be able to change governments but not policies. In protest, voters in Italy and Greece are turning to radical parties such as Alexis Tsipras's Syriza party in Greece and Beppe Grillo's Five Star Movement in Italy. These parties, however, could become part of the solution by forcing member states to revisit the strict austerity programs and go further in mutualizing debt across Europe—which they must ultimately do. So yes, European politics have a legitimacy problem; the solution is more likely to come from policy change rather than, say, giving yet more power to the European Parliament. Never mind what the skeptics say—it already has plenty.

"Europe Has a Democratic Deficit."

"Europe Is About to Fall off a Demographic Cliff."

So Is Nearly Everybody Else

The EU does have a serious demographic problem. Unlike the United States—whose population is projected to increase to 400 million by 2050—the EU's population is projected to increase from 504 million now to 525 million in 2035, but thereafter to decline gradually to 517 million in 2060, according to Europe's official statistical office. The problem is particularly acute in Germany, today the EU's largest member state, which has one of the world's lowest birth rates. Under current projections, its population could fall from 82 million to 65 million by 2060.

Europe's population is also aging. This year, the EU's working-age population will start falling from 308 million and is projected to drop to 265 million in 2060. That's expected to increase the old-age dependency ratio (the number of over-65s as a proportion of the total working-age population) from 28 percent in 2010 to 58 percent in 2060. Such figures can lead to absurd predictions of civilizational extinction. As one Guardian pundit put it, "With each generation reproducing only half its number, this looks like the start of a continent-wide collapse in numbers. Some predict wipe-out by 2100."

Demographic woes are not, however, something unique to Europe. In fact, nearly all the world's major powers are aging—and some more dramatically than Europe. China is projected to go from a population with a median age of 35 to 43 by 2030, and Japan will go from 45 to 52. Germany will go from 44 to 49. But Britain will go from 40 to just 42—a rate of aging comparable to that of the United States, one of the powers with the best demographic prospects.

So sure, demography will be a major headache for Europe. But the continent's most imperiled countries have much that's hopeful to learn from elsewhere in Europe. France and Sweden, for example, have reversed their falling birth rates by promoting maternity (and paternity) rights and child-care facilities. In the short term, the politics may be complicated, but immigration offers the possibility of mitigating both the aging and shrinking of Europe's population—so-called decline aside, there is no shortage of young people who want to come to Europe. In the medium term, member states could also increase the retirement age—another heavy political lift but one that many are now facing. In the long term, smart family-friendly policies such as child payments, tax credits, and state-supported day care could encourage Europeans to have more children. But arguably, Europe is already ahead of the rest of the world in developing solutions to the problem of an aging society. The graying Chinese should take note.

"Europe Is Irrelevant in Asia."

No

It is often said—most often and loudly by Singapore's Mahbubani—that though the EU may remain relevant in its neighborhood, it is irrelevant in Asia, the region that will matter most in the 21st century. Last November, then-Secretary of State Hillary Clinton proclaimed that the U.S. "pivot" to Asia was "not a pivot away from Europe" and said the United States wants Europe to "engage more in Asia along with us."

But Europe is already there. It is China's biggest trading partner, India's second-biggest, the Association of Southeast Asian Nations (ASEAN)'S second-biggest, Japan's third-biggest, and Indonesia's fourth-biggest. It has negotiated free trade areas with Singapore and South Korea and has begun separate talks with ASEAN, India, Japan, Malaysia, Thailand, and Vietnam. These economic relationships are already forming the basis for close political relationships in Asia. Germany even holds a regular government-to-government consultation—in effect a joint cabinet meeting—with China. If the United States can claim to be a Pacific power, Europe is already a Pacific economy and is starting to flex its political muscles there too.

Europe played a key role in imposing sanctions against Burma—and in lifting them after the military junta began to reform. Europe helped resolve conflicts in Aceh, Indonesia, and is mediating in Mindanao in the Philippines. While Europe may not have a 7th Fleet in Japan, some member states already play a role in security in Asia: The British have military facilities in Brunei, Nepal, and Diego Garcia, and the French have a naval base in Tahiti. And those kinds of ties are growing. For example, Japanese Prime Minister Shinzo Abe, who is trying to diversify Japan's security relationships, has said he wants to join the Five Power Defense Arrangements, a security treaty that includes Britain. European Union member states also supply advanced weaponry such as fighter jets and frigates to democratic countries like India and Indonesia. That's hardly irrelevance.

"Europe Will Fall Apart."

Too Soon to Say

The danger of European disintegration is real. The most benign scenario is the emergence of a three-tier Europe consisting of a eurozone core, "pre-ins" such as Poland that are committed to joining the euro, and "opt-outs" such as Britain that have no intention of joining the single currency. In a more malign scenario, some eurozone countries such as Cyprus or Greece will be forced to leave the single currency, and some EU member states such as Britain may leave the EU completely—with huge implications for the EU's resources and its image in the world. It would be a tragedy if an attempt to save the eurozone led to a breakup of the European Union.

But Europeans are aware of this danger, and there is political will to prevent it. Germany does not want Greece to leave the single currency, not least due to a fear of contagion. A British withdrawal is possible but unlikely and in any case some way off: Prime Minister David Cameron would have to win an overall majority in the next election, and British citizens would have to vote to leave in a referendum. In short, it's premature to predict an EU breakup.

This is not to say it will never happen. The ending of the long story of Europe remains very much unwritten. It is not a simple choice between greater integration and disintegration. The key will be whether Europe can save the euro without splitting the European Union. Simply by its creation, the EU is already an unprecedented phenomenon in the history of international relations—and a much more perfect union than the declinists will admit. If its member states can pool their resources, they will find their rightful place alongside Washington and Beijing in shaping the world in the 21st century. As columnist Charles Krauthammer famously said in relation to America, "Decline is a choice." It is for Europe too.

Europe's economic model is not unsustainable simply because its GDP growth has slowed of late.

The EU is already an unprecedented phenomenon in the history of international relations-and a much more perfect union than the declinists will admit.

Critical Thinking

1. How are demographic changes likely to affect the role of Europe in international affairs?

2. Is Europe irrelevant to Asia?

3. How likely is it that an integrated Europe will fall apart?

Create Central

www.mhhe.com/createcentral

Internet References

World Economic Forum
    www.weforum.org
European Commission
    http://ec.europa.eu/index_en.htm
Organization for Economic Co-operation and Development
    www.oecd.org

MARK LEONARD is director and **HANS KUNDNANI** is editorial director of the European Council on Foreign Relations.

Article   Prepared by: Robert Weiner, *University of Massachusetts, Boston*

Broken BRICs

RUCHIR SHARMA

Learning Outcomes

After reading this article, you will be able to:

- Identify and describe the major emerging market economies.

- Offer specific examples of economic and social differences between the so-called BRIC countries.

- Explain what are some of the fundamental problems with long-range economic forecasts.

Why the Rest Stopped Rising

Over the past several years, the most talked-about trend in the global economy has been the so-called rise of the rest, which saw the economies of many developing countries swiftly converging with those of their more developed peers. The primary engines behind this phenomenon were the four major emerging-market countries, known as the BRICS: Brazil, Russia, India, and China. The world was witnessing a once-in-a-lifetime shift, the argument went, in which the major players in the developing world were catching up to or even surpassing their counterparts in the developed world.

These forecasts typically took the developing world's high growth rates from the middle of the last decade and extended them straight into the future, juxtaposing them against predicted sluggish growth in the United States and other advanced industrial countries. Such exercises supposedly proved that, for example, China was on the verge of overtaking the United States as the world's largest economy—a point that Americans clearly took to heart, as over 50 percent of them, according to a Gallup poll conducted this year, said they think that China is already the world's "leading" economy, even though the U.S. economy is still more than twice as large (and with a per capita income seven times as high).

As with previous straight-line projections of economic trends, however—such as forecasts in the 1980s that Japan would soon be number one economically—later returns are throwing cold water on the extravagant predictions. With the world economy heading for its worst year since 2009, Chinese growth is slowing sharply, from double digits down to seven percent or even less. And the rest of the BRICS are tumbling, too: since 2008, Brazil's annual growth has dropped from 4.5 percent to two percent; Russia's, from seven percent to 3.5 percent; and India's, from nine percent to six percent.

None of this should be surprising, because it is hard to sustain rapid growth for more than a decade. The unusual circumstances of the last decade made it look easy: coming off the crisis-ridden 1990s and fueled by a global flood of easy money, the emerging markets took off in a mass upward swing that made virtually every economy a winner. By 2007, when only three countries in the world suffered negative growth, recessions had all but disappeared from the international scene. But now, there is a lot less foreign money flowing into emerging markets. The global economy is returning to its normal state of churn, with many laggards and just a few winners rising in unexpected places. The implications of this shift are striking, because economic momentum is power, and thus the flow of money to rising stars will reshape the global balance of power.

Forever Emerging

The notion of wide-ranging convergence between the developing and the developed worlds is a myth. Of the roughly 180 countries in the world tracked by the International Monetary Fund, only 35 are developed. The markets of the rest are emerging—and most of them have been emerging for many decades and will continue to do so for many more. The Harvard economist Dani Rodrik captures this reality well. He has shown that before 2000, the performance of the emerging markets as a whole did not converge with that of the developed world at all. In fact, the per capita income gap between the advanced and the developing economies steadily widened from 1950 until 2000. There were a few pockets of countries that did catch up with the West, but they were limited to oil states in the Gulf, the nations of southern Europe after World War II, and the economic "tigers" of East Asia. It was only after 2000 that the emerging markets as a whole started to catch up; nevertheless, as of 2011, the difference in per capita incomes between the rich and the developing nations was back to where it was in the 1950s.

This is not a negative read on emerging markets so much as it is simple historical reality. Over the course of any given decade since 1950, on average, only a third of the emerging markets have been able to grow at an annual rate of five percent or more. Less than one-fourth have kept up that pace for two decades, and one-tenth, for three decades. Only Malaysia, Singapore, South Korea, Taiwan, Thailand, and Hong Kong have maintained this growth rate for four decades. So even before the current signs of a slowdown in the BRICS, the odds were against Brazil experiencing a full decade of growth above five percent, or Russia, its second in a row.

Meanwhile, scores of emerging markets have failed to gain any momentum for sustained growth, and still others have seen their progress stall after reaching middle-income status. Malaysia and Thailand appeared to be on course to emerge as rich countries until crony capitalism, excessive debts, and overpriced currencies caused the Asian financial meltdown of 1997–98. Their growth has disappointed ever since. In the late 1960s, Burma (now officially called Myanmar), the Philippines, and Sri Lanka were billed as the next Asian tigers, only to falter badly well before they could even reach the middle-class average income of about $5,000 in current dollar terms. Failure to sustain growth has been the general rule, and that rule is likely to reassert itself in the coming decade.

In the opening decade of the twenty-first century, emerging markets became such a celebrated pillar of the global economy that it is easy to forget how new the concept of emerging markets is in the financial world. The first coming of the emerging markets dates to the mid-1980s, when Wall Street started tracking them as a distinct asset class. Initially labeled as "exotic," many emerging-market countries were then opening up their stock markets to foreigners for the first time: Taiwan opened its up in 1991; India, in 1992; South Korea, in 1993; and Russia, in 1995. Foreign investors rushed in, unleashing a 600 percent boom in emerging-market stock prices (measured in dollar terms) between 1987 and 1994. Over this period, the amount of money invested in emerging markets rose from less than one percent to nearly eight percent of the global stock-market total.

This phase ended with the economic crises that struck from Mexico to Turkey between 1994 and 2002. The stock markets of developing countries lost almost half their value and shrank to four percent of the global total. From 1987 to 2002, developing countries' share of global GDP actually fell, from 23 percent to 20 percent. The exception was China, which saw its share double, to 4.5 percent. The story of the hot emerging markets, in other words, was really about one country.

The second coming began with the global boom in 2003, when emerging markets really started to take off as a group. Their share of global GDP began a rapid climb, from 20 percent to the 34 percent that they represent today (attributable in part to the rising value of their currencies), and their share of the global stock-market total rose from less than four percent to more than ten percent. The huge losses suffered during the global financial crash of 2008 were mostly recovered in 2009, but since then, it has been slow going.

The third coming, an era that will be defined by moderate growth in the developing world, the return of the boom-bust cycle, and the breakup of herd behavior on the part of emerging-market countries, is just beginning. Without the easy money and the blue-sky optimism that fueled investment in the last decade, the stock markets of developing countries are likely to deliver more measured and uneven returns. Gains that averaged 37 percent a year between 2003 and 2007 are likely to slow to, at best, ten percent over the coming decade, as earnings growth and exchange-rate values in large emerging markets have limited scope for additional improvement after last decade's strong performance.

Past its Sell-by Date

No idea has done more to muddle thinking about the global economy than that of the BRICS. Other than being the largest economies in their respective regions, the big four emerging markets never had much in common. They generate growth in different and often competing ways—Brazil and Russia, for example, are major energy producers that benefit from high energy prices, whereas India, as a major energy consumer, suffers from them. Except in highly unusual circumstances, such as those of the last decade, they are unlikely to grow in unison. China apart, they have limited trade ties with one another, and they have few political or foreign policy interests in common.

A problem with thinking in acronyms is that once one catches on, it tends to lock analysts into a worldview that may soon be outdated. In recent years, Russia's economy and stock market have been among the weakest of the emerging markets, dominated by an oil-rich class of billionaires whose assets equal 20 percent of GDP, by far the largest share held by the superrich in any major economy. Although deeply out of balance, Russia remains a member of the BRICS, if only because the term sounds better with an R. Whether or not pundits continue using the acronym, sensible analysts and investors need to stay flexible; historically, flashy countries that grow at five percent or more for a decade—such as Venezuela in the 1950s, Pakistan in the 1960s, or Iraq in the 1970s—are usually tripped up by one threat or another (war, financial crisis, complacency, bad leadership) before they can post a second decade of strong growth.

The current fad in economic forecasting is to project so far into the future that no one will be around to hold you accountable. This approach looks back to, say, the seventeenth century, when China and India accounted for perhaps half of global GDP, and then forward to a coming "Asian century," in which such preeminence is reasserted. In fact, the longest period over which one can find clear patterns in the global economic cycle is around a decade. The typical business cycle lasts about five years, from the bottom of one downturn to the bottom of the next, and most practical investors limit their perspectives to one or two business cycles. Beyond that, forecasts are often rendered obsolete by the unanticipated appearance of new competitors, new political environments, or new technologies. Most CEOS and major investors still limit their strategic visions to three, five, or at most seven years, and they judge results on the same time frame.

The New and Old Economic Order

In the decade to come, the United States, Europe, and Japan are likely to grow slowly. Their sluggishness, however, will look less worrisome compared with the even bigger story in the global economy, which will be the three to four percent slowdown in China, which is already under way, with a possibly deeper slowdown in store as the economy continues to mature. China's population is simply too big and aging too quickly for its economy to continue growing as rapidly as it has. With over 50 percent of its people now living in cities, China is nearing what economists call "the Lewis turning point": the point at which a country's surplus labor from rural areas has been largely exhausted. This is the result of both heavy migration to cities over the past two decades and the shrinking work force that the one-child policy has produced. In due time, the sense of many Americans today that Asian juggernauts are swiftly overtaking the U.S. economy will be remembered as one of the country's periodic bouts of paranoia, akin to the hype that accompanied Japan's ascent in the 1980s.

As growth slows in China and in the advanced industrial world, these countries will buy less from their export-driven counterparts, such as Brazil, Malaysia, Mexico, Russia, and Taiwan. During the boom of the last decade, the average trade balance in emerging markets nearly tripled as a share of GDP, to six percent. But since 2008, trade has fallen back to its old share of under two percent. Export-driven emerging markets will need to find new ways to achieve strong growth, and investors recognize that many will probably fail to do so: in the first half of 2012, the spread between the value of the best-performing and the value of the worst-performing major emerging stock markets shot up from ten percent to

35 percent. Over the next few years, therefore, the new normal in emerging markets will be much like the old normal of the 1950s and 1960s, when growth averaged around five percent and the race left many behind. This does not imply a reemergence of the 1970s-era Third World, consisting of uniformly underdeveloped nations. Even in those days, some emerging markets, such as South Korea and Taiwan, were starting to boom, but their success was overshadowed by the misery in larger countries, such as India. But it does mean that the economic performance of the emerging-market countries will be highly differentiated.

The uneven rise of the emerging markets will impact global politics in a number of ways. For starters, it will revive the self-confidence of the West and dim the economic and diplomatic glow of recent stars, such as Brazil and Russia (not to mention the petro-dictatorships in Africa, Latin America, and the Middle East). One casualty will be the notion that China's success demonstrates the superiority of authoritarian, state-run capitalism. Of the 124 emerging-market countries that have managed to sustain a five percent growth rate for a full decade since 1980, 52 percent were democracies and 48 percent were authoritarian. At least over the short to medium term, what matters is not the type of political system a country has but rather the presence of leaders who understand and can implement the reforms required for growth.

Another casualty will be the notion of the so-called demographic dividend.

Because China's boom was driven in part by a large generation of young people entering the work force, consultants now scour census data looking for similar population bulges as an indicator of the next big economic miracle. But such demographic determinism assumes that the resulting workers will have the necessary skills to compete in the global market and that governments will set the right policies to create jobs. In the world of the last decade, when a rising tide lifted all economies, the concept of a demographic dividend briefly made sense. But that world is gone.

The economic role models of recent times will give way to new models or perhaps no models, as growth trajectories splinter off in many directions. In the past, Asian states tended to look to Japan as a paradigm, nations from the Baltics to the Balkans looked to the European Union, and nearly all countries to some extent looked to the United States. But the crisis of 2008 has undermined the credibility of all these role models. Tokyo's recent mistakes have made South Korea, which is still rising as a manufacturing powerhouse, a much more appealing Asian model than Japan. Countries that once were clamoring to enter the euro-zone, such as the Czech Republic, Poland, and Turkey, now wonder if they want to join a club with so many members struggling to stay afloat. And as for the United States, the 1990s-era Washington consensus—which called for poor countries to restrain their spending and liberalize their economies—is a hard sell when even Washington can't agree to cut its own huge deficit.

Because it is easier to grow rapidly from a low starting point, it makes no sense to compare countries in different income classes. The rare breakout nations will be those that outstrip rivals in their own income class and exceed broad expectations for that class. Such expectations, moreover, will need to come back to earth. The last decade was unusual in terms of the wide scope and rapid pace of global growth, and anyone who counts on that happy situation returning soon is likely to be disappointed.

Among countries with per capita incomes in the $20,000 to $25,000 range, only two have a good chance of matching or exceeding three percent annual growth over the next decade: the Czech Republic and South Korea. Among the large group with average incomes in the $10,000 to $15,000 range, only one country—Turkey—has a good shot at matching or exceeding four to five percent growth, although Poland also has a chance. In the $5,000 to $10,000 income class, Thailand seems to be the only country with a real shot at outperforming significantly. To the extent that there will be a new crop of emerging-market stars in the coming years, therefore, it is likely to feature countries whose per capita incomes are under $5,000, such as Indonesia, Nigeria, the Philippines, Sri Lanka, and various contenders in East Africa.

Although the world can expect more breakout nations to emerge from the bottom income tier, at the top and the middle, the new global economic order will probably look more like the old one than most observers predict. The rest may continue to rise, but they will rise more slowly and unevenly than many experts are anticipating. And precious few will ever reach the income levels of the developed world.

Critical Thinking

1. Why is the convergence between developing and the developed worlds a myth?

2. Why don't these four countries have much in common, i.e., what are their differences?

3. Which one of the four countries do you predict will continue to develop the fastest and why?

Create Central

www.mhhe.com/createcentral

Internet References

Organization for Economic Co-operation and Development
www.oecd.org

International Monetary Fund
www.imf.org

Inter-American Development Bank
www.iadb.org

Morgan Stanley Investment Management
www.morganstanley.com

RUCHIR SHARMA is head of Emerging Markets and Global Macro at Morgan Stanley Investment Management and the author of *Breakout Nations: In Pursuit of the Next Economic Miracles.*

Article

Prepared by: Robert Weiner, *University of Massachusetts, Boston*

The Roadblock

If the West Doesn't Shape Up, the Rest of the World Will Just Go around It.

MOHAMED A. EL-ERIAN

Learning Outcomes

After reading this article, you will be able to:

- Understand the dependency of the developing world on the financial and trading system of the developed world.

- Gain insights into how the International monetary system works.

O n My Travels Around The world this fall and at international meetings of economists and policymakers, one thing has become crystal clear: A growing number of developing-country officials are increasingly worried that "irresponsible" political behavior in the United States risks undermining the well-being of their citizens. Indeed, the 16-day government shutdown this October and the congressional brinkmanship over the debt ceiling that threatened a payments default are just two points in a seemingly endless series of strange developments that risk fueling unnecessary financial and economic instability in the rest of the world. And with Europe still struggling to regain a more robust economic footing, the developing world takes little comfort in operating in a global economic system that is constructed on the assumption of a stable, rational, and responsive West.

Being on the receiving end of Western-induced economic disruptions is not a new phenomenon for developing countries. Only 5 years ago, they felt the full impact of a financial crisis that peaked in the United States with the disorderly collapse of Lehman Brothers. With the frightening fragility of the Western banking system fully exposed, developing countries struggled to counter the collapse in global GDP and trade. Fortunately,

and to the surprise of many, emerging economies bounced back from the global financial crisis much better—and much faster—than most analysts had predicted. Moreover, they have handily and consistently outperformed the West in terms of growth and job creation.

But that may not last—and that would primarily be the West's fault. The West's current phase of economic policy inconsistency has a lot to do with the difficulties that democracies with short election cycles face in dealing with the consequences of low growth and persistently high unemployment. Five years after the 2008 financial crisis, and after billions of dollars poured into recapitalizing banks, Europe and the United States have not yet been able to kick-start their economies into escape velocity and return to sustainable high-growth rates and proper job creation. And disappointments have a habit of generating even greater disappointments. Rather than step up to the challenges, the political system in general, and the U.S. Congress in particular, has become much more susceptible to paralyzing polarization. In the process, policymaking has become more fragmented, and countries have become more insular and notably less open to holistic policy approaches that break the hold of prolonged downturns.

But while Americans can gripe and moan about their political dysfunction—and yes, eventually force a change through the ballot box—the rest of the world has no choice but to frown and bear it. The implications go beyond bracing for the occasional government shutdown and threats of technical default by the issuer of the world's reserve currency. They also affect the very manner in which the global economy functions.

Economic and financial resilience is key if developing countries are to navigate what is, to borrow an elegant phrase from outgoing U.S. Federal Reserve Chairman Ben Bernanke, an "unusually uncertain" outlook. And here, developing countries have to come to terms quickly with—and respond to—four increasingly entrenched realities: The global economic system anchored by the West will remain volatile going forward; multilateral reform is essentially stuck, eroding the already thin possibility that coordinated policies could improve the common good; developing countries have fewer economic and financial defenses at their disposal today as compared with 5 years ago; and they face greater temptation to return to bad old habits or remain in denial.

Like a poorly equipped car on a frustratingly long and bumpy journey, developing countries must weather the potholes created by the West with fewer spare tires. Financial cushions are less robust this time around; lower international reserves and greater corporate and household debt have left many countries less resilient to the shocks caused by American congressional dithering. Ironically, crisis managers in key countries seem to have become a little more complacent, either denying the growing potential for financial instability (Turkey) or engaging in pointless blame games (Brazil).

Like a poorly equipped car on a frustratingly long and bumpy journey, developing countries must weather the potholes created by the West with fewer spare tires.

Even the more agile policymakers in Asia and Latin America convey a sense of frustration, if not fatalism and helplessness, when it comes to an obvious and inconvenient truth: Developing countries are structurally wired into a global system—be it trade, finance, regulation, or multilateral governance—that is anchored by an increasingly insular and less predictable West. The post-World War II system is based on the assumption that the core will act rationally and responsibly when it comes to its global economic and financial functions. And lately, this has not been the case.

Developing countries are powerless to rewire the system quickly, especially as there is no other economic and financial superpower to replace the United States at the core. No wonder China expressed annoyance at U.S. congressional behavior, particularly as it threatened the estimated $1.3 trillion Beijing holds in bonds issued by the U.S. government. Yet even Beijing can do little save complain. Like other developing countries, China can't bail on the global economic system. It can only advocate better policymaking in the West, while at the margins tweaking some "south-south" trade and financial relationships.

But this impotence is not a permanent condition. If the United States and Europe can't figure out how to limit the damage that subpar politics is doing to their economies, the developing world will begin to seriously experiment with bolder approaches that sidestep the tired and obstructionist core of the global financial system. It might not happen today or tomorrow, perhaps not even this decade, but it will happen. And the effect could be material and irreversible. Indeed, the resulting fragmentation could well end up making the global economy less efficient, undermining both actual and potential global growth and making it more prone to cross-border tensions.

We should certainly all hope that it's just a matter of time before the West returns to being a more responsible and consistent steward of the international monetary system. We should also all hope that institutions like the International Monetary Fund will soon be empowered to fill the void that national governments have created. But all these things are just that—hopes. They speak to what needs to happen, not what likely will based on current realities.

As much as we should all hope for a better-functioning global system, developing countries will continue to be exposed to an unusual degree of Western economic malfunction, and U.S. congressional dysfunction in particular. If I were a policymaker in Latin America or Asia, I'd be packing a few more spare tires and planning for quite a bumpy road.

Critical Thinking

1. How can the developing world extricate itself from an international financial system dominated by the West?

2. How can the global international economic system be effectively reformed?

3. Why is the United States accused of behaving in an irresponsible fashion in the international financial and economic system?

Create Central

www.mhhe.com/createcentral

Internet References

BRICS
  http://www.bricsforum.org
Group of 20
  https://www.g20.org
Millennium Development Goals
  http://www.un.org/millennium goals
Third World Network
  http://www.twnside.org/sg

Contributing editor **MOHAMED A. EL-ERIAN** is CEO and co-chief investment officer of global investment management firm Pimco and author of When Markets Collide.

Article
Prepared by: Robert Weiner, *University of Massachusetts, Boston*

New World Order: Labor, Capital, and Ideas in the Power Law Economy

Erik Brynjolfsson, Andrew McAfee, and Michael Spence

Learning Outcomes

After reading this article, you will be able to:

- Understand the relationship between globalization and digital technology.

- Understand what the authors mean by power law.

Recent advances in technology have created an increasingly-unified global marketplace for labor and capital. The ability of both to flow to their highest-value uses, regardless of their location, is equalizing their prices across the globe. In recent years, this broad factor-price equalization has benefited nations with abundant low-cost labor and those with access to cheap capital. Some have argued that the current era of rapid technological progress serves labor, and some have argued that it serves capital. What both camps have slighted is the fact that technology is not only integrating existing sources of labor and capital but also creating new ones.

Machines are substituting for more types of human labor than ever before. As they replicate themselves, they are also creating more capital. This means that the real winners of the future will not be the providers of cheap labor or the owners of ordinary capital, both of whom will be increasingly squeezed by automation. Fortune will instead favor a third group: those who can innovate and create new products, services, and business models.

The distribution of income for this creative class typically takes the form of a power law, with a small number of winners capturing most of the rewards and a long tail consisting of the rest of the participants. So in the future, ideas will be the real scarce inputs in the world—scarcer than both labor and capital—and the few who provide good ideas will reap huge rewards. Assuring an acceptable standard of living for the rest and building inclusive economies and societies will become increasingly important challenges in the years to come.

Labor Pains

Turn over your iPhone and you can read an eight-word business plan that has served Apple well: "Designed by Apple in California. Assembled in China." With a market capitalization of over $500 billion, Apple has become the most valuable company in the world. Variants of this strategy have worked not only for Apple and other large global enterprises but also for medium-sized firms and even "micro-multinationals." More and more companies have been riding the two great forces of our era—technology and globalization—to profits.

Technology has sped globalization forward, dramatically lowering communication and transaction costs and moving the world much closer to a single, large global market for labor, capital, and other inputs to production. Even though labor is not fully mobile, the other factors increasingly are. As a result, the various components of global supply chains can move to labor's location with little friction or cost. About one-third of the goods and services in advanced economies are tradable, and the figure is rising. And the effect of global competition spills over to the nontradable part of the economy, in both advanced and developing economies.

All of this creates opportunities for not only greater efficiencies and profits but also enormous dislocations. If a worker in China or India can do the same work as one in the United States, then the laws of economics dictate that they will end up earning similar wages (adjusted for some other differences in national productivity). That's good news for overall economic efficiency, for consumers, and for workers in developing countries—but not for workers in developed countries who now face low-cost competition. Research indicates that the tradable sectors of advanced industrial countries have not been net employment generators for two decades. That means job creation now takes place almost exclusively within the large nontradable sector, whose wages are held down by increasing competition from workers displaced from the tradable sector.

Even as the globalization story continues, however, an even bigger one is starting to unfold: the story of automation, including artificial intelligence, robotics, 3D printing, and so on. And this second story is surpassing the first, with some of its greatest effects destined to hit relatively unskilled workers in developing nations.

Visit a factory in China's Guangdong Province, for example, and you will see thousands of young people working day in and day out on routine, repetitive tasks, such as connecting two parts of a keyboard. Such jobs are rarely, if ever, seen anymore in the United States or the rest of the rich world. But they may not exist for long in China and the rest of the developing world either, for they involve exactly the type of tasks that are easy for robots to do. As intelligent machines become cheaper and more capable, they will increasingly replace human labor, especially in relatively structured environments such as factories and especially for the most routine and repetitive tasks. To put it another way, offshoring is often only a way station on the road to automation.

This will happen even where labor costs are low. Indeed, Foxconn, the Chinese company that assembles iPhones and iPads, employs more than a million low-income workers, but now, it is supplementing and replacing them with a growing army of robots. So after many manufacturing jobs moved from the United States to China, they appear to be vanishing from China as well. (Reliable data on this transition are hard to come by. Official Chinese figures report a decline of 30 million manufacturing jobs since 1996, or 25 percent of the total, even as manufacturing output has soared by over 70 percent, but part of that drop may reflect revisions in the methods of gathering data.) As work stops chasing cheap labor, moreover, it will gravitate toward wherever the final market is, since that will add value by shortening delivery times, reducing inventory costs, and the like. The growing capabilities of automation threaten one of the most reliable strategies that poor countries have used to attract outside investment: offering low wages to compensate for low productivity and skill levels. And the trend will extend beyond manufacturing.

Interactive voice response systems, for example, are reducing the requirement for direct person-to-person interaction, spelling trouble for call centers in the developing world. Similarly, increasingly reliable computer programs will cut into transcription work now often done in the developing world. In more and more domains, the most cost-effective source of "labor" is becoming intelligent and flexible machines as opposed to low-wage humans in other countries.

Capital Punishment

If cheap, abundant labor is no longer a clear path to economic progress, then what is? One school of thought points to the growing contributions of capital: the physical and intangible assets that combine with labor to produce the goods and services in an economy (think of equipment, buildings, patents, brands, and so on). As the economist Thomas Piketty argues in his best-selling book *Capital in the Twenty-first Century,* capital's share of the economy tends to grow when the rate of return on it is greater than the general rate of economic growth, a condition he predicts for the future. The "capital deepening" of economies that Piketty forecasts will be accelerated further as robots, computers, and software (all of which are forms of capital) increasingly substitute for human workers. Evidence indicates that just such a form of capital-based technological change is taking place in the United States and around the world.

In the past decade, the historically consistent division in the United States between the share of total national income going to labor and that going to physical capital seems to have changed significantly. As the economists Susan Fleck, John Glaser, and Shawn Sprague noted in the U.S. Bureau of Labor Statistics' Monthly Labor Review in 2011, "Labor share averaged 64.3 percent from 1947 to 2000. Labor share has declined over the past decade, falling to its lowest point in the third quarter of 2010, 57.8 percent." Recent moves to "re-shore" production from overseas, including Apple's decision to produce its new Mac Pro computer in Texas, will do little to reverse this trend. For in order to be economically viable, these new domestic manufacturing facilities will need to be highly automated.

Other countries are witnessing similar trends. Economists Loukas Karabarbounis and Brent Neiman have documented significant declines in labor's share of GDP in 42 of the 59 countries they studied, including China, India, and Mexico. In describing their findings, Karabarbounis and Neiman are explicit that progress in digital technologies is an important driver of this phenomenon: "The decrease in the relative price of investment goods, often attributed to advances in information technology and the computer age, induced firms to shift away from labor and toward capital. The lower price of investment goods explains roughly half of the observed decline in the labor share."

But if capital's share of national income has been growing, the continuation of such a trend into the future may be in jeopardy as a new challenge to capital emerges—not from a revived labor sector but from an increasingly important unit within its own ranks: digital capital.

In a free market, the biggest premiums go to the scarcest inputs needed for production. In a world where capital such as software and robots can be replicated cheaply, its marginal value will tend to fall, even if more of it is used in the aggregate. And as more capital is added cheaply at the margin, the value of existing capital will actually be driven down. Unlike, say, traditional factories, many types of digital capital can be added extremely cheaply. Software can be duplicated and distributed at almost zero incremental cost. And many elements of computer hardware, governed by variants of Moore's law, get quickly and consistently cheaper over time. Digital capital, in short, is abundant, has low marginal costs, and is increasingly important in almost every industry.

Even as production becomes more capital-intensive, therefore, the rewards earned by capitalists as a group may not necessarily continue to grow relative to labor. The shares will depend on the exact details of the production, distribution, and governance systems.

Most of all, the payoff will depend on which inputs to production are scarcest. If digital technologies create cheap substitutes for a growing set of jobs, then it is not a good time to be a laborer. But if digital technologies also increasingly substitute for capital, then all owners of capital should not expect to earn outsized returns, either.

Techcrunch Disrupt

What will be the scarcest, and hence the most valuable, resource in what two of us (Erik Brynjolfsson and Andrew McAfee) have called "the second machine age," an era driven by digital technologies and their associated economic characteristics? It will be neither ordinary labor nor ordinary capital but people who can create new ideas and innovations.

Such people have always been economically valuable, of course, and have often profited handsomely from their innovations as a result. But they had to share the returns on their ideas with the labor and capital that were necessary for bringing them into the marketplace. Digital technologies increasingly make both ordinary labor and ordinary capital commodities, and so a greater share of the rewards from ideas will go to the creators, innovators, and entrepreneurs. People with ideas, not workers or investors, will be the scarcest resource.

The most basic model economists use to explain technology's impact treats it as a simple multiplier for everything else, increasing overall productivity evenly for everyone. This model is used in most introductory economics classes and provides the foundation for the common—and, until recently, very sensible—intuition that a rising tide of technological progress will lift all boats equally, making all workers more productive and hence more valuable.

A slightly more complex and realistic model, however, allows for the possibility that technology may not affect all inputs equally but instead favor some more than others. Skill-based technical change, for example, plays to the advantage of more skilled workers relative to less skilled ones, and capital-based technical change favors capital relative to labor. Both of those types of technical change have been important in the past, but increasingly, a third type—what we call superstar-based technical change—is upending the global economy.

Today, it is possible to take many important goods, services, and processes and codify them. Once codified, they can be digitized, and once digitized, they can be replicated. Digital copies can be made at virtually zero cost and transmitted anywhere in the world almost instantaneously, each an exact replica of the original. The combination of these three characteristics—extremely low cost, rapid ubiquity, and perfect fidelity—leads to some weird and wonderful economics. It can create abundance where there had been scarcity, not only for consumer goods, such as music videos, but also for economic inputs, such as certain types of labor and capital.

The returns in such markets typically follow a distinct pattern—a power law, or Pareto curve, in which a small number of players reap a disproportionate share of the rewards. Network effects, whereby a product becomes more valuable the more users it has, can also generate these kinds of winner-take-all or winner-take-most markets. Consider Instagram, the photo-sharing platform, as an example of the economics of the digital, networked economy. The 14 people who created the company didn't need a lot of unskilled human helpers to do so, nor did they need much physical capital. They built a digital product that benefited from network effects, and when it caught on quickly, they were able to sell it after only a year and a half for nearly three-quarters of a billion dollars—ironically, months after the bankruptcy of another photography company, Kodak, that at its peak had employed some 145,000 people and held billions of dollars in capital assets.

Instagram is an extreme example of a more general rule. More often than not, when improvements in digital technologies make it more attractive to digitize a product or process, superstars see a boost in their incomes, whereas second bests, second movers, and latecomers have a harder time competing. The top performers in music, sports, and other areas have also seen their reach and incomes grow since the 1980s, directly or indirectly riding the same trends upward.

But it is not only software and media that are being transformed. Digitization and networks are becoming more pervasive in every industry and function across the economy, from retail and financial services to manufacturing and marketing. That means superstar economics are affecting more goods, services, and people than ever before.

Even top executives have started earning rock-star compensation. In 1990, CEO pay in the United States was, on average, 70 times as large as the salaries of other workers; in 2005, it was 300 times as large. Executive compensation more generally has been going in the same direction globally, albeit with considerable variation from country to country. Many forces are at work here, including tax and policy changes, evolving cultural and organizational norms, and plain luck. But as research by one of us (Brynjolfsson) and Heekyung Kim has shown, a portion of the growth is linked to the greater use of information technology. Technology expands the potential reach, scale, and monitoring capacity of a decision-maker, increasing the value of a good decision-maker by magnifying the potential consequences of his or her choices. Direct management via digital technologies makes a good manager more valuable than in earlier times, when executives had to share control with long chains of subordinates and could affect only a smaller range of activities. Today, the larger the market value of a company, the more compelling the argument for trying to get the very best executives to lead it.

When income is distributed according to a power law, most people will be below the average, and as national economies writ large are increasingly subject to such dynamics, that pattern will play itself out on the national level. And sure enough, the United States today features one of the world's highest levels of real GDP per capita—even as its median income has essentially stagnated for two decades.

Preparing for the Permanent Revolution

The forces at work in the second machine age are powerful, interactive, and complex. It is impossible to look far into the future and predict with any precision what their ultimate impact will be. If individuals, businesses, and governments understand what is going on, however, they can at least try to adjust and adapt.

The United States, for example, stands to win back some business as the second sentence of Apple's eight-word business plan is overturned because its technology and manufacturing operations are once again performed inside U.S. borders. But the first sentence of the plan will become more important than ever, and here, concern, rather than complacency, is in order. For unfortunately, the dynamism and creativity that have made the United States the most innovative nation in the world may be faltering.

Thanks to the ever-onrushing digital revolution, design and innovation have now become part of the tradable sector of the global economy and will face the same sort of competition that has already transformed manufacturing. Leadership in design depends on an educated work force and an entrepreneurial culture, and the traditional American advantage in these areas is declining. Although the United States once led the world in the share of graduates in the work force with at least an associate's degree, it has now fallen to 12th place. And despite the buzz about entrepreneurship in places such as Silicon Valley, data show that since 1996, the number of U.S. start-ups employing more than one person has declined by over 20 percent.

If the trends under discussion are global, their local effects will be shaped, in part, by the social policies and investments that countries choose to make, both in the education sector specifically and in fostering innovation and economic dynamism more generally. For over a century, the U.S. educational system was the envy of the world, with universal K-12 schooling and world-class universities propelling sustained economic growth. But in recent decades, U.S. primary and secondary schooling have become increasingly uneven, with their quality based on neighborhood income levels and often a continued emphasis on rote learning.

Fortunately, the same digital revolution that is transforming product and labor markets can help transform education as well. Online learning can provide students with access to the best teachers, content, and methods regardless of their location, and new data-driven approaches to the field can make it easier to measure students' strengths, weaknesses, and progress. This should create opportunities for personalized learning programs and continuous improvement, using some of the feedback techniques that have already transformed scientific discovery, retail, and manufacturing.

Globalization and technological change may increase the wealth and economic efficiency of nations and the world at large, but they will not work to everybody's advantage, at least in the short to medium term. Ordinary workers, in particular, will continue to bear the brunt of the changes, benefiting as consumers but not necessarily as producers. This means that without further intervention, economic inequality is likely to continue to increase, posing a variety of problems. Unequal incomes can lead to unequal opportunities, depriving nations of access to talent and undermining the social contract. Political power, meanwhile, often follows economic power, in this case undermining democracy.

These challenges can and need to be addressed through the public provision of high-quality basic services, including education, health care, and retirement security. Such services will be crucial for creating genuine equality of opportunity in a rapidly changing economic environment and increasing intergenerational mobility in income, wealth, and future prospects.

As for spurring economic growth in general, there is a near consensus among serious economists about many of the policies that are necessary. The basic strategy is intellectually simple, if politically difficult: boost public-sector investment over the short and medium term while making such investment more efficient and putting in place a fiscal consolidation plan over the longer term. Public investments are known to yield high returns in basic research in health, science, and technology; in education; and in infrastructure spending on roads, airports, public water and sanitation systems, and energy and communications grids. Increased government spending in these areas would boost economic growth now even as it created real wealth for subsequent generations later.

Should the digital revolution continue to be as powerful in the future as it has been in recent years, the structure of the modern economy and the role of work itself may need to be rethought. As a group, our descendants may work fewer hours and live better— but both the work and the rewards could be spread even more unequally, with a variety of unpleasant consequences. Creating sustainable, equitable, and inclusive growth will require more than business as usual. The place to start is with a proper understanding of just how fast and far things are evolving.

Critical Thinking

1. What is the relationship between globalization and technology in bringing about more innovation?

2. Why is the relationship between labor, capital, and ideas important?

3. Why is what the authors call the second machine age important?

Create Central

www.mhhe.com/createcentral

Internet References

Commission on Digital Economy
http://www.iccw60.org/about-icc/policy-commissions/digital-economy

Embracing Digital technology
http://sloanreview.mit.edu/projects/embracing-digital-technology

MIT Center for Digital Business
http://digital.mit.edu

ERIK BRYNJOLFSSON is Schussel Family Professor of Management Science at the MIT Sloan School of Management and Co-Founder of MIT's Initiative on the Digital Economy. **ANDREW MCAFEE** is a Principal Research Scientist at the MIT Center for Digital Business at the MIT Sloan School of Management and Co-Founder of MIT's Initiative on the Digital Economy. **MICHAEL SPENCE** is **WILLIAM R. BERKLEY** Professor in Economics and Business at the NYU Stern School of Business.

Article   Prepared by: Robert Weiner, *University of Massachusetts, Boston*

As Objects Go Online: The Promise (and Pitfalls) of the Internet of Things

Neil Gershenfeld and J. P. Vasseur

Learning Outcomes

After reading this article, you will be able to:

- Understand what the Internet of Things is.

- Understand why the Internet of Things may represent a third industrial revolution.

Since 1969, when the first bit of data was transmitted over what would come to be known as the Internet, that global network has evolved from linking mainframe computers to connecting personal computers and now mobile devices. By 2010, the number of computers on the Internet had surpassed the number of people on earth.

Yet that impressive growth is about to be overshadowed as the things around us start going online as well, part of what is called "the Internet of Things." Thanks to advances in circuits and software, it is now possible to make a Web server that fits on (or in) a fingertip for $1. When embedded in everyday objects, these small computers can send and receive information via the Internet so that a coffeemaker can turn on when a person gets out of bed and turn off when a cup is loaded into a dishwasher, a stoplight can communicate with roads to route cars around traffic, a building can operate more efficiently by knowing where people are and what they're doing, and even the health of the whole planet can be monitored in real time by aggregating the data from all such devices.

Linking the digital and physical worlds in these ways will have profound implications for both. But this future won't be realized unless the Internet of Things learns from the history of the Internet. The open standards and decentralized design of the Internet won out over competing proprietary systems and centralized control by offering fewer obstacles to innovation and growth. This battle has resurfaced with the proliferation of conflicting visions of how

devices should communicate. The challenge is primarily organizational, rather then technological, a contest between command-and-control technology and distributed solutions. The Internet of Things demands the latter, and openness will eventually triumph.

The Connected Life

The Internet of Things is not just science fiction; it has already arrived. Some of the things currently networked together send data over the public Internet, and some communicate over secure private networks, but all share common protocols that allow them to interoperate to help solve profound problems.

Take energy inefficiency. Buildings account for three-quarters of all electricity use in the United States, and of that, about one-third is wasted. Lights stay on when there is natural light available, and air is cooled even when the weather outside is more comfortable or a room is unoccupied. Sometimes fans move air in the wrong direction or heating and cooling systems are operated simultaneously. This enormous amount of waste persists because the behavior of thermostats and light bulbs are set when buildings are constructed; the wiring is fixed and the controllers are inaccessible. Only when the infrastructure itself becomes intelligent, with networked sensors and actuators, can the efficiency of a building be improved over the course of its lifetime.

Health care is another area of huge promise. The mismanagement of medication, for example, costs the health-care system billions of dollars per year. Shelves and pill bottles connected to the Internet can alert a forgetful patient when to take a pill, a pharmacist to make a refill, and a doctor when a dose is missed. Floors can call for help if a senior citizen has fallen, helping the elderly live independently. Wearable sensors could monitor one's activity throughout the day and serve as personal coaches, improving health and saving costs.

Countless futuristic "smart houses" have yet to generate much interest in living in them. But the Internet of Things succeeds to the extent that it is invisible. A refrigerator could communicate with a grocery store to reorder food, with a bathroom scale to monitor a diet, with a power utility to lower electricity consumption during peak demand, and with its manufacturer when maintenance is needed. Switches and lights in a house could adapt to how spaces are used and to the time of day. Thermostats with access to calendars, beds, and cars could plan heating and cooling based on the location of the house's occupants. Utilities today provide power and plumbing; these new services would provide safety, comfort, and convenience.

In cities, the Internet of Things will collect a wealth of new data. Understanding the flow of vehicles, utilities, and people is essential to maximizing the productivity of each, but traditionally, this has been measured poorly, if at all. If every street lamp, fire hydrant, bus, and crosswalk were connected to the Internet, then a city could generate real-time readouts of what's working and what's not. Rather than keeping this information internally, city hall could share open-source data sets with developers, as some cities are already doing.

Weather, agricultural inputs, and pollution levels all change with more local variation than can be captured by point measurements and remote sensing. But when the cost of an Internet connection falls far enough, these phenomena can all be measured precisely. Networking nature can help conserve animate, as well as inanimate, resources; an emerging "interspecies Internet" is linking elephants, dolphins, great apes, and other animals for the purposes of enrichment, research, and preservation.

The ultimate realization of the Internet of Things will be to transmit actual things through the Internet. Users can already send descriptions of objects that can be made with personal digital fabrication tools, such as 3D printers and laser cutters. As data turn into things and things into data, long manufacturing supply chains can be replaced by a process of shipping data over the Internet to local production facilities that would make objects on demand, where and when they were needed.

Back to the Future

To understand how the Internet of Things works, it is helpful to understand how the Internet itself works, and why. The first secret of the Internet's success is its architecture. At the time the Internet was being developed, in the 1960s and 1970s, telephones were wired to central office switchboards. That setup was analogous to a city in which every road goes through one traffic circle; it makes it easy to give directions but causes traffic jams at the central hub. To avoid such problems, the Internet's developers created a distributed network, analogous to the web of streets that vehicles navigate in a real city. This design lets data bypass traffic jams and lets managers add capacity where needed.

The second key insight in the Internet's development was the importance of breaking data down into individual chunks that could be reassembled after their online journey. "Packet switching," as this process is called, is like a railway system in which each railcar travels independently. Cars with different destinations share the same tracks, instead of having to wait for one long train to pass, and those going to the same place do not all have to take the same route. As long as each car has an address and each junction indicates where the tracks lead, the cars can be combined on arrival. By transmitting data in this way, packet switching has made the Internet more reliable, robust, and efficient.

The third crucial decision was to make it possible for data to flow over different types of networks, so that a message can travel through the wires in a building, into a fiber-optic cable that carries it across a city, and then to a satellite that sends it to another continent. To allow that, computer scientists developed the Internet Protocol, or IP, which standardized the way that packets of data were addressed. The equivalent development in railroads was the introduction of a standard track gauge, which allowed trains to cross international borders. The IP standard allows many different types of data to travel over a common protocol.

The fourth crucial choice was to have the functions of the Internet reside at the ends of the network, rather than at the intermediate nodes, which are reserved for routing traffic. Known as the "end-to-end principle," this design allows new applications to be invented and added without having to upgrade the whole network. The capabilities of a traditional telephone were only as advanced as the central office switch it was connected to, and those changed infrequently. But the layered architecture of the Internet avoids this problem. Online messaging, audio and video streaming, e-commerce, search engines, and social media were all developed on top of a system designed decades earlier, and new applications can be created from these components.

These principles may sound intuitive, but until recently, they were not shared by the systems that linked things other than computers. Instead, each industry, from heating and cooling to consumer electronics, created its own networking standards, which specified not only how their devices communicated with one another but also what they could communicate. This closed model may work within a fixed domain, but unlike the model used for the Internet, it limits future possibilities to what its creators originally anticipated. Moreover, each of these standards has struggled with the same problems the Internet has already solved: how to assign network names to devices, how to route messages between networks, how to manage the flow of traffic, and how to secure communications.

Although it might seem logical now to use the Internet to link things rather than reinvent the networking wheel for each industry, that has not been the norm so far. One reason is that manufacturers have wanted to establish proprietary control. The Internet does not have tollbooths, but if a vendor can control the communications standards used by the devices in a given industry, it can charge companies to use them.

Compounding this problem was the belief that special purpose solutions would perform better than the general-purpose Internet. In reality, these alternatives were less well developed and lacked the Internet's economies of scale and reliability. Their designers overvalued optimal functionality at the expense of interoperability. For any given purpose, the networking standards of the Internet are not ideal, but for almost anything, they are good enough. Not only do proprietary networks entail the high cost of maintaining multiple, incompatible standards; they have also been less secure. Decades of attacks on the Internet have led a large community of researchers and vendors to continually refine its defenses, which can now be applied to securing communications among things.

Finally, there was the problem of cost. The Internet relied at first on large computers that cost hundreds of thousands of dollars and then on $1,000 personal computers. The economics of the Internet were so far removed from the economics of light bulbs and doorknobs that developers never thought it would be commercially viable to put such objects online; the market for $1,000 light switches is limited. And so, for many decades, objects remained offline.

Big Things in Small Packages

But no longer do economic or technological barriers stand in the way of the Internet of Things. The unsung hero that has made this possible is the microcontroller, which consists of a simple processor packaged with a small amount of memory and peripheral parts. Microcontrollers measure just millimeters across, cost just pennies to manufacture, and use just milliwatts of electricity, so that they can run for years on a battery or a small solar cell. Unlike a personal computer, which now boasts billions of bytes of memory, a microcontroller may contain only thousands of bytes. That's not enough to run today's desktop programs, but it matches the capabilities of the computers used to develop the Internet.

Around 1995, we and our colleagues based at mit began using these parts to simplify Internet connections. That project grew into a collaboration with a group of the Internet's original architects, starting with the computer scientist Danny Cohen, to extend the Internet into things. Since "Internet2" had already been used to refer to the project for a higher-speed Internet, we chose to call this slower and simpler Internet "Internet 0."

The goal of Internet 0 was to bring IP to the smallest devices. By networking a smart light bulb and a smart light switch directly, we could enable these devices to turn themselves on and off rather than their having to communicate with a controller connected to the Internet. That way, new applications could be developed to communicate with the light and the switch, and without being limited by the capabilities of a controller.

Giving objects access to the Internet simplifies hard problems. Consider the Electronic Product Code (the successor to the familiar bar code), which retailers are starting to use in radio-frequency identification tags on their products. With great effort, the developers of the EPC have attempted to enumerate all possible products and track them centrally. Instead, the information in these tags could be replaced with packets of Internet data, so that objects could contain instructions that varied with the context: at the checkout counter in a store, a tag on a medicine bottle could communicate with a merchandise database; in a hospital, it could link to a patient's records.

Along with simplifying Internet connections, the Internet 0 project also simplified the networks that things link to. The quest for ever-faster networks has led to very different standards for each medium used to transmit data, with each requiring its own special precautions. But Morse code looks the same whether it is transmitted using flags or flashing lights, and in the same way, Internet 0 packages data in a way that is independent of the medium. Like IP, that's not optimal, but it trades speed for cheapness and simplicity. That makes sense, because high speed is not essential: light bulbs, after all, don't watch broadband movies.

Another innovation allowing the Internet to reach things is the ongoing transition from the previous version of IP to a new one. When the designers of the original standard, called IPv4, launched it in 1981, they used 32 bits (each either a zero or a one) to store each IP address, the unique identifiers assigned to every device connected to the Internet—allowing for over four billion IP addresses in total. That seemed like an enormous number at the time, but it is less than one address for every person on the planet. IPv4 has run out of addresses, and it is now being replaced with a new version, IPv6. The new standard uses 128-bit IP addresses, creating more possible identifiers than there are stars in the universe. With IPv6, everything can now get its own unique address.

But IPv6 still needs to cope with the unique requirements of the Internet of Things. Along with having limitations involving memory, speed, and power, devices can appear and disappear on the network intermittently, either to save energy or because they are on the move. And in big enough numbers, even simple sensors can quickly overwhelm existing network infrastructure; a city might contain millions of power meters and billions of electrical outlets. So in collaboration with our colleagues, we are developing extensions of the Internet protocols to handle these demands.

The Inevitable Internet

Although the Internet of Things is now technologically possible, its adoption is limited by a new version of an old conflict. During the 1980s, the Internet competed with a network called Bitnet, a centralized system that linked mainframe computers. Buying a mainframe was expensive, and so Bitnet's growth was limited; connecting personal computers to the Internet made more sense. The Internet won out, and by the early 1990s, Bitnet had fallen out of use. Today, a similar battle is emerging between the Internet of Things and what could be called the Bitnet of Things. The key distinction is where information resides:

in a smart device with its own IP address or in a dumb device wired to a proprietary controller with an Internet connection. Confusingly, the latter setup is itself frequently characterized as part of the Internet of Things. As with the Internet and bitnet, the difference between the two models is far from semantic. Extending IP to the ends of a network enables innovation at its edges; linking devices to the Internet indirectly erects barriers to their use.

The same conflicting meanings appear in use of the term "smart grid," which refers to networking everything that generates, controls, and consumes electricity. Smart grids promise to reduce the need for power plants by intelligently managing loads during peak demand, varying pricing dynamically to provide incentives for energy efficiency, and feeding power back into the grid from many small renewable sources. In the not-so-smart, utility-centric approach, these functions would all be centrally controlled. In the competing, Internet-centric approach, they would not, and its dispersed character would allow for a marketplace for developers to design power-saving applications.

Putting the power grid online raises obvious cybersecurity concerns, but centralized control would only magnify these problems. The history of the Internet has shown that security through obscurity doesn't work. Systems that have kept their inner workings a secret in the name of security have consistently proved more vulnerable than those that have allowed themselves to be examined—and challenged—by outsiders. The open protocols and programs used to protect Internet communications are the result of ongoing development and testing by a large expert community.

Another historical lesson is that people, not technology, are the most common weakness when it comes to security. No matter how secure a system is, someone who has access to it can always be corrupted, wittingly or otherwise. Centralized control introduces a point of vulnerability that is not present in a distributed system.

The flip side of security is privacy; eavesdropping takes on an entirely new meaning when actual eaves can do it. But privacy can be protected on the Internet of Things. Today, privacy on the rest of the Internet is safeguarded through cryptography, and it works: recent mass thefts of personal information have happened because firms failed to encrypt their customers' data, not because the hackers broke through strong protections. By extending cryptography down to the level of individual devices, the owners of those devices would gain a new kind of control over their personal information. Rather than maintaining secrecy as an absolute good, it could be priced based on the value of sharing. Users could set up a firewall to keep private the Internet traffic coming from the things in their homes—or they could share that data with, for example, a utility

that gave a discount for their operating their dishwasher only during off-peak hours or a health insurance provider that offered lower rates in return for their making healthier lifestyle choices.

The size and speed of the Internet have grown by nine orders of magnitude since the time it was invented. This expansion vastly exceeds what its developers anticipated, but that the Internet could get so far is a testament to their insight and vision. The uses the Internet has been put to that have driven this growth are even more surprising; they were not part of any original plan. But they are the result of an open architecture that left room for the unexpected. Likewise, today's vision for the Internet of Things is sure to be eclipsed by the reality of how it is actually used. But the history of the Internet provides principles to guide this development in ways that are scalable, robust, secure, and encouraging of innovation.

The Internet's defining attribute is its interoperability; information can cross geographic and technological boundaries. With the Internet of Things, it can now leap out of the desktop and data center and merge with the rest of the world. As the technology becomes more finely integrated into daily life, it will become, paradoxically, less visible. The future of the Internet is to literally disappear into the woodwork.

Critical Thinking

1. What civil rights and security issues may be raised by the Internet of Things?
2. How may the Internet of Things transform the world?

Create Central

www.mhhe.com/createcentral

Internet References

European Research Center on the Internet of Things
http://www.internet-of-things-research.eu/documents.htm

Goldman Sachs
http://www.goldmansachs.com/our-thinking/outlook/internet-ofthings/index
.html?cid=PS_01_89_07_00_00_OIM

The Internet of Things Council
http://www.theinternetofthings.eu/

Pew Research Internet Project
http://www.pewinternet.org

NEIL GERSHENFELD is a Professor at the Massachusetts Institute of Technology and directs MIT's Center for Bits and Atoms. **JP VASSEUR** is a Cisco Fellow and Chief Architect of the Internet of Things at Cisco Systems.

Neil Gershenfeld & J. P. Vasseur, "As Objects Go Online: The Promise (and Pitfalls) of the Internet of Things," *Foreign Affairs*, March/April 2014, pp. 60–67. Copyright © 2014 by Foreign Affairs. All rights reserved. Used with permission.

Article   Prepared by: Robert Weiner, *University of Massachusetts, Boston*

Britain and Europe: The End of the Affair?

Matthias Matthijs

Learning Outcomes

After reading this article, you will be able to:

- Be familiar with the history of the relationship between the United Kingdom and Europe.

- Understand the relationship between the British system of government and the European Union.

After a tumultuous professional marriage of just over 40 years, Britain and Europe are facing the possibility of divorce. In January 2013, Prime Minister David Cameron decided to celebrate Britain's 40th anniversary as a member of the European Union by pledging a fundamental renegotiation of his country's terms of membership. Cameron further promised to submit any renegotiated deal to a clear "in-or-out" referendum in 2017 on whether or not to leave the EU, assuming his own Conservative Party wins a majority in the next general election in May 2015. Egged on by his party's growing ranks of restive Euroskeptics and trying to fight off a challenge on his right flank from populist Nigel Farage's UK Independence Party (UKIP), Cameron rolled the dice. He hoped to settle once and for all the Europe question, which has so often cast a dark shadow over the political debate in Westminster and Whitehall.

Renegotiating international treaties is extremely difficult, given that such pacts usually result from carefully crafted compromises among multiple states. Undoing one element could quickly unravel the whole construction. Additionally, the 27 other members of the EU—emerging cautiously from the existential angst of the euro crisis and visibly frustrated with Britain's increasingly obstructive attitude toward Brussels—are in no mood to permit substantial steps in the direction of à la carte membership. Allowing such flexibility for Britain would open the door to renegotiations for other members as well. While there is undoubtedly some sympathy for Britain's qualms from like-minded northern member states such as Sweden, the Netherlands, and Germany, any new deal that Cameron can negotiate will likely fall well short of his party's Euroskeptic bottom line.

A Conservative majority is still a distant prospect for next year's general election—at the time of writing, another hung Parliament seems the most likely outcome—but it is certainly within the realm of possibility, especially if growth picks up, living standards start to improve, and the economy recovers from 5 years of stagnation. As a result, Britain today is as close as it has ever been to actually leaving the EU, and at risk of turning inward to embrace not-so-splendid isolation.

How did it come to this? Cameron is not the first occupant of 10 Downing Street to struggle with former US Secretary of State Dean Acheson's famous thesis, expounded in a 1962 speech at West Point, that Britain had "lost an empire and has not yet found a role." Ever since World War II, British prime ministers—Edward Heath being the one notable exception—have tried to deny their country's European destiny.

Resisting the calls for unity from Brussels in favor of a rather vague notion of a "global" Britain, free from continental chains, most British leaders either have been seduced by the mirage of being America's junior partner or have fallen prey to the legacy of an empire on which the sun never set. However, since Heath achieved accession to the European Economic Community (EEC) in 1973, every British leader has been unable to stop the momentum behind European integration. They have found their country—for better or worse—tied closer to Europe and its supranational institutions than they were ever willing to admit.

Since the advent of the euro crisis, though, the dynamic of European integration has qualitatively changed. The pace of integration has dramatically picked up, and the direction Europe is now taking toward more supranational oversight of economic and financial policy is increasingly at odds with how Britain has defined its national interests. The City, London's financial district, is worried about a barrage of restrictive regulations from Brussels. With the UK unlikely to join the euro and, with continental Europeans determined to do whatever it takes to save the common currency—including surrendering ever more sovereignty to Brussels to build a more genuine Economic and Monetary Union (EMU)—London has started to wonder whether its EU game is still worth the candle.

Postwar Fog

V-E Day—May 8, 1945—marked the end of European hostilities in World War II and put Britain in the unique position of being the only European power that had not been occupied or defeated. This fact alone made the country of Winston Churchill the natural leader of Europe. The small island nation had stood alone against Nazi Germany for 18 long months. Aside from an upsurge of patriotic fervor, the other legacy of war was that it left Britain financially vulnerable, if not bankrupt. Britain managed to stay afloat during the war thanks to America's Lend-Lease Act, but when that funding was abruptly cut off in the summer of 1945, it left the new Labour Party government of Clement Attlee scrambling.

At the same time, Britain was quickly exposed as a power in decline, suffering from "imperial overstretch" (as the historian Paul Kennedy put it). It faced turmoil in India, a relentless drain of US dollars to pay for national reconstruction, the mounting cost of building a universal welfare state at home, and the need to maintain the British garrison in defeated Germany. India—the jewel in the crown of the British Empire—became an independent country in 1947. It was not until Marshall Plan aid reached Britain in 1948 and a 30 percent devaluation of sterling in 1949 that Britain's economy started to make a full recovery.

The Cold War was under way, and it was clear to many observers at the time (though to almost no one in Britain) that the world was increasingly turning bipolar, with America in the West and the Soviet Union in the East fighting for global supremacy. Britain was relegated to second-power status, occupied with "the orderly management of decline." The first three postwar prime ministers—Attlee (1945–51), Churchill (1951–55), and Anthony Eden (1955–57)—all preferred to ignore reality and deliberately kept their foreign policy focus away from Europe, toward the wider world.

Defeated and humiliated, France realized that any future peace in Europe could only be secured through some kind of pragmatic reconciliation with its archenemy Germany. Britain had initially resisted taking part in the continental endeavors

of what quickly became "the Six" (France, West Germany, Italy, and the Benelux countries), starting with French Foreign Minister Robert Schuman's call for a European Coal and Steel Community in 1950. Britain was also notably absent in Messina, Italy, in 1955 when the idea of a European common market first took hold. An aging Churchill, back in office in 1951, showed no interest. Nor did his successor Eden, whose chancellor of the exchequer, R.A. Butler, derisively referred to the Messina talks as "archaeological excavations." But the Six went ahead, and the Treaty of Rome was signed in May 1957 without Britain's participation.

In October of that same year, a London *Times* headline famously read: "Heavy Fog in Channel—Continent Cut Off." Nothing summed up better the British state of mind regarding Europe than the idea that the world still evolved around "Great" Britain—that the continent could somehow be "cut off" from the island, rather than the other way around. The "heavy fog" in the channel was an apt metaphor for the enduring and often willful British misreading of what exactly those continentals in Brussels were up to.

When Harold Macmillan became prime minister in 1957, Britain's attitude toward Europe slowly started to change. While he was himself very much a Conservative politician in the mold of Churchill, periodically musing that postwar Britain could play the role of an older and wiser Greece to America's increasingly imperial Rome, Macmillan was also a realist and a pragmatist. Not only did he observe the "winds of change" of national independence movements all over British Africa, he also saw the continental economies systematically outperform Britain's during the 1950s.

Macmillan eventually submitted a half-hearted application to Brussels in the early 1960s but went out of his way to emphasize that this was merely to find out whether "favorable membership conditions" could be established. French President Charles de Gaulle was having none of it, seeing Britain's application as an American Trojan horse, and proclaimed an unequivocal *"Non"* at a January 1963 press conference.

After Labour came back to power in 1964, Prime Minister Harold Wilson eventually decided to reapply in 1967, but for the second time de Gaulle issued a veto. A few months later, Wilson announced the withdrawal of all British military forces from "east of Suez." Britain had reached the limit of its global pretentions and needed to retrench. The Europe question still loomed, with the fog in the Channel thicker than ever.

Even though Britain may decide to leave Europe, Europe will never leave Britain.

Bold Statesmanship

"A week is a long time in politics," Wilson once remarked to an aide. Two years after his second veto of Britain, de Gaulle left the French political scene and was replaced by Georges Pompidou, himself a Gaullist but much less intransigent than his predecessor. One year later, in 1970, Edward Heath's Tories surprised everyone by beating Wilson's Labour in a general election. Suddenly the Europe question took on a renewed sense of urgency. Heath, who had experienced the carnage of World War II firsthand in both France and the Low Countries, was a true man of Europe. Having participated in the British liberation of Antwerp in September 1944, he was part of the "never again" generation of Europeans who passionately believed that reconstruction and reconciliation had to go hand in hand with greater political unity.

While his government's official reasons for reapplying to join the EEC in the early 1970s were mainly economic, Heath always emphasized the broader political significance of Britain's fully belonging to Europe. The negotiations were relatively swift, even though some difficult issues like Britain's future budgetary contribution would have to be resolved later. Britain's entry in January 1973—alongside Denmark and Ireland—constituted a bold act of statesmanship and a personal triumph for Heath. At last, the Europe question received an unambiguous response. The fog had cleared.

The love affair would be short-lived. By early 1974, after yet another miners' strike, nationwide power cuts, and the imposition of a three-day workweek to conserve electricity, the British people answered the "Who Governs Britain?" general election slogan of Heath's Conservatives with "Not you." Wilson and Labour returned to power with a fragile minority government, and now had to honor their pledge to put Britain's EEC membership up for a nationwide referendum—the first in the country's history. Labour's left-wingers, led by Tony Benn and Barbara Castle, opposed the common market, which they saw as a free-market plot to undermine socialist planning and erode workers' rights.

To everyone's surprise, the 1975 referendum delivered a 2-to-1 endorsement of membership. While the Labour leadership, together with the opposition Conservatives, had campaigned in favor of staying in, Wilson and his successor James Callaghan did so only reluctantly. The same was true for the leader of the opposition, Margaret Thatcher, who lacked the European zeal of her predecessor Heath.

While the Europe question was settled for the moment, the second half of the 1970s and the early 1980s were spent dealing with economic crises at home and saw few steps toward further integration. An exception was the establishment of the European Monetary System of fixed exchange rates in 1979, but Britain, led by Callaghan, refused to join.

Thatcher and the Superstate

The Europe question regained prominence during the mid-1980s with Thatcher in her second term as prime minister, her big economic battles at home decisively won. The first issue on her European agenda was to renegotiate Britain's budgetary contribution. Since Britain had a relatively efficient agricultural sector, it received comparatively small subsidies from the EEC's Common Agricultural Policy. At the same time, being a trading nation with a long tradition of commerce with its former colonies, it also paid disproportionately more into the EEC budget than other members due to the common external tariff. Thatcher had made it clear that she wanted "our own money back" from Brussels.

During a June 1984 European summit in Fontainebleau, she bargained hard with French President François Mitterrand. They finally agreed that Britain would receive a 66 percent rebate of the amount it was "overpaying." The deal was hailed as a decisive victory for Thatcher back home, but her confrontational method of negotiation would soon reach its limits. Mitterrand and German Chancellor Helmut Kohl, together with Jacques Delors—Mitterrand's former finance minister and the new European Commission president—were determined to pursue further integration, despite Thatcher's stubborn opposition.

Delors's relaunch of Europe after 10 years of relatively little progress came in 1985, with a new intergovernmental conference on completing the common market. This led to the signing of the Single European Act in 1986, the first major revision of the Treaty of Rome. Thatcher eagerly signed on to the treaty because of its liberalizing, deregulating, and market-freeing potential and its overall sound economic rationale. However, the price she had to pay was an increase in qualified majority voting in the European Council, where more decisions concerning the common market would no longer require unanimity. The Single Act sailed through the House of Commons in six days, requiring little debate.

Britain today is as close as it has ever been to actually leaving the EU.

Thatcher exultantly claimed to have exported her free market revolution to the European continent. But that was not how Delors viewed the Single Act, which he favored because it made both political and economic sense, given the ascendancy of free market ideas at the time. For Delors, a *dirigiste* French socialist, the new treaty was but one necessary step toward a closer federal political union. Increased majority voting in the Council was a key part of that strategy.

On September 8, 1988, Delors received a hero's welcome at the annual meeting of Britain's Trades Union Congress, when thousands of Labour activists belted out "Frère Jacques"—most likely the only French tune they knew—marking Labour's shift away from its knee-jerk Euroskepticism toward an embrace of Delors's strategy. Delors became their brother in arms against Thatcherism's assault on union rights. Thatcher felt betrayed: This was not the Europe she had signed up for. Twelve days later, in a speech in Bruges, she attacked Delors's vision of Europe, declaring, "We have not successfully rolled back the frontiers of the state in Britain only to see them reimposed at a European level, with a European superstate exercising a new dominance from Brussels."

But the European train had already left the station. Thatcher found herself increasingly at odds with her two most faithful cabinet lieutenants, Nigel Lawson at the Treasury and Geoffrey Howe at the Foreign Office, over whether to join Europe's Exchange Rate Mechanism (ERM) to fight inflation at home—a strategy Lawson favored—and over her intransigence toward European integration, an attitude Howe began to despise. After Lawson and Howe resigned, Thatcher's animosity toward Europe only intensified as her reign drew to a close. Michael Heseltine, lamenting the disastrous state of Britain's relations with Europe because of "one woman's prejudice," openly challenged Thatcher's party leadership. John Major beat Heseltine in the Tory contest to succeed her.

Maastricht's Aftermath

After Thatcher's defenestration in November 1990, the Europe issue turned toxic in the Conservative Party. While Major could by no means be classified as a Europhile, the Maastricht negotiations in December 1991 would test his diplomatic skills to the limit. Although Britain finally joined the ERM right before Thatcher's resignation, it clearly would not take part in any early stage of Economic and Monetary Union.

The idea behind EMU was not new, harkening back to the the late 1960s. It gained new momentum after the Berlin Wall came down and German reunification became a geopolitical fact. EMU would incorporate a reunified Germany into an irreversible union with a single currency and tie Berlin's fate to the rest of Europe through a common monetary policy. France was particularly keen on this, attracted by Germany's hard-won reputation for price stability.

Moreover, European elites widely shared the view that the forces of globalization, evident in rapidly rising international trade and capital flows, meant a substantial hollowing out of the traditional nation-state, and hence would require an answer at the supranational level. EMU was to serve as the vehicle that would enable Europe to compete as a unified economic bloc

with a rising Japan, a nascent North American free trade area, and other emerging giants, mainly in Asia.

Major's Conservative government, with some exceptions such as Kenneth Clarke and Heseltine, did not share that view. There was no majority in the House of Commons for transferring so much sovereignty to an independent European Central Bank, and most policy makers in Britain agreed that it would be unwise to permanently give up its national monetary policy authority. Still, Kohl and Mitterrand were adamant in pursuing monetary union.

Aware that vetoing the Maastricht Treaty would leave Britain isolated in Europe, Major painstakingly negotiated hard opt-outs from the single currency (as well as from the Social Chapter, which concerned issues such as employment conditions and social security) before signing the treaty in February 1992. At the time, the general feeling in Britain was that Major had gotten a good deal for the country. Major himself might have exaggerated when he claimed "game, set, and match," but even Thatcher admitted that her successor had negotiated well. Two months later, in April 1992, Major unexpectedly led the Tories to another general election victory.

Soon after, open warfare broke out in the Conservative Party over the ratification of the Maastricht Treaty. A combative Thatcher, now in the House of Lords, tore the treaty to pieces and declared she would have never signed it. In May 1992, Major carried the narrowest of votes in favor of Maastricht in the House of Commons, but the wounds within his party were deep. Three months later, a humiliating exit from the ERM came on "Black Wednesday," as currency speculators forced the Bank of England's hand, leading to a significant devaluation of the pound. In one day, the Conservatives had lost their electoral trump card of economic competence.

Maastricht was the harbinger of new developments in British politics in the 1990s: the founding of UKIP in 1993, the electoral suicide of the Conservative Party over Europe in the mid-1990s after years of cabinet infighting, and New Labour's rise to power in 1997 after promising "five economic tests" to join the single currency. The euro eventually came into circulation in January 2002 without Britain's participation. The pro-European Prime Minister Tony Blair promised to join "when the time was right," but was held back by a much more skeptical Chancellor of the Exchequer Gordon Brown.

French and Dutch "no" votes in 2005 referenda put Europe's constitutional dreams on ice. The substitute was the much more modest Treaty of Lisbon, which kept most of the constitutional treaty's substance and aimed to make a much-enlarged union function better. Blair and Brown thus avoided the risk of a "no" vote in a referendum of their own.

Meanwhile, the Conservative Party started to emerge from the political wilderness after its third consecutive defeat at the polls by choosing David Cameron as its new leader. After the global financial crisis and the ensuing Great Recession led to the downfall of Gordon Brown and Labour in May 2010, the Tories—now more Euroskeptic than ever—returned to office in an awkward and unnatural coalition with the pro-European Liberal Democrats.

Cameron's Dilemma

Cameron was barely installed as prime minister when he found himself in the midst of the European sovereign debt crisis. As many analysts had been pointing out since Maastricht, the EMU was only a half-built house: It had a common monetary policy, but lacked the elements of a real "economic government," including a fiscal union, a common debt instrument, a banking union, or the legitimacy of a political union. In order to save the euro, the euro zone members would now have to complete the unfinished tasks.

However, the logic of building a genuine EMU could only mean a further transfer of national powers to Brussels and Frankfurt—a clear red line for Cameron's government. The crisis laid bare the contradictions of a continent caught between the centripetal demands of making a supranational currency union function and the centrifugal force of more than 25 domestic political agendas. And for better or worse, democratic legitimacy still mainly lies with the nation-state, as Euroskeptic Britons know all too well.

With the UK out of the euro zone, and continental Europeans committed to completing their unfinished monetary union, Cameron faces a dilemma. On the one hand, he would like to see the euro succeed without Britain. However, that is only possible with much more centralized powers in Brussels and Frankfurt, which will have to implement a host of new regulations affecting all members of the common market, including those who are currently not members of the euro zone. This would do particular harm to Britain's powerful financial industry.

On the other hand, Cameron wants to maintain maximum sovereignty over his country's economic future, while retaining the ability to influence European decisions concerning the common market that directly affect Britain. It now looks increasingly as though Britain, if it wants a real say in the EU's future institutional infrastructure, will have to join the euro itself. Yet the depth and duration of the sovereign debt crisis have definitely not helped the case for euro entry.

Penny Wise

There is no denying that the case for Britain to leave the EU altogether has become stronger since the euro crisis, if it wants to keep the pound. As Wolfgang Münchau of the *Financial Times* has argued, the euro zone is likely to supersede the common market as the main organizing principle for the EU, which weakens the case for Britain to stay. Radek Sikorski, Poland's foreign minister, unwittingly made the case for a British exit from the EU by arguing recently that the euro zone is the "real" EU. He committed his country to joining the single currency by 2020, since he feels that the euro is now the true political heart of Europe, and he wants Poland to play a central role in it.

Every British leader has been unable to stop the momentum behind European integration.

Nigel Lawson, speaking for much of Euroskeptic Britain, argues that the costs imposed by harmful EU regulations cancel out the benefits from opening Europe's markets, especially in financial, legal, and consulting services, where Britain has a clear comparative advantage. Furthermore, Lawson points out, trade with the EU has reached a plateau, whereas growth potential lies with the emerging economies in Asia and Latin America. EU membership, the logic goes, holds Britain back from accessing those lucrative markets.

But that does not mean the benefits from leaving the EU would outweigh the costs, especially in the short and medium term. In a recent report for the Center for European Reform, a London-based think tank, John Springford and Simon Tilford point out that Britain has precious little to gain, but a lot to lose. Any new arrangement with the EU after quitting—either as a member of the European Economic Area like Norway, a customs union à la Turkey, or a free trade agreement like Switzerland's—implies a loss of influence in negotiating nontariff barriers such as product and safety standards and environmental regulations. Springford and Tilford also note that Britain stands to gain the most from further liberalization of the services industry in Europe. The best guarantee for this *not* to happen would be for Britain to turn its back on the EU.

Leaving the EU would also mean a dramatic loss of influence on the world stage, not just in negotiations within the World Trade Organization, which are dominated by the United States, China, and the EU, but also in foreign affairs. Losing influence in Brussels will be equated to an overall loss of British power from the vantage point of Washington, Moscow, Beijing, or New Delhi.

Growing Tension

Objectively, the case for staying in the EU remains stronger than the case for leaving it. However, if we can believe the opinion polls, that is not how a majority of British voters currently sees it—a trend perhaps encouraged by chronic misinformation from the ferociously Euroskeptic tabloid press. From Cameron's point of view, the best way out is to create a different Europe. Not having been present at the creation always meant that the UK would have to join the club on Europe's terms, rather than its own.

Through a renegotiation of its own fundamental membership terms, Britain wants to reform Europe from within—but by staying out of the euro it refuses to be at the core of European policy making. In the words of Foreign Secretary William Hague, Britain wants to be "in Europe, but not run by Europe." After 3 years of the euro crisis, it is not clear how a country can remain in Europe without being subject to its laws and regulations, of which there will only be more in the future.

A sign of things to come is the growing tension between London and Brussels on the subject of the free movement of labor, one of the "four freedoms" that form the bedrock of the Treaty of Rome. After the December 2013 EU summit, Cameron—under huge pressure from his party to defy Brussels and maintain labor restrictions against Bulgarians and Romanians—threatened to veto any new EU member accessions from the Balkans if welfare "benefit tourism" is not stamped out. This shows that Cameron's Conservatives are concerned not only by fiscal and financial regulation, but also by basic questions of national sovereignty.

It might be too late for Britain to have its cake and eat it too. The reforms that London wants Europe to undertake, including structural measures to increase competitiveness and austerity budgets to put its fiscal house in order, will simply not be enough to save the euro and the European project in the long term. The single currency can only work if it is part of a broader political project. If Britain decides that it wants no part of such a future, it may well choose the exit option. But before it comes to that, the Scots first have to decide in September 2014 whether they want to remain in the UK. Most opinion polls show that a clear majority would like to stay, so the main issue remains Britain's relationship with Europe.

The irony is that even though Britain may decide to leave Europe, Europe will never leave Britain. If Britain leaves the EU, it will find that it is still "run by Europe" to some extent, as Switzerland and Norway can attest. The continent will never be truly "cut off." But the heavy fog in the Channel is unlikely to clear anytime soon.

Critical Thinking

1. Will the United Kingdom withdraw from the European Union? Why or why not?

2. What will be the effect of the next set of general elections of 2015 on British membership in the European Union?

3. Why would the Conservative Party hold a referendum on British membership in the European Union?

Create Central

www.mhhe.com/createcentral

Internet References

European Parliament Information Office in the United Kingdom
http://www.europarl.org.uk/en/your-meps.html

Lisbon Treaty
europea.eu/Lisbon_treaty/full-text

UK Independence Party
http://www.ukip.org

UK Representation to the EU
https://www.gov.uk-government/world/uk-representation-to-the-eu

---

MATTHIAS MATTHIJS is an assistant professor of international political economy at Johns Hopkins University's School of Advanced International Studies and the author of *Ideas and Economic Crises in Britain from Attlee to Blair (1945–2005)* (Routledge, 2010).

Article

Prepared by: Robert Weiner, *University of Massachusetts, Boston*

Is Africa's Land Up for Grabs?

Foreign Acquisitions: Some Opportunities, but Many See Threats.

ROY LAISHLEY

Learning Outcomes

After reading this article, you will be able to:

- Discuss a new form of neo-colonialism in Africa.
- Discuss what can be done to counter the negative effects of land grabs in Africa.

An apparent surge in the purchase of African land by foreign companies and governments to grow food and other crops for export has set alarm bells ringing on and off the continent. The headlines have been strident: "The Second Scramble for Africa Starts," "Quest for Food Security Breeds Neo-Colonists," "Food Security or Economic Slavery?"

2.5 mn hectares of African farmland allocated to foreign-owned entities between 2004 and 2009.

The outcries reflect the continuing impact of the continent's history, when as recently as the last century colonial powers and foreign settler populations arbitrarily seized African land and displaced those who lived on it, lending considerable emotion to the current volatile issue. Some agricultural experts have wondered whether such land deals could lead to a form of "neo-colonialism". But immediate, practical concerns are also prominent. "This is a worrisome trend," noted Akinwumi Adesina, the then vice president of the advocacy group Alliance

for a Green Revolution in Africa (AGRA). Such foreign land acquisitions, he argued, have the potential to hurt domestic efforts to raise food production and could limit broad-based economic growth. Many deals have little oversight, transparency or regulation, have no environmental safeguards and fail to protect smallholder farmers from losing their customary rights to use land, added Mr. Adesina, now Nigeria's minister for agriculture.

The sheer size of some of the land agreements has added to the alarm. A deal to allow South Korea's Daewoo Corporation to lease 1.3 million hectares was a key factor in building support for the ouster of Madagascar's President Marc Ravalomanana in March 2009. In Kenya the government struggled to overcome local opposition to a proposal to give Qatar and others rights over some 40,000 hectares in the Tana River Valley in return for building a deep-sea port.

A number of international organizations reacted to this development. The Food and Agriculture Organization (FAO) and the World Bank commissioned studies into so-called "land grabs." At the 2009 summit of the Group of Eight (G-8) industrialized countries in Italy, Japan pushed for a code of conduct to govern such schemes. Any code of conduct is going to be difficult to negotiate, and it will be even more difficult for industrialized countries to apply to deals that are primarily worked out between countries in the South, the UN's Special Rapporteur on the Right to Food, Olivier De Schutter, told *Africa Renewal*.

In a report titled "Large-scale Land Acquisitions and Leases," Mr. De Schutter wrote that while such investments provide certain development opportunities, they also represent a threat to food security and other core human rights. "The stakes are huge," he told *Africa Renewal*. Unfortunately, "the deals as they have been concluded up to now are very meagre as far as the obligations of the investors are concerned." He also noted that agreements concerning thousands of hectares of farmland are sometimes just three or four pages long.

Yet for African countries agreeing to such deals, the possible advantages are also attractive. While African agriculture rarely attracts significant investments or external aid—and the current global economic downturn has made external financing even more scarce—leasing unused land to foreign governments and companies for large-scale cultivation can seem like a way to boost an underdeveloped sector and create new job opportunities.

A study by the International Institute for Environment and Development (IIED), a research group based in the UK, estimated that nearly 2.5 mn hectares of African farmland had been allocated to foreign-owned entities between 2004 and 2009 in just five countries (Ethiopia, Ghana, Madagascar, Mali and Sudan) it studied in depth. The sheer scale of many leases is unprecedented, said the IIED report, *Land Grab or Development Opportunity?*, which was prepared for the FAO and the UN's International Fund for Agricultural Development.

The surge in interest in African land has been driven by a number of factors. On the side of investors, those include a desire for food security back home and to a lesser extent rising demand for biofuels. Behind both is the expectation of rising costs of land and water as world demand for food and other crops continues to expand.

Many of the government-to-government deals are aimed at meeting food needs, especially in the states of the Arab Gulf and in South Korea. Indian companies, backed in part by their government, have invested millions of dollars in Ethiopia to meet rising domestic food and animal feed demand. Commercial enterprises, many of them European, as well as Chinese companies, have been in the lead in cultivating jatropha, sorghum and other biofuels in countries such as Madagascar, Mozambique and Tanzania.

Africa is a particular focus for this investment explosion because of the perception that there is plenty of cheap land and labour available, as well as a favourable climate, Mr. De Schutter points out. In Mozambique, Tanzania and Zambia, for example, only some 12% of arable land is actually cultivated.

Africa so far has been able to mobilize only limited financing to develop its arable land. Despite persistent calls for increased domestic investment, agriculture has lagged well behind other sectors. The African Union has urged governments to devote 10% of their spending to agriculture, but not many have actually met that target. Donor countries and institutions have also failed to play their part, with agriculture's share of aid tending to fall.

With land apparently in abundance, but money not, the offer by foreign investors to develop agricultural land appears very attractive. But with much of the land not as unused as it might seem and with actual returns on agricultural investment far lower

than presented in initial feasibility studies, the political and economic reality for African governments can be very sobering.

"Governments are sitting on a box of dynamite," Namanga Ngongi, former president of AGRA, initiated by former UN Secretary-General Kofi Annan, told the media.

Towards a Strategic Approach

Recent assessments by IIED, FAO, the World Bank and the Washington-based International Food Policy Research Institute (IFPRI) all confirm the shortcomings and potential dangers. These include the risks of undermining domestic efforts to increase food production, the danger that agricultural projects aimed exclusively at foreign markets may do little to stimulate domestic economic activities, and the potential loss of land rights for local farmers.

Many of the studies also point to possible benefits for a sector strapped for cash. These include the creation of jobs, the introduction of new technologies, improvements in the quality of agricultural production and opportunities to develop higher-value agricultural processing activities. There might even be "an increase in food supplies for the domestic market and for export," the FAO says.

To reap the benefits of this new trend, says an IFPRI study, *"Land Grabbing" by Foreign Investors in Developing Countries: Risks and Opportunities*, governments need to develop the capacity to negotiate sound contracts and to exercise oversight. This can help create "a win-win scenario for both local communities and foreign investors." The studies advise African governments to be strategic in their approach. In his report, Mr. De Schutter puts forward a number of recommendations to guide such land deals. These include the free, prior and full participation and agreement of all local communities concerned—not just their leaders, the protection of the environment, based on thorough impact assessments that demonstrate a project's sustainability, full transparency, with clear and enforceable obligations for investors, backed by specified sanctions and legislation, as necessary and measures to protect human rights, labour rights, land rights and the right to food and development. Such comprehensive deals would be in the long-term interest of investors and local communities alike, IFPRI notes, pointing out that land disputes can become violent, and governments may quickly find themselves with no alternative but to change or rescind contractual arrangements.

Land Rights

Land ownership is a core issue. Only a relatively small portion of land in Africa is subject to individual titling. Much land is community-owned, and in some countries state-owned. Even land that is officially categorized as un- or under-utilized may in fact be subject to complex patterns of "customary" usage. "Better systems to recognize land rights are urgently needed," the FAO argues in a policy brief, From Land Grab to Win-Win.

The World Bank points to the importance of international bodies helping African governments develop land registry systems. The IIED study stresses that such schemes must allow for collective registration of community lands that protect "customary" land rights.

Mr. De Schutter argues that internationally agreed-upon human rights instruments can be used to protect such rights, including those of livestock herders and indigenous forest dwellers.

According to the IIED study, the bulk of recent large-scale land acquisitions in Africa have been based on the leasing of land to foreign entities with the intent of using labour to work the land. The study argues the need for governments to include clauses ensuring the use of local labour in contracts for such schemes. "Agreements to lease or cede large areas of land in no circumstance should be allowed to trump the human rights obligations of the states concerned," Mr. De Schutter argues.

Proposals for such ideal agreements, backed by necessary national legislation and enforcement principles, are being put forward. But, as the IIED study points out, there is already a large gulf between contractual provisions and their enforcement. The gap between the statute books and the reality on the ground may entail serious costs for local communities.

A code of conduct for host governments and foreign investors could help ensure that land deals are a "win-win" arrangement for investor and local communities alike, IFPRI suggests. It cites the Extractive Industries Transparency Initiative, which binds participating governments and companies to certain standards in mining and oil activities, as one possible model for large-scale land deals.

Mr. De Schutter is sceptical that such a code can be negotiated or enforced. He instead emphasizes the existing body of human rights laws, which can be applied to large-scale land acquisitions and used to get governments to meet their obligations to citizens.

Either way, experts agree that African governments must have the will and the ability to apply laws. "Strengthening the negotiation capacity is vital," Mr. De Schutter argues. And that capacity cannot be of governments alone, he says. Local communities must also be empowered and national parliaments must be involved. Achieving that, many fear, may be the most difficult gap to bridge.

Critical Thinking

1. Why are the "land grabs" in Africa considered harmful to African economic interests?

2. What advantages do African countries gain from land deals?

3. Why is Africa the focus of land investment by such countries as India and South Africa?

Internet References

African Development Bank
http://www.afdb.org
Food and Agriculture Organization
www.fao.org
UN Economic Commission for Africa
www.uneca.org

Article   Prepared by: Robert Weiner, *University of Massachusetts, Boston*

The Blood Cries Out

Burundi is about the size of Maryland but holds nearly twice as many people. Brothers are now killing brothers over mere acres of earth. Could africa's next civil war erupt over land?

JILLIAN KEENAN

Learning Outcomes

After reading this article, you will be able to:

- Understand why Burundi may be on the edge of civil conflict.
- Discuss the relationship between land and civil conflict.

When Pierre Gahungu thinks about the small farm in the Burundian hills where he grew up and started a family, he remembers the soil—rich and red, perfect for growing beans, sweet potatoes, and bananas. He used to bend over and scoop up a handful of the earth just to savor its moist feel. To Gahungu, now in his 70s, the farm was everything: his home, his livelihood, and his hope. After he was gone, he had always believed, the land would sustain his eventual heirs.

But then, in an instant, his dreams were thrown into jeopardy. On a dusky evening in 1984, Gahungu was walking home when he heard a noise behind him. He turned and found himself face to face with Alphonse, the son of a cousin. For months, Alphonse had been begging Gahungu, whom he called "uncle," for a portion of the farm. Alphonse's polygamous father had many sons—more than 20, Gahungu says—which meant each one would get just a tiny plot of his land. (In Burundi, generally only men may inherit property.) Alphonse wanted more space, a rapidly shrinking commodity, on which to build a house and a life. Gahungu had a much smaller family—ultimately, he and his wife would have three children, but only one boy, named Lionel—so he had plenty of land to share, Alphonse reasoned. Why shouldn't he get a piece of it? Gahungu, however, had refused repeatedly. When he saw Alphonse that night on the road, he assumed they were in for another round of the same exhausting refrain.

Alphonse, however, had not come to talk. Without saying a word, he raised a machete and brought it down onto his uncle's skull. Gahungu remembers feeling a flash of pain and hearing a bone crunch before everything went black.

"I was terrified," Gahungu says through an interpreter. He woke up wounded and later saw a doctor. He began recovering from his injury, but he feared that his farm would never be safe in his hands. Gahungu decided that if Alphonse couldn't kill him, the land's legal owner, his son's inheritance would be safe. So he left his family behind and moved alone to the nearby city of Muramvya, where he worked at a tailoring shop downtown.

Before long, more problems arose, but not with Alphonse (who, Gahungu says, died in a car accident). One of Alphonse's brothers built two houses on Gahungu's land without his uncle's permission. In 1991, on the eve of a brutal, 12-year civil war that pitted Burundi's two main ethnic groups, Hutu and Tutsi, against one another, Gahungu took the man to a local court, which ruled in the owner's favor. But winning the legal battle did nothing to change Gahungu's situation: To this day, he says, Alphonse's brother, the brother's family, and the two houses illegally occupy his farm.

Gahungu has tried to go back to Burundi's backlogged courts for help, but he doesn't have the money to pay for a case. Tragedy, too, has continued to follow him: Lionel died at just 19. As his own life draws toward an inevitable end, Gahungu lives alone in Muramvya. He now fears he will die before ever getting his beloved land back. "It was the perfect farm because it was my farm," Gahungu recalls. "It was my whole life."

Gahungu's experience mirrors other stories familiar to Burundi for decades—stories that are multiplying and worsening as the country copes with a veritable explosion of people. At 10,745 square miles, Burundi is slightly smaller than the U.S. state of Maryland, but it holds nearly twice as many people: about 10 million, according to the U.N. Development Programme, or roughly 40 percent more than a decade ago. The population growth rate is 2.5 percent per year, more than twice the average global pace, and the average Burundian woman has 6.3 children, nearly triple the international fertility rate. Moreover, roughly half a million refugees who fled the country's 1993–2005 civil war or previous ethnic violence had come back as of late 2014. Another 7,000 are expected to arrive this year.

The vast majority of Burundians rely on subsistence farming, but under the weight of a booming population and in the long-standing absence of coherent policies governing land ownership, many people barely have enough earth to sustain themselves. Steve McDonald, who has worked on a reconciliation project in Burundi with the Woodrow Wilson International Center for Scholars, estimates that in 1970 the average farm was probably between 9 and 12 acres. Today, that number has shrunk to just over one acre. The consequence is remarkable scarcity: In the 2013 Global Hunger Index, Burundi had the severest hunger and malnourishment rates of all 120 countries ranked. "As the land gets chopped into smaller and smaller pieces," McDonald says, "the pressure intensifies."

This pressure has led many people who want land, like Alphonse years ago, to take matters into their own hands—at times violently. The United Nations estimates that roughly 85 percent of disputes pending in Burundian courts pertain to land. Between 2013 and 2014, incidents of arson and attempted murder related to land conflict rose 19 percent and 36 percent, respectively. Violence sometimes occurs within families, but it also can play out between ethnic groups: Most returning refugees are Hutu, but the land they left behind has often been purchased by Tutsis. "The land issue comes into politics when parties say, 'I promise to return to you what is rightfully yours,'" says Thierry Uwamahoro, a Burundian political analyst based in the Washington, D.C. area.

Against this fragile backdrop, the Institute for Security Studies, a South African-based think tank, has warned that "attempts to politicise land management . . . risks reigniting ethnic tensions" before national elections scheduled for May and June. Many locals, however, fear that an even bigger disaster is looming. "The next civil war in Burundi will absolutely be over land," says a communications consultant in Bujumbura, the capital, who works for U.N. agencies and asked not to be named for security reasons. "If there is no new land policy, we won't last a decade."

There are no easy solutions to Burundi's mounting land crisis, but stories like Gahungu's offer a glimpse of what might happen if this ticking time bomb is not diffused. "In the past, this situation didn't exist," Gahungu says, standing outside the tailoring shop where he still works, cleaning and ironing clothes. "There was land for all, but not anymore. I fled because

I feared that what happened to me before could happen again. It happens to someone every day now."

Before European colonizers arrived in Burundi, farmers cultivated the country's arable hilltops, while less desirable, low-lying swamplands went largely unclaimed. An aristocratic class, known as the Ganwa, technically owned the land, but farmers' access was administered at the local level by a network of "land chiefs," many of whom were Hutu. The chiefs also resolved land conflicts, according to Timothy Longman, director of Boston University's African Studies Center.

Under Belgian rule, which lasted from 1916 to 1962, this all began to change. The king, the head of the Ganwa, kept control of the highlands. (According to scholar Dominik Kohlhagen, the king was seen as the land's spiritual guardian.) But the state assumed ownership of the lowlands and began to encourage their cultivation. The colonial government also concentrated political power among the Tutsi minority, which comprised about 14 percent of the population, giving the group a near monopoly on Burundi's government, military, and economy. Among other actions, this consolidation involved gradually stripping the Hutu land chiefs of their authority. More broadly, too, it sowed the seeds of dangerous ethnic polarization.

Under colonialism, official land deeds and titles were few and far between, which meant that Burundians often could not prove that they owned acreage. In the early 1960s, as independence loomed, the government began offering land registration to parties that requested it, which, for the most part, were foreign businesses such as hotels. Families also had the option to register their land, but because the centralized system was inaccessible for most farmers and required a huge tax payment, few did. So land plots quickly fell into two categories: those with boundaries recognized by the state, and those with borders determined by custom—that is to say, residents understood trees, rocks, paths, creeks, and huts to mark de facto property lines.

The Catholic Church was among the institutions that benefited from the colonial approach to land. Missionaries, known as "White Fathers," began arriving in the late 19th century, and over several decades, the king gave them large tracts of land, which they used to establish churches, schools, hospitals, and farms. After colonialism ended, the self-sufficiency that land provided the church helped it retain influence, even as its relationship with the newly independent government grew fraught. Most notably, Jean-Baptiste Bagaza, a military leader who in the mid-1970s seized Burundi's presidency in a bloodless coup, saw the church as an extension of colonial power and a rival to his own, so he limited the hours in which congregations could gather, shut down a Catholic radio station, and used visa nonrenewals and expulsions to decrease the number of missionaries in the country. Nevertheless, the church retained millions of

Burundian followers, along with plenty of land, though no one, it seems, knew exactly how much.

The Catholic Church was also complicit in nurturing Burundi's ethnic divisions; Catholic schools, for instance, were largely reserved for "elite" children, meaning Tutsis. Intensifying schisms led to various outbreaks of ethnic violence, and in 1972, the Tutsi-dominated military launched a series of pogroms targeting Hutus. More than 300,000 Hutus fled the country in under a year, leaving behind their land. Bujumbura took advantage of some of this newly vacated property and extended agricultural schemes called *paysannats* (derived from the French word for "peasantry"): The state leased the land to farmers, who would grow cotton, tobacco, and coffee and then sell these crops back to the government, the only legal buyer. Officials in Bujumbura hoped to boost Burundi's weak economy by reselling the crops on the international market.

But the paysannat system failed miserably due to corruption, inefficient government bureaucracy, and variations in global commodity prices. Seeking bigger profits than they were able to get in Burundi, farmers began to smuggle their harvests over the country's borders, and state-run agricultural buying programs floundered in the mid-1980s. Paysannats also ignored and often destroyed the physical markers that had defined traditional land boundaries. Along with the pervasive lack of legal documents showing land ownership, this made it impossible for most returning refugees to reclaim their lost acreage. (Today, the Hutus who left in 1972, some of whom have never come home or are only just doing so, are called "old-caseload" refugees.)

The government set up two commissions, in 1977 and 1991, to resolve land disputes, but they proved largely ineffective. Ethnic tensions continued to mount, coming to a head in 1993. That October, Tutsi extremists assassinated Burundi's first democratically elected Hutu president, Melchior Ndadaye, and civil war erupted as Hutu peasants responded by murdering Tutsis. In just the first year of conflict, tens of thousands of people were killed; by the time the war ended more than a decade later, some 300,000 Burundians, most of them civilians, had died. The war also produced a new wave of roughly 687,000 refugees.

When the dust settled, the effects of mass death and displacement were exacerbated by widespread poverty, food insecurity, and a host of other post-conflict challenges, all of which persist today. In 2014, the World Bank estimated Burundi's GDP growth rate at 4.0 percent, below the average of 4.5 percent for countries in sub-Saharan Africa. The bank forecasts that this gap will only widen in 2015, with Burundi's rate declining to 3.7 percent and the region's climbing slightly. The issue of land, meanwhile, has become a casualty in its own right, thrown into greater flux than ever before.

Today, there are dozens of scenarios under which people claim land, and the same plot, no matter how tiny, is often the subject of competing claims. Some families still say they own acreage because of paysannat leases; seeking to make a profit, tenants have even sold their land over the years, despite the fact that it is technically state-owned. According to the International Crisis Group, some 95 percent of Burundian land still falls under customary law: A family says it purchased its farm from neighbors before the war, but holds no formal deed, while another claims village elders approved the purchase of a few acres after the war, and so on. A centralized registration system does exist, and according to the country's land code (which was revised in 2011 for the first time since 1986), any person who owns property must hold a land certificate. The bureaucratic system, however, is complicated, and the government has done little in terms of enforcement. According to Kohlhagen, offices that issue certificates exist in only three cities, and as of 2008, only about 1 percent of the country's surface area was registered.

Complicating matters further is the continuous flow of refugees who return home to find their land occupied by new owners. In some cases, a Hutu farmer who fled the 1972 pogroms may come back to find two other people claiming his property: whoever lived on it up until 1993, and whoever claimed it after the civil war. The last resident may have purchased the land legally, even from the government itself, and may have been paying off mortgages for decades. The question then becomes, who should get to live on the land now—and how should the claimants who can't have it be compensated?

The government has no clear answer. "It's tricky to say what land policy is today, because there is not a uniform dispute-resolution strategy," says Mike Jobbins, a senior program manager at Search for Common Ground, a conflict-prevention NGO that works in Burundi. "Every case is decided on its own merits."

The Burundian courts and *bashingantahe,* or traditional panels comprising senior men in villages, are empowered to settle land disputes. But while courts issue legally binding rulings, cases are time-consuming and often prohibitively expensive. The bashingantahe, meanwhile, are free, yet operate according to customary law. A third body would seem to offer a more promising option: The National Commission for Land and Property, known by its French acronym, CNTB, was established in 2006 to resolve arguments over who owns land vacated by refugees. Its 50 members are required to be 60 percent Hutu and 40 percent Tutsi, and since its creation, the CNTB has processed nearly 40,000 cases.

But the CNTB has struggled to adopt a consistent approach to its judgments. At first, its preference was to divide land between valid claimants, both past and current owners. Since 2011, however, it has begun returning land to its original owners, usually Hutu refugees displaced in 1972. In some cases, it has even revised decisions on previously closed disputes. This has led to angry claims that the government of President Pierre Nkurunziza, a Hutu, is trying to curry support from Burundi's predominantly Hutu electorate.

The government did nothing to quiet these concerns when, in December 2013, it expanded the CNTB's mandate to review cases that predate even 1972, made it a criminal offense to obstruct the commission's actions, and allowed rulings to be appealed to a new "land court" that can issue binding decisions. This move to boost the CNTB's power created suspicion among critics of the government that the commission's biases would only become more firmly entrenched. "Far from uniting Burundians or reconciling them, this new law on the CNTB will divide them," Charles Nditije, then leader of an opposition political party, told the media at the time.

Other criticisms surround the government's failure to establish a compensation fund for people who do not win land disputes. The 2000 Arusha Peace and Reconciliation Agreement for Burundi, which outlined a plan for the country's postwar peace process, guaranteed compensation, but according to Thierry Vircoulon, the International Crisis Group's project director for central Africa, the state seems to have "completely ignored" this detail. Some government detractors say this is a deliberate, ethnically driven decision by Nkurunziza's administration, because many people eligible for compensation would be Tutsi.

Murky, controversial land policies have at times led to inter-ethnic violence. In 2013, riots broke out in Bujumbura when the police tried to evict a Tutsi family from the house it had owned for 40 years in order to give the house back to its previous Hutu residents. "We are here to oppose injustice, to oppose the CNTB, which is undermining reconciliation in Burundi society," a protester was quoted by Agence France-Presse as shouting. Over a six-hour standoff, more than a dozen people were reportedly hurt, and 20 were arrested.

Ethnic tensions, however, are only part of the puzzle in Burundi's land crisis. Poor farming families are straining the country's limited ground space. About two-thirds of Burundians live in poverty, and families often have several male heirs who are forced to share plots of earth that barely fit a home and a few rows of crops. As a result, according to research conducted by land-rights consultant Kelsey Jones-Casey, "[T]he most destructive conflicts experienced by rural people in Burundi are intra-family disputes, most of which manifest over the issue of inheritance." Violence sometimes occurs within polygamous families, with sons born by different mothers fighting for finite land. In Muramvya, people speak in low voices about a woman who slit her husband's throat to accelerate a land inheritance for her son.

Violence over land is roiling a country that already clings to an uneasy peace. Nkurunziza's government has been accused of ordering convictions, murders, and disappearances of political adversaries, among other abuses. In January, the state claimed to have killed nearly 100 rebels who had crossed the border between Burundi and the Democratic Republic of the Congo with the intention of destabilizing the country and the upcoming national elections, in which Nkurunziza is expected to seek a third term despite a constitutional limit of two. Vital Nshimirimana of the Forum for the Strengthening of Civil Society in Burundi told Voice of America that officials had suggested the country's political opposition supported the rebels: "This is what leads us to think that it might be a fake explanation to actually take advantage of the same to arrest opposition leaders or some civil society [members]," he said. A few days later, youth leader Patrick Nkurunziza (no relation to the president) was arrested for his alleged connections to the rebels; at the same time, the government sentenced opposition leader and former Vice President Frédéric Bamvuginyumvira to five years in prison for bribery.

Violence over land is roiling a country that already clings to an uneasy peace.

Many Burundians fear that land could be the detail that pushes swelling political tension into something far worse. "Land is the blood and the flesh of any human being," says Placide Hakizimana, a judge in Muramvya, who notes that 80 percent of the cases he adjudicates pertain to property disputes. "Without land, we are condemned to death. No one will accept that. [Families] will fight. We prefer to die rather than live without land."

Policy reform may be a dead end or, at least, one that is too rife with corruption and partisan battles to ever solve the land crisis. This thinking is driving some people to focus on restricting population growth. "Family planning is the only exit point to the land problem," says Norbert Ndihokubwayo, a member of Burundi's parliament and president of a legislative commission on social and health issues. "No other solution is possible."

In 2011, the government approved a national development strategy called "Vision Burundi 2025" with ambitious demographic goals: to reduce national growth from its current rate, which would cause the population to double every 28 years, to 2 percent over the next decade, and to slash the birthrate in half. To hit these numbers, the government said it would partner with civil society to "stress . . . information and education on family planning and reproductive health." Ndihokubwayo says the government is also "absolutely" considering a law that would limit the number of children each family can have.

Many international donors are helping to expand access to family-planning services. The Netherlands chose Burundi in 2011 as one of 15 "partner countries" in which to emphasize programs that promote peace and stability, and according to Jolke Oppewal, the Dutch ambassador to Burundi, his country now donates 8 million euros annually to programs promoting sexual and reproductive health, among other human rights. In a 2005–2013 contract, the German government-owned development bank KfW dedicated more than 1.4 million euros to "strengthening and reorganizing [Burundi's] reproductive health and family planning services."

Some medical professionals are keenly aware of the role they are meant to play in keeping population growth in check. Christine Nimbona is a nurse at a secondary health clinic in Kayanza province, which, with nearly 1,500 inhabitants per square mile, is one of Burundi's most overpopulated regions. One day in August, several women waiting outside the clinic where Nimbona works nursed babies; dozens of children played nearby. Nimbona says that of the roughly 30 patients she sees each week, "almost all" cite fears about land resources and potential inheritance conflicts as their reasons for seeking family planning. "I know that by what I am doing, I am lighting the escalation of violence in my country," Nimbona says.

It's an uphill battle, littered with enormous, deep-seated obstacles. According to the United Nations, modern contraceptive use among females between the ages of 15 and 49 was just 18.9 percent in 2010. In Burundi's male-dominated society, women are often powerless to convince their husbands to use birth control. Then there is the Catholic Church: In addition to claiming an estimated 60 percent of Burundians as followers, the church has affiliations with roughly 30 percent of national health clinics, which are forbidden from distributing or discussing condoms, the pill, and other medical contraceptives. "Catholic teachings against birth control are very resonant with Burundian culture, which says that children are wealth," explains Longman, of Boston University. "Because the Catholic Church is so powerful and controls so much of the health sector, it creates a huge stumbling block for family-planning practice."

Bujumbura insists that the Catholic Church is a collaborative partner on land issues. The president even appointed a Catholic bishop, Sérapion Bambonanire, as head of the CNTB in 2011. But cracks do exist between church and state. In 2012, the Ministry of Public Health launched a series of "secondary health posts," which offer medical contraceptives; sometimes these clinics, including Nimbona's, are built right next door to existing Catholic ones.

There is also tension over a variable with unknown dimensions: how much of the land the Catholic Church held onto after colonialism it still owns today. "The Catholic Church can't keep owning all the land while Christians are starving," says a regional government employee in Kayanza, who spoke on condition of anonymity out of concern for his safety. According to him, in 2013 the government quietly launched a mapping program to determine, among other things, how much land the church controls. "National politics don't allow us to focus on the Catholic Church," he says, referring to the fact that the church's followers are also voters. "So the government thinks this indirect method is best."

Ndihokubwayo says a land-mapping program does exist, but won't confirm or deny whether it was created specifically to find out how much property the Catholic Church possesses. "This is a very delicate issue," he explains. "I'm not sure whether we'll ever find out how much land the church owns, but we'll keep trying." (Cara Jones, an assistant professor at Mary Baldwin College who studies Burundi, pointed out that the program would also give President Nkurunziza's ruling party information about how much land its political opponents own.)

Some religious leaders are on board with the push for family planning. Pastor Andre Florian, a priest in Burundi's Anglican Church, which has an estimated 900,000 followers, says he used to be part of the problem. From the pulpit of his small stone church in Kayanza, he once railed against the evils of contraception. Family planning, he told his congregation, was best left to God. Yet Florian watched with grave concern as members of his flock struggled to feed their babies. One day, he looked at a child with dull orange hair, a clear sign of advanced malnutrition, and asked himself: Was this really God's plan? Shaken, Florian isolated himself for three months, studying scripture and praying. "When I returned from my research, I realized that I had done wrong," Florian says. "If nothing happens, if we just keep doing what we're doing, tomorrow is not certain. We will see families killing each other. We will see chaos in the country. The day after tomorrow will disappear."

Other Burundians, however, fear that support for family planning is too little too late. Joaquim Sinzobatohana, a father of four in Ngozi province, says he first learned about vasectomies on a radio program. (Eighty percent of Burundians have a radio, and U.N.-funded songs and soap operas now dramatize stories of families that have suffered the burden of many pregnancies but are saved by family planning.) He decided to get the operation, a simple outpatient procedure, because he and his wife, Clautilde, are "very scared" that their small plot will provoke conflict among their children, and more offspring would only increase the chances of violence. But Sinzobatohana admits that even a demographic freeze might not save his family, or his neighbors. The numbers just don't add up: Already, too many people are squeezed onto too little land.

"It's unfortunate that these contraceptive programs came after we already had too many kids," Sinzobatohana says. "The damage has been done. Now we wait."

International experts say a comprehensive approach to Burundi's land crisis is necessary—one that combines policy reform, better dispute-resolution options, family planning, and new economic opportunities that will ensure fewer Burundians rely solely on the earth for survival. "People need to have economic opportunities besides agriculture, to incorporate people into other kinds of jobs and trades, so that not everyone is dependent on farming for their livelihoods," says Jobbins of Search for Common Ground. "Without some prospect for economic growth within the context of the region and the East African community, land scarcity will continue to be a stressor."

But the land problem is infinitely complex, with roots that run deep into Burundi's history. The resources and political will to deal with it are scarce. And whether in a new law or a family's decades-long story, there will always be critical details that go overlooked—details that could become matters of life and death.

In 1999, Emmanuel Hatungimana, an elderly farmer in northern Ngozi, could feel his body slipping away. Death was very close. So he gathered his family—two wives and 14 children—around his bedside. It was time to divide his farm.

At 37.5 acres, Hatungimana's lush plot of land was a decent size. An equitable division would have left his eight sons with roughly 4.7 acres each. The eldest sons of Hatungimana's two wives stepped up to represent his part of the family: Pascal Hatungimana for the four sons of the first wife, and Prudence Ndikuryayo for the four sons of the second wife. Hatungimana gave exactly half of his land to Pascal and his brothers, and the other half to Prudence and his brothers. The patriarch was satisfied, according to Pascal; his family's future was secure. A few days later, he died at peace.

But no one had considered the road.

One side of Hatungimana's land runs alongside a paved road, a very desirable quality because access to that lane makes it significantly easier to bring supplies in and out of the property.

In dividing his land right down the middle, however, Hatungimana had ensured that only four of his sons could claim the road as a border.

Today, more than 15 years after Hatungimana's death, his family teeters on the brink of violence. Prudence says it's unfair that Pascal and his brothers have the better land, and that he is willing to fight to get what he deserves. But Pascal doesn't want to give up anything. So they've brought their case to a local bashingantahe. If the panel can't resolve the dispute, both brothers say they don't know what will happen.

Standing on the land outside his brother's house, Prudence looks left, over his shoulder, at Pascal, who listens nearby with his arms crossed. A dark expression falls over Prudence's face. "Around Burundi, brothers are killing brothers. Sons are killing fathers. And it's all for land," he says. "Hopefully our family won't reach that stage. But if something doesn't happen, everything will fall apart."

Critical Thinking

1. What is the cause of ethnic tension between the Hutus and the Tutsis in Burundi?
2. Why is less land available for farming in Burundi?
3. What is the legacy of colonial rule in Burundi?

Internet References

African Development Bank Burundi
www.afdb.org/en/countries/east-africa/Burundi/
Burundi Government
https://www.cia.gov/library/publications/the-world-factbook/geos/bu.html
IFAD's Strategy in Burundi
operations.ifad.org/web/ifad/operations/country/home/tags/burundi

JILLIAN KEENAN (@JillianKeenan) is a writer based in New York. She is working on a book about Shakespeare and global sexuality. A grant from the United Nations Population Fund supported research for this article.

Article   Prepared by: Robert Weiner, *University of Massachusetts, Boston*

Can a Post-Crisis Country Survive in the Time of Ebola?

Issues Arising with Liberia's Post-War Recovery.

Jordan Ryan

Learning Outcomes

After reading this article, you will be able to:

- Learn about the responses to infectious diseases in a post-conflict country.
- Learn about the focus on development in post-Ebola Liberia.

We suspect that the first case of the current outbreak of Ebola Virus Disease (EVD) began with the illness of a two-year old child who died at the end of December 2013. This occurred in Guéckédou prefecture, Guinea, located in the sub-region adjoining Liberia and Sierra Leone. This area is well known for its porous borders and peoples who share ethnic and tribal identities, and it has been a cauldron for the brutal conflict that enveloped the area for well over 15 years.

We may never see the face or know the name of "Patient Si." That infant's death and the thousands of others since the EVD outbreak provoked the near collapse of the health systems in these three countries. It is a catastrophe that demands more than an emergency response from the world. Now at stake are health systems, their scope, quality, and impact—and more broadly, governance, policy choices, and progress.

At a more fundamental level, the EVD outbreak requires nothing less than a wholesale reordering of our priorities and the way in which we respond to crisis and to the emerging threats, including infectious diseases—especially in areas emerging from violent conflict.

This article will consider some of these issues in broad terms, drawing on my personal experience and association with Liberia's progress over the past nine years. Following a review of progress and challenges, it will provide a perspective on the lessons and priorities for doing development differently in post-Ebola Liberia and the neighboring countries. One point is clear: building on the creative energies of the Liberian people, the international system needs to learn to act in a proactive manner, rather than wait until a global crisis arises.

Some Starting Points

First, to be absolutely clear, there can be no doubt of the paramount need to support the all-out effort to stop the spread of EVD now. Although the initial response was unfortunately marked with far too much hesitation, there is now a robust Security Council approved mission, the UN Mission for Ebola Emergency Response (UNMEER). Now is the time for UNMEER, with all concerned national authorities, bilateral and other partners, including philanthropies and the private sector, to act in concert with one single goal. Donors need to provide immediate and generous support now, not next year when it will be too late.

Second, it will be important to look carefully at why the epidemic flourished and what factors allowed it to do so. This outbreak must serve as a wake-up call to the international community, for certainly this situation is and will not be an isolated incident. We are witnessing the dawn of a much more complicated world to come: one which is regularly challenged by upheavals that may at first sight appear local, but because of the nature of globalization, can have a dramatic impact upon populations living far away.

Finally, the unleashing of Ebola in the 21st century is not a case of science fiction coming true. It is instead the result of a series of failures: failures to invest in a timely manner in the right infrastructure; failures to build accountable systems; failures to concentrate on resilience; and failures to put an end to the taxing, time-consuming bureaucracy which saps the ability of people to focus on what matters most—making a difference in the lives of people, especially the poorest.

Liberia: A Story of Hope and Work

I arrived in Liberia in November 2005 to serve as the deputy special representative of the UN secretary-general (DSRSG) in the peacekeeping mission, UN Mission in Liberia (UNMIL), which had been established under a Security Council mandate to support a successful peace process.

Within the first days, I had the thrill of witnessing a former UN colleague, Ellen Johnson-Sirleaf, win the run-off presidential election to become Africa's first democratically elected female president. In the mid 1990s, she had held senior positions within the United Nations Development Programme (UNDP), and prior to that, the World Bank.

In her inaugural address on January 6, 2006, President Sirleaf called on her fellow compatriots to "break with the past," declaring: "The future belongs to us because we have taken charge of it. We have the resources. We have the resourcefulness. Now, we have the right government. And we have good friends who want to work with us."

Each year since then, the international community has invested over US$1.5 billion to support the government of Liberia's five-pillar strategy of security, economic revitalization, basic services, infrastructure, and good governance. The international community embraced this strategy and its aim to direct assistance to support tangible development gains for rural Liberians.

My primary task as the DSRSG for UNMIL was to coordinate the provision of life-saving humanitarian aid as well as the longer-range development assistance of the United Nations and international partners. Working closely with the Special Representative, I had the vast logistical, technical, and military resources of UNMIL at my disposal. I could also call on the UN agencies which were collectively known as the UN Country Team, a number of which—including UNDP, the UN Refugee Agency, UNICEF, and the World Food Programme—had been working in Liberia for many years.

My job was to harness and integrate the different strengths and activities of the peacekeeping mission and the UN Country Team in a "one UN" effort to provide relief assistance, strengthen the capacity of Liberian institutions to govern and deliver basic services (e.g., security, justice, policing, and social services like health and education), revive economic activity (particularly in rural areas), and, ultimately, to foster national healing and reconciliation.

As I realized during my first visit with the minister of health (whose office was unreliably lit by a single electric bulb dangling from a wire), the war had left the Liberian health sector (and many others) in shambles. The minister told me that the primary provider of health services, both during and after the war, had been international NGOs. The country's health facilities had been completely looted and vandalized during the war, and medical supplies were simply unavailable. A country of over 4 million people had only 26 practicing doctors. In most parts of rural Liberia, health services and referral systems (including any kind of maternal or reproductive health care services and information) simply did not exist. It was clear that in the health sector, as in several other sectors, the work of strengthening institutions would be a case of rebuilding them virtually from scratch.

The Liberian government and its international partners took several key steps to rebuild these institutions. We understood from the past that the concession economy and the politics of elite capture and bribery had mutually reinforced one another, creating the dynamics that led to civil war and the deprivation, suffering, and traumatization of the Liberian people.

This led to the initial institution-building focus on building up much needed capacities and systems within government institutions, with the aim of developing the systems for accountable and transparent financial management, budgeting, and procurement. Steps were taken to put in place accountability mechanisms to reduce corruption and increase transparency, introduce a cash management system, devise a new procurement commission, and establish a general auditing commission.

Another critical task was to restore trust and public confidence between the Liberian people and the government. This needed to begin early and with quick support from Sweden and the UNDP/UN Country Team, which simultaneously began the process of decentralizing governance by supporting local development initiatives in each of the 15 administrative regions. Steps were taken to foster citizen involvement to build peace and re-establish trust between the government and the general population.

Transparency International (TI) ranked the transparency of Liberia's government as the third best in Africa, citing the independence of the General Auditing Commission, support for the establishment of the Liberia Anti-Corruption Commission, the promotion of transparent financial management, public procurement and budget processes, and the establishment of a national law to ensure Liberia's compliance with the Extractive Industries Transparency Initiative. The economy was also growing at an impressive rate. Trade, production, commerce, and construction expanded rapidly.

A series of government plans were issued to outline the way that the Johnson Sirleaf Administration would capitalize upon this post-conflict economic boom, starting with her 150-day Action Plan to jump-start economic recovery. This was followed by an 18-month Interim Poverty Reduction Strategy, and in 2008, the government completed its first Poverty Reduction Strategy (PRS), whose formulation drew on countywide consultations and citizen engagement.

> **"Within months, the disease dislocated the institutional fabric of Liberia, disrupting not just the health system, but also the entire system of governance. Of the 10, 129 reported cases globally as of October 23, 2014, 4,665 are in Liberia and 2,705 have died."**

Framed around the five pillars mentioned above, the "Lift Liberia" strategy was specifically designed to promote rapid, shared growth. Officials at all levels sought to assure the Liberian population that "growth without development," which prior to the war had generated extreme inequality and deprivation, was gone forever. Promoting shared growth entailed the provision of quality public services (especially education and health) and the revival of small-scale agriculture and rural livelihoods supported by the expansion of infrastructure (roads, bridges, water, and sanitation) throughout the whole country.

This was music to the international community's ears. Aid continued to pour in. As a result, investor confidence rose dramatically. Starting with the rubber plantations, the concession economy (iron ore, rubber, and timber) began attracting large-scale international investors. Private investment increased rapidly in residential and commercial property, telecommunications, and transport.

Growth Without Development: A Return of Despair?

When I left Liberia in 2009, the story was still one of hope, and there was still widespread confidence in the national leadership. In fact, the nation's narrative of peace, stability, and recovery was heralded as a prized example of post-conflict stability, reconstruction, and development. Up until the outbreak, Liberia had experienced a sustained peace, two successful democratic elections, improved access to justice and human rights, a restoration of public services, and a reemergence of private

sector activity. In conjunction were unprecedented growth rates, showcasing Liberia's considerable strides since the August 2003 Comprehensive Peace Accord and the profound chaos and disorder the country found itself in at that time.

Since my departure, the robust transparency and accountability architecture that led to the country being ranked favorably by TI and various international watchdog groups have disintegrated as quickly as they rose. The 2010 TI Global Corruption Barometer graded Liberia as among the world's most corrupt countries, especially in the area of citizens who need to pay bribes to public servants.

The forward march and the bright future to which the president called her compatriots at her first inauguration appear to have stalled. Instead of building on a promising foundation of public hope that greeted Liberia's post-war government, its performance began to erode rather than continue to build trust.

It has become evident that the old, tired pursuit of "growth without development," as well as its perennial companion, the politics of greed, have indeed begun to settle in Liberia—as far too often the case in many resource-rich countries. While the economy continued to grow, its impact on the lives of ordinary Liberians has been limited. The original promise of broad-based engagement with citizens in order to foster countrywide development faded as government's attention turned to the revival of the concession economy. A rail line from the port to the mines was rebuilt and the iron ore mines were reopened, while after a corruption-tainted start, large-scale timber concessions were granted and the rubber plantations were rehabilitated and expanded. With the discovery of oil along the Gulf of Guinea (especially in Ghana), Liberian officials began contemplating and preparing for the emergence of a petroleum industry.

The lofty mission to "Lift Liberia" as captured in the PRS, especially as it related to small-scale agriculture, rural infrastructure, and strengthening rural public services, has failed. Rural poverty has remained high, exacerbating the low levels of health, poor standards of education, and food insecurity. Sadly it appears that Liberia has firmly moved onto a trajectory that it has already been [on] before: once more growing but not developing.

Ebola in a Time of Crisis

It is against this background—one of governance failures—that the EVD outbreak began. Within months, the disease dislocated the institutional fabric of Liberia, disrupting not just the health system, but also the entire system of governance. Of the 10,129 reported cases globally as of October 23, 2014, 4,665 people are in Liberia and 2,705 have died. The estimated figures will be more alarming if the epidemic is not brought under control.

Many health centers have shut down as health workers abandoned their posts for fear of contracting the virus, leaving hundreds of Liberians without access to health services. These centers have done so due to poor conditions, and provide no protective equipment and incentive to perform the life-threatening work for which they were created. They have already seen over 95 of their fellow health workers die, and hundreds of others fighting for their lives.

Consequently, there are widespread reports that people with high blood pressure and diabetes are no longer cared for. Pregnant women have been turned away from hospitals. They are left to die or lose their babies before they are born. Desperate Liberians have abandoned neighbors or relatives suspected of having EVD to die slow and painful deaths. Both Liberia's society and culture are being challenged in many new and desperate ways. This is a different type of war now, not the civil war of the 1990s, but a war brought about in part by a health system unable to cope with the scale of the Ebola outbreak.

The Ebola epidemic is not just devastating the Liberian population. It is also severely crippling all sectors of the country's economy: notably health, trade and commerce, and education. The World Bank recently projected major reductions in the economy over the coming years—estimating that Liberia could see significant contractions of its growth. The impact of EVD has seen the original GDP growth projections revised downwards from the initial 8.7 percent, progressively to 5.9 percent, 2.5 percent, and most recently, to 1 percent. This will have a direct impact on the country's Human Development Index. With villages decimated by the disease and agricultural fields being abandoned, famine is becoming a reality. The prices of food have been rising due to shortages. Liberian professionals who hold foreign passports, many of whom returned with high hopes of contributing to the development of their motherland, are leaving the country. This will accentuate Liberia's deficiencies in human resources. Schools have closed, business has declined, and international connections (via air and sea transport) have been curtailed. In rural Liberia, communities shun many who contract the virus for bringing calamity upon their neighbors. This is further undermining the fragile social fabric that had been slowly rebuilding after the war.

Lessons for Re-engaging on Recovery in Liberia

There will be plenty of calls for lessons learned from future analysis of what happened in Guinea, Sierra Leone, and Liberia during the early months from the death of Si in December 2013. Looking even further back, the efforts and choices made by the government of Liberia, as well as other governments in the regional

and international community, in strengthening institutions appear to be either inappropriate, misguided, or too superficial to support the country's development. We need to learn from this so that we can build new and robust local, national, regional, and global architectures that can effectively respond to current and future epidemics. For me, the following lessons are worth heeding:

(1) Effective and accountable institutions remain key in post-conflict recovery and transition out of fragility.

It is already expected that climate change will shift where infectious diseases break out. We should expect both an increasing number of epidemics and mega multi-hazards. The WHO has warned that climate change will see the rise of infectious diseases. Many of these would likely originate in the so-called conflict-affected fragile states, so we must learn quickly to engage these states in ways that increase their resilience as the first line of defense in our emerging complex new world.

In my role as director for the former Bureau for Crisis Prevention and Recovery in UNDP, we supported the work of the g7+ (a group of self-identified fragile states) that organized themselves under what is called the New Deal for Peacebuilding and Statebuilding. The specific goal was to determine how the countries could transition out of fragility to become more resilient. At the heart of the New Deal is the building of institutions to promote inclusive politics, security, justice, revenue and jobs, and basic services. These countries recognize that until they build more resilient and participatory governance systems, their prospects for peace and sustained development are limited.

We cannot continue this firefight since the resources and the know-how are simply not available to respond in an ad-hoc manner to all of these mega-hazards. Resilient institutions are essential. According to the World Bank, these are institutions that "can sustain and enhance results overtime, can adapt to changing circumstances, anticipate new challenges, and cope with exogenous shocks." Building such institutions requires that they be embedded in the societal, political, and geographical contexts from where they derive meaning and legitimacy.

Creating such an institutional context means investing in education so that these countries can have the critical mass through which a supportive institutional environment can develop. While institution-building is for the long-term, this is a great opportunity to experiment with the concept of the use of the country system, national ownership, and the rebuilding of trust between government and society as well as governments and international partners. These are the core principles of the New Deal for Peacebuilding and Statebuilding.

(2) Timely, targeted, coordinated, and coherent result-oriented response.

The Ebola crisis is not just a health emergency; it is a multidimensional social and humanitarian crisis. It requires a complex, multi-pronged response involving health, aid-coordination, personal security, food security, appropriate budgetary decision-making, and responsive governance, among others. It is a whole of government challenge. While this point cannot be ducked we in the international community regularly develop "whole of government" approaches in ways that overextend the agendas of already fragile countries well beyond their capacities to respond. We often call on the government to act in a coordinated manner, but as international partners we can be disorganized and consequently fail to act in a unified or coordinated manner ourselves.

Rolling back an epidemic is not the time for long complicated layers of bureaucracies and agency-driven interests. We need targeted and efficient responses that produce results rapidly. In their immediate response, these countries need enough ambulances to quickly collect the sick and the dead. They need health workers including infectious disease control doctors on the ground in all affected parts of the countries. They need funds to pay health workers adequately for undertaking such dangerous work. Most importantly, they need the international community to accompany them by nurturing the use of their respective country systems. This includes the training programs at the local and regional level that will continuously build capacity to stay abreast with medical science, technology, and innovation.

In the medium-term, the network of health workers across the countries must be strengthened to exchange experiences and build practices on a regular basis. But much more is needed if the countries are to rebound. They need considerable support to revitalize the productivity of their agricultural sectors, and they need innovative ways to open the schools. Early recovery activities should be prioritized, including cash transfers targeting not only the directly impacted, but also the affected households; enterprise recovery must be a key component as well.

These are concrete tasks and should be carried out without being subject to the typical bogs of bureaucracy and complexity. How can this be done differently? Where is the venture-capitalist mindset behind all the Silicon Valley startups for all of West Africa? We need to adopt modern methods of training rural health workers, the young women and men who are ready to stay in the provinces and counties and who are willing to provide real services to their fellow citizens, in exchange for being paid real salaries on time.

(3) Limit coordination layers.

There are multiple actors who are returning to these countries to help. They must be coordinated and the governments must be at the center of these coordination platforms, but these should not result in multiple and burdensome transaction costs for coordination. Coordination at the center of government is one of the core functions that needs to be strengthened, particularly in countries where such systems are still not fully consolidated. There is absolutely no time for competing layers of coordination.

As director of the former Bureau for Crisis Response and Recovery in UNDP, I saw firsthand how effective a network of actors across the government can be. In fact, it is critical. As the former UN resident coordinator in Vietnam, I witnessed firsthand that nation's response to SARS and the avian flu. The remarkable success in that country was primarily due to the cohesive response of the government, as well as its clarity of purpose and its decisiveness.

(4) From knee-jerk international mobilization to global solidarity to shared security.

We all hope the current epidemic will be brought under control as rapidly as possible. Soon we will need to face the next challenge: rebuilding the affected countries. Yes, there will be calls to build back better. But it will take much more than slogans this time. The world will be challenged to make the right investments. It cannot be business as usual nor can we allow ourselves to slip simply into old comfortable patterns of working. We should be measured as to whether we are doing the right things. Who are the best judges? The people on the ground are. Do they see an improvement in the education system, the delivery of health, and access to clean water, all of which make life livable?

It is no longer a cliche to say that our security and existence are intertwined even with that of remote villages and impoverished fragile states. It is no longer a world of them and us. Whatever support we give to affected countries is not an act of a good Samaritan. It is for our very own personal safety and well-being.

In our affluent and technologically sophisticated world, complacency is not an option. We cannot glibly dismiss seemingly faraway threats as problems of the poor and remote parts of the world. As the Ebola epidemic in West Africa has revealed in just a matter of months, it is in our personal interest to address those problems at their source before they escalate. This will reduce the tragic impact of the epidemic locally and avoid having it become a global crisis.

With 2015 approaching, and its world of conferences and goals, now would be [the] time to take decisive action that fundamentally reinvigorates the ability of the international system to work in a more effective and cohesive manner with human, physical, and financial resources upfront to support national response plans as well as those that transcend national boundaries in the way modern threats do. Whether this requires fine-tuning or a complete overhaul, now is the time for action, and hopefully for something a bit more ambitious than just making the United Nations "fit for purpose" which seems to mean simply "good enough to get the job done".

Critical Thinking

1. What strategy was followed by the UN team to promote the postwar economic development of Liberia?
2. What was the effect of the Ebola virus on the Liberian economy?
3. What lessons were learned from the Ebola epidemic in Liberia?

Internet References

Centers for Disease Control and Prevention

www.cdc.gov/

Liberia

https://www.cia.gov/library/publications/the-world-factbook/geos/li.html

UN Mission in Liberia

www.un.org/en/peacekeeping/missions/unmil/

WHO Ebola virus Disease outbreak

www.who.int/csr/disease/ebola/en/

JORDAN RYAN headed the UN Development Programme's Bureau for Crisis Prevention and Recovery from 2009 until retiring this September. He has also served as the UN Secretary-General's Deputy Special Representative in Liberia from 2005 to 2009. Ryan has worked as a lawyer in Saudi Arabia and China and was a Visiting Fellow at the Harvard Kennedy School in 2001.

Article   Prepared by: Robert Weiner, *University of Massachusetts, Boston*

Can Africa Turn from Recovery to Development?

"For more than a decade, African policy making was limited to a narrow space prescribed by the Washington Consensus. Things are changing now, facilitated by the collapse of that doctrine."

THANDIKA MKANDAWIRE

Learning Outcomes

After reading this article, you will be able to:

- Understand what is meant by the Washington consensus.

- Understand what the factors behind the Washington consensus are.

During the last decade or so, Africa, once labeled by the *Economist* as the "Hopeless Continent," has been rebranded by the same magazine as "Africa Rising." Described by then—British Prime Minister Tony Blair in 2001 as "a scar on our consciences," Africa has become the home of "roaring lions" and the "fastest billion"—contrasting with the image of the world's most impoverished "bottom billion," in the words of the economist Paul Collier. These new monikers and the ebullient optimism they reflect are a welcome change. They have replaced a costly "Afropessimism" that reigned in Western media and academic circles during much of the 1980s and 1990s. The costs of the negative stereotypes of that period were felt not only in terms of Africa's self-esteem but also financially: They depicted Africa as economically much riskier than it ever was and dampened the animal spirits of investors.

Afropessimism never caught on in Africa itself. With 70 percent of its population under the age of 20, the continent is perhaps too youthful to indulge in despair. Now the threat to sound reflection on the future is "Afro-euphoria." But opinion surveys by Afrobarometer suggest that Africans may also be immune to the new fad.

Despite serious doubts about the reliability of official data, African economies have been growing fairly rapidly during the past decade or so, following the two "lost decades" that nourished negative images. African cities now exude a new vibrancy after years of depressing decay. Western media, which have often turned a blind eye to the continent, are beginning to notice the change. And so the "Africa Rising" narrative is understandable.

However, as the graph on the next page suggests, we should bear in mind that this is not the first time that postindependence African economies have grown rapidly. The first decade of independence saw growth rates that exceeded the current ones. Africa then endured the lost decades. It took close to two decades of growth to regain the peak income levels of the 1970s: Much of this has simply involved recovery from the consequences of the adjustment debacle that resulted from policies imposed in the 1980s and 1990s by the International Monetary Fund and the World Bank.

There is a difference between recovery and catching up with the rest of the world. Recovery basically puts to use existing, underutilized capacities to get back to earlier levels of development. Catching up involves the creation of new capacities and is thus an inherently more demanding task. The real challenge for Africa is not merely recovery but "accelerated development"—the unfulfilled promise made by the World Bank to Africa in its landmark 1981 Berg Report.

Given this postcolonial experience, there are three tasks that should preoccupy policy makers. The first task is simply trying to understand the factors behind the economic recovery, separating fortuitous windfall gains from the more durable factors that can be harnessed to sustain the current boom and make the recovery stronger and more inclusive. The second task is to assess the magnitude of the growth and its adequacy for addressing the severe underdevelopment and poverty that afflict the African continent. The third is to deal with the legacy of the structural adjustment debacle.

Paternity Claims

Claims to paternity of the African economic recovery have come from many quarters. Incumbent politicians claim that their wisdom, foresight, and astuteness produced the "miracles" that have taken place under their watch. Yet most leaders would be hard-pressed to explain what particular policies or acts of theirs account for the high growth in their respective economies. It can, however, be attributed to greatly improved governance in Africa, thanks to a broad trend of democratization.

International financial institutions (IFIs) such as the World Bank and the IMF have argued that the adjustment and stabilization policies that they imposed in Africa are finally bearing fruit. These claims are rather disingenuous for a number of reasons. For one, at no time did these institutions indicate that their policies would have such a long gestation period. Since there were no clear indications of the time lag between policy and recovery, they could claim success for any recovery occurring at any time.

Nevertheless, one can surmise from earlier pronouncements that the envisaged time lag was 3 to 5 years. Indeed, within a few years of initiating its policies in the 1980s, the World Bank began producing numerous tables of "good adjusters." The interesting thing about these tables was how briefly countries featured on them before they were relegated to the group of "poor adjusters." Much of the supposed success had to do with

the funding from IFIs and other donors that often temporarily relaxed foreign exchange constraints on production and led to improved capacity utilization. Generally, this had a one-off effect, as austerity measures imposed by the IFIs as conditions for the funding caused countries to fall back onto the low-growth path. The donors themselves were at great pains to deny that their money had anything do with the recoveries, insisting instead that their policies accounted for the turnaround.

There was a considerable academic literature, some of it produced within these financial institutions themselves, suggesting that their policies led to poor performance—or at best had produced few, if any, positive outcomes. In the mid-1980s, the IFIs and economists associated with them began to suggest a whole range of explanations for the poor results of their policies. By 1989 the World Bank had identified bad governance as the problem. This was soon followed by a call for "getting institutions right," building on the work of the American economist Douglass North, which was easy to link to a market reform agenda because it focused on the protection of property rights. New property laws were enacted, central banks were given independence, and many other autonomous institutions were set up to reduce the discretionary role of politicians.

When these changes did not seem to lead to the expected result, new culprits were identified. There was unfortunate geology that nourished the "resource curse"—the idea that rents from natural resources destroy the export competitiveness of other industries, and make the state less accountable and more prone to corruption. There were borders that rendered many countries landlocked or ethnically diverse and conflict-prone. There was the lethal cultural brew of neopatrimonialism, a form of clientelistic rule in which political power is personalized and whose ingredients were African culture and Western rationality. And there was the colonial heritage that left Africa with only extractive institutions. Obviously, the more that is attributed to these exogenous factors, the less agency and policy are considered to play an important role in the performance of African economies.

Finally, we should recall that up until the sudden surge of African economies in the mid-1990s, advocates of the Washington Consensus (a set of ideas about the liberalization of markets, macro-economic policies, and the role of government that formed the basis for IMF and World Bank policies in the 1980s and 1990s) argued that the failure of their prescribed policies was due to recidivism and noncompliance by African governments. We would need to know the factors behind political leaders' supposed Damascene conversion to "good policies" for the story to make sense.

In any case, the Washington institutions themselves expressed doubts about the efficacy of their model and admitted that, through errors of omission or commission, their policies had done harm to African economies. The mistakes adduced in this

Sub-Saharan African GDP Per Capita, 1960–2012 (Constant 2000 US$)

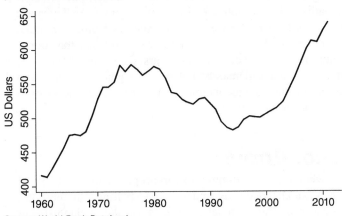

Source: World Bank Databank

mea culpa included neglect of infrastructure, assumption of "policy ownership" by donors, excessive retrenchment of the state, neglect of tertiary education, "one-size-fits-all" institutional reform, and wrongly sequenced privatization and financial sector reforms. There were attempts to revise the policy package into something known as the post–Washington Consensus. However, the dramatic recovery of African economies has allowed this mea culpa to be conveniently set aside, and African countries have been urged not to waste time looking at the recent past.

Recovery Factors

So what lies behind the recovery? Different factors have played out differently in various countries, but we can highlight a number of significant ones. The first is improved earnings from exports due to significantly improved terms of trade since the mid-1990s. A major reason for improved export performance is closer linkage with more dynamic economies. For much of the postcolonial period, African nations were intimately associated with the economies of their erstwhile colonial masters through arrangements that tended to reinforce inherited export structures. African economies were thus tethered to slow-growing partners. But during the last decade, the share of African exports going to Asia has more than doubled, to 27 percent of total exports, allowing African countries to benefit from Asia's high growth rates.

The current commodity boom has not been accompanied by systematic attempts to industrialize.

There is, however, a disturbing side to this improved export performance that raises questions about its long-term sustainability: The expansion in export earnings has been based on increased prices rather than increased production of export commodities. Furthermore, the export structure exhibits no diversification, rendering many countries just as prone to "monocropping" (relying on a single export commodity) as they were under colonial rule.

Another cause of the recovery is a revival of foreign direct investment (FDI). Whereas in the early 1990s FDI was directly associated with privatization policies that naturally tapered off, much of the new investment is driven by factors unrelated to policy. Two sectors have proved especially attractive to FDI: mining, and information and communications technology. The latter attracted more than 50 percent of FDI investment into Africa from 1996 to 2006.

During the period of structural adjustment there was little public investment in infrastructure; the belief was that the private sector would step in to provide public services in a more efficient way. In the event, private-sector infrastructure investment was not forthcoming, apart from telecommunications and mining infrastructure.

Public investment, especially in infrastructure, has picked up recently. Governments have financed this investment mainly with their own resources and loans from other emerging economies. Some of the more resource-rich countries have also been able to raise money in financial markets to invest in infrastructure. For the least developed countries, debt relief has positively affected public investment, but policies accompanying that relief have attenuated the full benefits by limiting state capacity and by inducing greater consumption rather than saving.

A case can also be made that political changes have had a positive impact on economies. The end of militarism and the greater democratization of African countries have placed economic performance at the core of states' sources of legitimation. The success of a leader, even in the remaining authoritarian strongholds, is no longer measured by the longevity of his reign, and even less so by the number of self-awarded medals on his chest, but by the performance of the economy and the stability of the political order.

The urgency of development is strongly felt by a young population that is aware of Africa's lagging behind and of the economic achievements in other parts of the world. In addition, democratization—bringing greater accountability to local constituencies—over the years has made it harder for external actors to impose their preferred policies. These political changes are no small matter, given the fact that Africa has had many leaders whose political aspirations never rose beyond satisfying local clients and the external masters who underwrote their rule.

The Washington Consensus lost its intellectual bearings following the 2008 financial crisis, which became a crisis of neoliberalism in the West. The increased foreign exchange reserves of many African countries have tended to undermine the conditionality-enforced policy regime, creating more policy space for African leaders. New aid donors, such as China and India, have no particular commitment to the consensus that donor institutions have wielded over the years. China may have made certain diplomatic demands, but it is not generally known to interfere in the macroeconomic policies of aid recipients or to insist on a certain regime type.

Donor Errors

Even when the IFIs admitted to certain policy errors, they did so in a backhanded way and without spelling out their full implications. The cumulative effects of the maladjustment they caused pose serious problems for Africa's attempts to go beyond recovery and accelerate its catch-up pace.

One obvious error was the devastating erosion of state capacity produced by reckless downsizing and formulaic retrenchment, the demoralizing of the civil service, and neglect of the physical apparatus of the state. In addition, poorly coordinated donors' massive interference with and experimentation on local institutions, and their assumption of policy "ownership," undermined the legitimacy of local political and bureaucratic actors by reducing their effectiveness through one-size-fits-all reform. The view of the IFIs was that African countries had bloated civil services; Africa now has the lowest number of civil servants per 100 citizens, making it the least governed continent.

There is a difference between recovery and catching up with the rest of the world.

Much of the institutional reform focused on enhancing the restraining arms (independent central banks, courts, police, accounting tribunals), rather than the transformative arms of the state (the so-called spending ministries in charge of social services, industry, agriculture, infrastructure, and so on). The state was effectively removed from the development policy arena, which was occupied by peripatetic experts, fencing off key institutions from local political oversight. The extent to which the donors controlled African economies became an embarrassment to the donors themselves, and they began to fret about "ownership" and "partnership."

Africa has substantial human resources and is moving toward a more favorable demographic profile, which can facilitate stronger economic growth. But this depends on whether the population is educated and provided opportunities for employment. One costly error of the adjustment policies was the neglect of tertiary education, based on the dubious argument that rates of return were better for primary education than for the secondary and tertiary levels, leading to dramatic shifts of resources from higher education. This is hampering Africa's capacity to exploit current opportunities, including the "demographic dividend" of a younger population. The lack of an educated workforce sets limits to economic growth and transformation.

Investment Needed

Compared with the importance of mining and telecommunications in the revival of FDI, investment in other sectors, such as agriculture and manufacturing, has not recovered. Investment rates in the early 1960s averaged between 7 percent and 8 percent of GDP, rising to a high point of about 13 percent during 1975–80, before falling back to about 7.5 percent during 1990–95. Starting in the second half of the 1990s, investment rates rose slowly. They are currently close to 10 percent—too low to sustain the structural transformation of Africa's economies.

The low investment is related to low savings. Africa's growth path has been consumption-intensive, at least when compared with the investment-driven East Asian model. Wal-Mart goes to Africa not to buy manufactured goods for the US market, as it does in China, but to sell goods to the new middle class. The proliferation of shopping malls in Africa is the reverse side of deindustrialization—cheap imported goods have undermined local industry.

The IFIs pressed African states to refrain from investment in basic public goods. One effect of this jaundiced viewed toward publicly provided infrastructure has been poor responsiveness to incentives, and especially to the opportunities of the current boom. The results can be seen in traffic congestion, electricity blackouts, and overall high transaction costs. Lack of infrastructure is proving to be one of the major constraints on diversifying African economies and placing them on a sustainable course.

Donors also erred with financial reforms whose failures they initially misinterpreted as a matter of incorrect sequencing, since the reforms took effect before fiscal consolidation. Later, they admitted that financial liberalization before the establishment of proper regulatory institutions led to fragmentation, financial chaos (including the collapse of a number of banks), and high levels of noncompetitive behavior due to the wide gap between interest rates paid to savers and those paid by borrowers. This gap often reflects the low levels of competition among banks in Africa, though the banks attribute it to high transaction costs. The reformed financial sector does not mobilize savings or allocate deposits productively. Much of the credit it extended went to speculative real estate investments and into consumption, symbolized by the shopping malls that have mushroomed all over Africa.

The donors also demanded that African countries liberalize their capital accounts. An immediate effect of this was a drastic increase in economic volatility. In turn, this volatility prompted an obsessive focus on foreign exchange reserves, leading in many countries to excessive accumulation of reserves.

Missed Opportunities

One widespread consequence of these failed policies was the deindustrialization of the continent. Quite remarkably, the current commodity boom has not been accompanied by systematic attempts to industrialize. This is in sharp contrast to the situation in the 1960s and 1970s, when initial attempts at industrialization were made—though not always successfully. Now there is a low level of diversification of African exports, particularly with respect to manufactured goods.

With no industrial policies or financial institutions to underwrite industrialization, African economies have not been able to enhance the interface between raw material production and manufacturing. As a 2014 United Nations report observes, "If Africa does not capitalize on its opportunities to diversify and add value to these presently lucrative activities, it may miss the opportunity presented by the commodity boom."

Up until the mid-1990s poverty rates in Africa were increasing. There is evidence suggesting that poverty has fallen since then. However, most of the data is for absolute levels of poverty, based on a one-dollar-a-day limit. This focus ignores the issue of inequality in Africa, which is politically more salient. The Africa Progress Panel led by former UN Secretary General Kofi Annan reported in 2012 that little progress had been made in addressing inequality. In the sub-Saharan region, inequality is now higher than in all other regions of the world except for Latin America and the Caribbean. This can be partly attributed to the fact that economic growth is not creating jobs in the formal sector, and large numbers of people must resort to casual and precarious work.

A major promise of adjustment policies was to reverse Africa's agricultural production decline and, even more significantly, to bring about a reversal in reliance on imported food. But per capita food availability remains way below the average levels of the years before the 2008 financial crisis. The agricultural transformation did not take place. This failure can be attributed to the collapse of rural infrastructure, the disappearance of marketing boards that have not been adequately replaced by the private sector, and low levels of investment in research and extension services, an outcome of the state's retrenchment.

Reasons for hope

One of the most significant conceptual transformations that took place in Africa during the era of structural adjustment (1980–95) was the recognition of the importance of markets as institutions for the exchange and allocation of resources. The adjustment debacle underscores the importance of regulating markets and of making them socially acceptable and politically viable. An important challenge for Africa is to go beyond the conflation of "pro-market" policy with capitalism and the "pro-business" regimes it entails. This requires that states conduct a proactive relationship with business, applying both carrots and sticks.

Countries need space not only to craft policies that are appropriate to their circumstances, but also for experimentation. For

more than a decade, African policy making was limited to a narrow space prescribed by the Washington Consensus. Things are changing now, facilitated by the collapse of that doctrine. The IFIs, with their tarnished brand, have retreated from several once-firmly held positions about the role of the state in development and the nature of market failure in developing countries. In addition, the more favorable foreign exchange positions of African governments have reduced the leverage of donors and their ability to impose prescriptive conditions attached to loans, dramatically increasing the policy space of those governments. This is the paradox of IMF programs: When countries are obliged to (wastefully) accumulate reserves, they become less pliable.

The urgency of development is strongly felt by a young population that is aware of Africa's lagging behind.

For latecomers, the developmental role of institutions is central. After years of touting Asian economies as proving the effectiveness of policies advocated by the IFIs, in 1993 the World Bank finally accepted the overwhelming evidence that the state had played a central role in the developmental experiences of these countries. Yet even when it admitted that industrial policy had proved effective in Asia and, indeed, in virtually every instance of economic catch-up, this reversal was still accompanied by caveats about why Africa could not replicate the Asian experience. The continent was said to be laboring under various forms of culturally bound forms of government that made it impossible to think about the collective good.

In more recent years, this dogma has softened, and perceptions of the African capacity to pursue industrialization strategies have changed. Even so, unbridled one-size-fits-all experimentation has left the continent without adequate institutions to manage the structural changes it needs: institutions for mobilizing financial resources and allocating them with a long-term vision, and institutions for drawing up and implementing industrial policy. During the era of adjustment, the expression "industrial policy" was taboo in policy circles, associated with the much maligned "import substitution strategies" that involved state protection of infant industries.

Africa has plentiful natural resources that, if harnessed, could enhance its development potential. This has been obscured by a debate focused on the concept of the resource curse, founded on a rather tendentious deployment of data. If there has been a resource curse, it is to be found in the low earnings of African countries due to poor terms of trade or the rapacity of foreign mining interests and their local collaborators. The most disturbing outcome of this trend has been the failure to capture mineral rents during the current commodities boom. Many other parts of the world have addressed these problems through various forms of resource nationalism. Africa may need a dose of that. There are now heated debates on the continent about what share of resource rents should go to the state, and calls for renegotiating a number of shadowy deals that African governments entered into.

African economies are generally recovering and growing fairly rapidly. But a number of factors in the recovery, such as the commodities boom, are one-off events unlikely to be repeated in the immediate future. One challenge is to identify internal drivers that can sustain growth into the future. These will include improved and prudent mobilization of human, material, and financial capital, which entails making the most of the continent's vast resources through increased technological mastery in order to achieve socially inclusive (and therefore politically sustainable) growth. Another challenge is to address the social problems that the new economic growth spawns, such as inequality—problems that are too often neglected by the Africa Rising narrative.

Critical Thinking

1. Do you agree that democratization has worked in Africa? Why or why not?

2. Why didn't the Washington consensus work in developing countries?

3. Why has "Afropessimism" been replaced by "Afroeuphoria"?

Create Central

www.mhhe.com/createcentral

Internet References

Millennium Development Goals
http://www.un.org/millenniumgoals.

The African Development Bank
http://www.afdb.org

The African Union
http://www.africa-union.org

UN Economic Commission for Africa
www.uneca.org

THANDIKA MKANDAWIRE is a professor of African development at the London School of Economics.

Article Prepared by: Robert Weiner, *University of Massachusetts, Boston*

The Early Days of the Group of 77

KARL P. SAUVANT

Learning Outcomes

After reading this article, you will be able to:

- Understand why the G-77 was created.

- Understand how the developing countries have organized themselves within the United Nations Conference on Trade and Development.

In December 1961, the United Nations General Assembly designated the 1960s as the "United Nations Development Decade."[1] At the same time, it also adopted a resolution on "International Trade as the Primary Instrument for Economic Development,"[2] in which the United Nations Secretary-General was asked to consult governments on the advisability of holding an international conference on international trade problems. These resolutions led to the United Nations Conference on Trade and Development (UNCTAD). Their underlying developmental model-trade as the motor of development-shaped the outlook and approach of the new institution.

After obtaining favourable reactions from most governments and strong support from a developing countries' Conference on the Problems of Economic Development held in Cairo in July 1962,[3] the United Nations General Assembly decided to convene the first session of UNCTAD.[4] A Preparatory Committee was established to consider the agenda of the Conference and to prepare the necessary documentation. During the deliberations of the Preparatory Committee—in identifying the relevant issues and problems, endeavouring to list proposals for action, and indicating lines along which solutions might be sought—the divergence of the interests of the developing countries from those of the developed countries began to emerge sharply. The distinctive interests of the Third World manifested themselves at the closing of the second session of the Preparatory Committee (21 May to 29 June 1963), when representatives of the developing countries submitted a "Joint Statement" to the Committee in which they

summarized the views, needs and aspirations of the Third World with regard to the impending UNCTAD session.[5] Later that year, this Statement was submitted to the General Assembly as a "Joint Declaration" on behalf of 75 developing countries that were members of the United Nations at that time.[6] This Declaration was the prelude to the establishment of the Group of 77 (G-77).

UNCTAD I met in Geneva from 23 March to 16 June 1964. It was the first major North-South conference on development questions. During the negotiations at that conference, economic interests clearly crystallized along geopolitical group lines, and the developing countries emerged as a group that was beginning to find its own. The "Joint Declaration of the Seventy-Seven," adopted on 15 June 1964, referred to UNCTAD I as "an event of historic significance"; it continued:

> The developing countries regard their own unity, the unity of the 75, as the outstanding feature of this Conference. This unity has sprung out of the fact that facing the basic problems of development they have a common interest in a new policy for international trade and development. They believe that it is this unity that has given clarity and coherence to the discussions of this Conference. Their solidarity has been tested in the course of the Conference and they have emerged from it with even greater unity and strength.

> The developing countries have a strong conviction that there is a vital need to maintain, and further strengthen, this unity in the years ahead. It is an indispensable instrument for securing the adoption of new attitudes and new approaches in the international economic field. This unity is also an instrument for enlarging the area of co-operative endeavour in the international field and for securing mutually beneficent relationships with the rest of the world. Finally, it is a necessary means for co-operation amongst the developing countries themselves.

> The 75 developing countries, on the occasion of this declaration, pledge themselves to maintain, foster and

strengthen this unity in the future. Towards this end they shall adopt all possible means to increase the contacts and consultations amongst themselves so as to determine common objectives and formulate joint programmes of action in international economic co-operation. They consider that measures for consolidating the unity achieved by the 75 countries during the Conference and the specific arrangements for contacts and consultations should be studied by government representatives during the nineteenth session of the United Nations General Assembly.[7]

Although the recommendations adopted by UNCTAD I were, to a large extent, inspired by the conceptual work undertaken in the preceding decade by the Economic Commission for Latin America—whose Executive Secretary, Raúl Prebisch, became the Secretary-General of UNCTAD I and stayed in that post as one of the principal promoters of Third World unity until 1969[8]—the conference was nonetheless a new departure: for the first time, the Third World as a whole had participated in the elaboration of a comprehensive set of measures.[9] Accordingly, "new" was the theme of the "Joint Declaration of the Seventy-Seven": UNCTAD I was recognized as a significant step towards "creating a new and just world economic order"; the basic premises of the "new order" were seen to involve "a new international division of labour" and "a new framework of international trade"; and the adoption of "a new and dynamic international policy for trade and development" was expected to facilitate the formulation of "new policies by the governments of both developed and developing countries in the context of a new awareness of the needs of developing countries." Finally, a "new machinery" was considered necessary to serve as an institutional focal point for the continuation of the work initiated by the conference. This machinery was established later that year, when the General Assembly decided to institutionalize UNCTAD as an organ of the General Assembly.[10] UNCTAD became the main forum for global development discussions, and—guided by the expectations voiced in 1964—it became the focal point of the activities of the G-77, which, by April 2014, counted 133 members[11] (United Nations membership totaled 193). During that period, the G-77 became an integral part of UNCTAD, was one of the most important agents for the socialization of the developing countries in matters relating to international political economy, and established itself firmly in all major relevant parts of the United Nations system as the Third World's principal organ for the articulation and aggregation of its collective economic interest and for its representation in the negotiations with the developed countries.[12]

No one has formulated the political point of departure of the Third World more succinctly than Julius K. Nyerere when he said in his address to the Fourth Ministerial Meeting of the G-77 in Arusha, in February 1979: "What we have in common is that we are all, in relation to the developed world, dependent—not interdependent—nations. Each of our economies has developed as a by-product and a subsidiary of development in the industrialized North, and it is externally oriented. We are not the prime movers of our own destiny. We are ashamed to admit it, but economically we are dependencies-semicolonies at best-not sovereign States."[13]

The objective is, therefore, quite naturally, "to complete the liberation of the Third World countries from external domination".[14]

Until the early 1970s, the G-77 thought to achieve this objective through improvements of the system, the high points being UNCTAD II (New Delhi, 1968) and UNCTAD III (Santiago, 1972) and the preparatory First (Algiers, 1967) and Second (Lima, 1971) Ministerial Meetings of the G-77, as well as UNIDO I (Vienna, 1971) and the adoption of the international development strategy for the Second United Nations Development Decade (1970). A number of changes were, in fact, made (witness, for instance, the Generalized System of Preferences), but many other negotiations (for instance, in the commodity sector) hardly made any progress, and no drastic improvements took place. On the contrary, the gap between the North and South widened, especially for the least developed among the developing countries.

The limitations of this approach naturally took time to become apparent. In addition, until the end of the 1960s, neither developed nor developing countries had fully realized the importance of economic development as a necessary complement to political independence. The development issue was regarded as "low politics," left to the technical ministries of planning, economics, commerce, finance and development. Attempts to politicize the issue therefore failed. The most prominent among these was the "Charter of Algiers," adopted by the First Ministerial Meeting of the G-77 in October 1967 in preparation of UNCTAD II. The intention of this first comprehensive declaration and programme of action of the G-77 was to give a new impetus to the North-South negotiations. For this purpose, the Ministerial Meeting even decided to send high-level "goodwill missions" to a number of developed countries (both those with centrally planned and those with market economies) to inform key governments about the conclusions of the meeting and to persuade them of the need for accelerated progress.[15]

At the beginning of the 1970s, however, several developments converged to produce a change in attitudes: the political decolonization process had largely run its course and the political independence of most of the new states had been consolidated; the political-military pressures of the Cold War were subsiding somewhat; the regional and international development efforts had shown disappointing results; and doubts had begun to be voiced about the prevailing development model.[16] As a result, more attention could be given to other important matters. For the developing countries, this meant that questions of economic development began to receive greater attention, and these countries became increasingly aware that the institutions of the international economic system had been established by the developed market economies to serve primarily their own purposes.[17] It was felt that the interests, needs and special conditions of the developing countries had largely been ignored, thus they remained in poverty and dependency. Hence, fundamental changes in the international economic system were required to establish a framework conducive to development and to create the economic basis of independence. In fact, the system itself had come under serious strain with the breakdown of the Bretton Woods system, the food and oil crises, payment imbalances, a general surge of inflation, world recessions, increasing protectionism, rising environmental concerns, and the spectre of the scarcity of raw materials. When the economic tranquility of the 1960s gave way to the turbulence of the 1970s, international economic matters could no longer be ignored.

The Non-Aligned Movement (NAM) offered the framework for this recognition to grow. Within a few years, development questions became "high politics"; they were elevated to the level of heads of state or government and were made a priority item on their agenda. Between 1970 and 1973, NAM evolved into a pressure group for the reorganization of the international economic system.[18] Since the Non-Aligned Countries (NAC) considered themselves to be playing a catalytic role within the G-77,[19] the politicization of the development issue had an important effect on the manner in which this issue was perceived, presented and pursued within North-South negotiations. Thus, the political clout and pressure of NAC, coupled with the Organization of the Petroleum Exporting Countries' (OPEC) forceful actions, led to the Sixth Special Session of the United Nations General Assembly which adopted, on 1 May 1974, the "Declaration and Programme of Action on the Establishment of a New International Economic Order."[20]

Hence, almost exactly one decade after the first session of UNCTAD, and after years of debates about improving the international economic system, the call for a new beginning was again taken up—this time, however, with a view towards a structural reorganization of the world economy. The establishment of a New International Economic Order (NIEO) became the main objective of the Third World. The concrete changes that the G-77 proposed in order to achieve this objective were spelled out in detail in the "Arusha Programme for Collective Self-Reliance and Framework for Negotiations," adopted by the Fourth Ministerial Meeting of the G-77 in Arusha, in February 1979.

While NAC played a key role in making the development issue a priority item on the international agenda, the G-77

became the principal organ of the Third World through which the concrete actions required for changing the international conditions for promoting development became to be negotiated within the framework of the United Nations system. This objective dominated UNCTAD IV (Nairobi, 1976) and UNCTAD V (Manila, 1979) and the preparatory Third (Manila, 1976) and Fourth (Arusha, 1979) Ministerial Meetings of the Group of 77; UNIDO II (Lima, 1975) and UNIDO III (New Delhi, 1980) and the preparatory meetings of the G-77 in Vienna (1974), Algiers (1975) and Havana (1979); the regional preparatory meetings convened for each of these UNCTAD and UNIDO conferences by the African, Arab (for UNIDO only), Asian, and Latin American members of the G-77; the 1976 Mexico City Conference on Economic Co-operation among Developing Countries; the 1975–1977 Paris Conference on International Economic Co-operation, in which the G-77 acted through the Group of 19; and a series of ministerial-level meetings of the G-77 (including meetings of ministers for foreign affairs) in preparation for sessions of the United Nations General Assembly. It also entered into the discussions in the International Monetary Fund (IMF) and the World Bank, where the G-77 had been acting through the Group of 24 since 1972.

There continued to be a considerable gap between the declaration of a new order and the action programmes formulated to establish it in the major areas of North–South interactions: commodities and trade, money and finance, research and development and technology, industrialization and transnational enterprises, and food and agriculture. In fact, an analysis of the contents of the NIEO programme showed that, although a number of additional problems had been identified, many of the concrete proposals under discussion remained the same since 1964—even if the emphasis on some of them (e.g., proposals concerning technology) had grown. This was an indication of the slow progress made in the past. The new proposals were aimed at creating new economic structures. And, to a greater extent than in the past, it became recognized that the various dimensions of North–South interactions are interrelated and hence have to be approached in a comprehensive and integrated manner. Over time, the gap between objectives and concrete proposals could have been closed through the elaboration of new policies or possibly even through changes in the underlying development model. At the time of the beginning of the G-77, however, the model continued to assume that development is best served by a close association of the developing with the developed countries.

A conceptual change, though, was in the offing with the concept of individual and collective self-reliance. In contradistinction to the prevailing associative development strategy with its orientation towards the world market and its heavy reliance on linkages with the developed countries for stimulating industrialization, the self-reliance concept sought greater selectivity in traditional linkages, accompanied by a greater mobilization of domestic and Third World resources and a greater reliance on domestic and Third World markets. It is these markets, rather than those in the developed world, which were expected to provide the principal stimulus for economic development.

The concept of self-reliance was introduced into the international development discussion by NAC in 1970, which were also responsible for most of the practical follow-up that was undertaken in the subsequent years.[21] Although self-reliance can be strengthened by international measures,[22] it requires primarily a strengthening of linkages among developing countries. For this reason, the G-77 which, as pointed out above, concentrates almost exclusively on North-South negotiations within the system of the United Nations had been slow in incorporating self-reliance into its own programme.

The first effort to do so was made at the 1976 Third Ministerial Meeting of the G-77, during which a resolution on economic cooperation among developing countries was adopted.[23] Through this resolution, which linked the work of the G-77 with that of NAC (whose pioneering work in this area was recognized in the resolution), it was decided to convene a meeting in Mexico City during the month of September 1976 to prepare a detailed programme on economic co-operation. Originally, it was planned to hold this meeting at the level of an intergovernmental working group; but at the subsequent UNCTAD IV session it was decided to hold it at the highest possible level.[24] Hence, the Conference on Economic Co-operation among Developing Countries was convened in Mexico City from 13 to 22 September 1976. Until the early 1980s, it was the only major conference of the G-77 that was not closely and directly related to an important impending activity within the United Nations system.

The full integration of this approach into the Groups conceptual mainstream came, however, only during the 1979 Arusha Fourth Ministerial Meeting in preparation of UNCTAD V, an event signaled by the very title of the final declaration of that meeting, the "Arusha Programme for Collective Self-Reliance and Framework for Negotiations." As this title indicates, the declaration consisted of two parts: a programme for self-reliance (even if this was formulated only in terms of economic cooperation among developing countries) and a programme for North–South negotiations. Thus, a shift in the United Nations orientation of the Group seemed to be taking place, and the G-77 (together with, or in addition to, NAC) was poised to make greater efforts towards stronger South–South co-operation.

Accordingly, the Arusha Ministerial Meeting strongly endorsed the recommendations of the Mexico City Conference regarding an institutional follow-up for economic cooperation among developing countries. As a consequence, a regionally prepared first meeting of Governmental Experts of Developing Countries on Economic Co-operation among Developing Countries was convened for March–April 1980. This interregional gathering, in turn, was fully supported by a March 1980 Ministerial Meeting of the G-77 in New York, during which economic cooperation among developing countries was a special item on the agenda and which decided to set up an open-ended ad hoc group "with the task of elaborating appropriate action-oriented recommendations for the early and effective implementation of the objectives of economic cooperation among developing countries."[25] This task was begun by the "Ad-hoc Intergovernmental Group of the G-77 on Economic Co-operation among Developing Countries in Continuation of the Ministerial Meeting of the G-77 held in New York in March 1980 in a session in Vienna in June 1980. Its conclusions and recommendations were considered a useful basis for further discussions by the 1980 Meeting of Ministers for Foreign Affairs of the G-77. Deliberating shortly after the 1980 Eleventh Special Session of the General Assembly, the failure of that Session led the Ministers to stress that economic co-operation among developing countries is "an indispensable element both of the accelerated development of developing countries and of strengthening their negotiating power in their relations with the rest of the world."[26] The Ministers decided, therefore, to convene a high-level conference on economic cooperation among developing countries for 1981, to expedite the implementation of various programmes and decisions relating to this subject matter.

The effort towards greater South-South cooperation was encouraged by the slow pace of progress in North-South negotiations and the frustrations created thereby, as well as the recognition of the limits of the prevailing associative development model. It was also facilitated by the bidimensional nature—recognized explicitly in the statement quoted in the preceding paragraph—of the self-reliance approach. One dimension, as described above, was seen as involving to bring about changes in the patterns of interaction between North and South that would allow for a more equitable sharing of the benefits of, and control over, international economic activities by developed and developing countries.

Besides being a part of the necessary structural change, self-reliance was also seen as an instrument for achieving it: self-reliance increases the individual and collective bargaining strength of the developing countries and, especially where it allows joint action, creates the countervailing power that is needed to negotiate the desired changes in the international system. In this respect, self-reliance meant strengthening the joint action capacity of the Third World.

In the end, however, developing countries returned to the associative development model. While economic cooperation among developing countries remained an important goal, the objective of self-reliance gave way in the following years to an export-oriented development strategy, partly driven by the spreading 1982 Mexican debt crisis and the quickening pace of globalization.

But the awareness of the weakness of each individual developing country in isolation was the genesis of the G-77 and remained its raison d'être. In this sense, then, the G-77 "is a kind of trade union of the poor," which is kept together by "a unity of nationalisms" and "a unity of opposition"—not by "the ideals of human brotherhood, or human equality, or love for each other" or, for that matter, a common ideology.[27] The unity of the G-77 is based on a shared historical experience, a shared continuing economic dependence, and a shared set of needs and aspirations.

Still, since the Group is by no means homogeneous, cohesiveness is not an easy matter to maintain. The immediate interests and specific negotiating priorities of many of its many members—the great number in itself makes it difficult to achieve consensus—are different from those of the others. The individual countries differ vastly from one another with respect to their cultural, ideological, political, and economic systems. No strong unifying institutional force existed: the G-77 had no long-term leadership, regular staff, headquarters, secretariat or, for that matter, any other permanent institution. In fact, the office of the coordinator rotated on an annual basis in New York and Vienna and on a three-monthly basis in Geneva. And although countries like Algeria, Argentina, Brazil, Egypt, India, Indonesia, Jamaica, Mexico, Nigeria, Pakistan, the Philippines, Sri Lanka, Venezuela, and Yugoslavia often played an important role on many issues, none of them dominated the Group. Very important also were and are great differences in the level of economic development, especially between the Latin American Group on the one hand and the African Group on the other. This cleavage was accentuated further by the exclusion of most Latin American countries from the preferential schemes of the Lome Convention. The individual weight of some countries could also complicate matters, especially when these countries were specially cultivated by developed countries and when occasions for separate bilateral deals arose. Similar complications could be created by the continuation of strong traditional links of some developing with some developed countries, e.g., some Central American countries with the United States or some African countries with France. Some of the special interests of the members of the Group also led to the formation of informal sub-groups of, for example, the most seriously affected, the least developed, the newly industrialized and, of course, the oil producing countries.[28] While the success of OPEC was welcomed by most developing countries, especially since it strengthened the bargaining power of the Third World as a whole, the balance of payments burden of the increased oil price introduced considerable strains into the G-77 (and, for that matter, into NAM). But since there existed no alternative for the oil importing developing countries, this experience, however painful, did not end the unity of the G-77.[29]

In the face of these factors it is a formidable task indeed to maintain the cohesiveness of the Group. But the strength of the basic common interests, the capacity to maintain consensus through acceptable trade-offs among the developing countries themselves, the recognition that separate deals bring only marginal and temporary concessions, and the resistance of the developed countries to enter into a broad range of detailed negotiations succeeded in overriding the pressures towards disunity. The maintenance and strengthening of the unity of the G-77 was, and therefore remains, a precondition for achieving the desired changes in the international economic system. To return to Nyerere's analogy and his evaluation of OPEC's "historic action" in 1973:

> But since then OPEC has learned, and we have learned once again, that however powerful it is, a single trade union which only covers one section of a total enterprise cannot change the fundamental relationship between employers and employees. . . . For the reality is that the unity of even the most powerful of the subgroups within the Third World is not sufficient to allow its members to become full actors, rather than reactors, in the world economic system. The unity of the entire Third World is necessary for the achievement of the fundamental change in the present world economic arrangements.[30]

These words spoken in 1979 encapsulate the challenges the G-77 faced throughout its existence, and continues to face today.

What we have in common is that we are all, in relation to the developed world, dependent—not interdependent nations. Each of our economies has developed as a by-product and a subsidiary of development in the industrialized North, and it is externally oriented. We are not the prime movers of our own destiny. We are ashamed to admit it, but economically we are dependencies semi-colonies at best— not sovereign States.

Notes

1. General Assembly resolution 1710 (XVI) of 19 December 1961.

2. General Assembly resolution 1707 (XVI) of 19 December 1961.

3. For the text of the "Cairo Declaration of Developing Countries," see Odette Jankowitsch and Karl P. Sauvant, eds., The Third World without Superpowers: The Collected Documents of the Non-Aligned Countries (Dobbs Ferry, NY: Oceana (now: Oxford University Press), 1978), vol. I, pp. 72–75, hereinafter cited as Jankowitsch and Sauvant. This meeting was the first attempt of the developing countries to coordinate their international development policies in the United Nations.

4. See Economic and Social Council resolution 917 (XXXIV) of 3 August 1962 and General Assembly resolution 1785 (XVII) of 8 December 1962.

5. Of the 32 members of the Preparatory Committee, 19 were developing countries (including Yugoslavia, which at that time played a key role in the Group of 77). Seventeen of the 19 supported the "Joint Statement." The other two, El Salvador and Uruguay, did so only when the "Joint Statement" became the "Joint Declaration."

6. Contained in Karl P. Sauvant, ed., The Third World without Superpowers, 2nd Ser., The Collected Documents of the Group of 77 (Dobbs Ferry, NY: Oceana (now: Oxford University Press), 1981), 20 vols., hereinafter cited as Sauvant. (The documents of the Group of 77 referred to below, as well as those of the United Nations meetings for which they were an input, are contained in these volumes; no specific reference is, therefore, made to them each time they are mentioned.) In 1963, 76 developing countries were members of the United Nations. Except for Cuba and the Ivory Coast, all developing countries, along with New Zealand, co-sponsored the Joint Declaration. Cuba was ostracized by the Latin American Group at that time and hence was not accepted as a co-sponsor of the resolution. (The principle of co-sponsorship requires that every sponsor of a given resolution has to accept any new co-sponsor.)

7. Ibid., document I.C.I.a. At the time of UNCTAD 177 developing countries were members of the United Nations. Of these, the Ivory Coast again did not join at that time and Cuba remained excluded until the 1971 Second Ministerial Meeting. Two others, the Republic of Korea and the Republic of Vietnam (which were not members of the United Nations but were the only other developing countries at UNCTAD I), did join after being accepted for membership by the Asian Group, so that 77 countries supported the "Joint Declaration of the Seventy-Seven." Since, however, the original membership of the group was 75—see the 1963 "Joint Declaration"—the resolution continued to refer to 75 countries. It was only with UNCTAD I that the Group acquired its present name.

8. Prebisch actively encouraged the developing countries during the preparations for UNCTAD I, the session itself, and the subsequent years to cooperate and to strengthen their unity in the framework of the Group of 77. His successor, Manuel Perez-Guerrero, continued this policy.

9. The resolution on the First United Nations Development Decade did not spell out a strategy.

10. Through resolution 1995 (XIX), contained in Sauvant. For the membership, principal functions, organization, etc., of UNCTAD, see that resolution.

11. Including the Palestine Liberation Organization, the only non-state member of the Group of 77.

12. The literature on the Group of 77 is scarce. One of the best analyses at that time was Branislav Gosovic, UNCTAD: Conflict and Compromise. The Third World's Quest for an Equitable World Economic Order through the United Nations (Leiden: Sijthoff, 1972). For another analysis of the Group of 77 and the nonaligned movement and their role in the main international economic conferences dealing with the New International Economic Order, see Robert A. Mortimer, The Third World Coalition in International Politics (New York: Praeger, 1980).

13. Infra, p. 133.

14. Ibid., p. 134.

15. See, Co-ordinating Committee, "Ministerial Mission" and First Ministerial Meeting of the Group of 77, "Charter of Algiers," Part III, in Sauvant, documents II.B.3 and II.D.7, respectively.

16. For an elaboration, see Karl P. Sauvant, "The Origins of the NIEO Discussions," in Karl P. Sauvant, ed., Changing Priorities on the International Agenda: The New International Economic Order (Elmsford, NY: Pergamon, 1981).

17. Apart from the Latin American states, only the following developing countries took part in the Bretton Woods Conference: Egypt, Ethiopia, India, Iran, Iraq, Liberia, and the Philippines.

18. See Odette Jankowitsch and Karl P. Sauvant, "The Initiating Role of the Non-Aligned Countries," in ibid. This observation should not be taken to slight the political purpose and function of the nonaligned movement; it is intended only to point out that the movement had also acquired an equally important economic function and that this change proved to be of crucial importance for making the development issue a priority item on the international agenda.

19. See, e.g., the "Final Communique" adopted at the 1978 Havana meeting of the Co-ordinating Bureau of the Non-Aligned Countries at the Ministerial Level, reprinted in Jankowitsch and Sauvant, vol. V.

20. General Assembly resolutions 3201 (S-VI) and 3202 (S-VI). Together with the "Charter of Economic Rights and Duties of States," adopted 12 December 1974 by the Twenty-Ninth Regular Session of the General Assembly as resolution 3281 (XXIX), and resolution 3362 (S-VII), entitled "Development and International Economic Cooperation," adopted on 16 September 1975 by the Seventh Special Session of the

General Assembly, these resolutions (which are contained in Sauvant) laid the foundations of the programme for a New International Economic Order.

21. Especially in the framework of the "Action Programme for Economic Cooperation," adopted by the 1972 Georgetown Third Conference of Ministers of Foreign Affairs of Non-Aligned Countries, as a consequence of which Coordinator Countries were designated for 18 fields of activity. Important also were a number of the follow-up activities to the Conference of Developing Countries on Raw Materials, which was held in Dakar from 4 to 8 February 1975; although the Dakar Conference was convened by the Non-Aligned Countries, it was explicitly designed to include all developing countries. For the relevant documents, see Jankowitsch and Sauvant.

22. The establishment, within UNCTAD and in pursuance of a resolution adopted at UNCTAD IV, of a Committee on Economic Co-operation among Developing Countries in 1976and the increased emphasis given since UNCTAD V to this approach were efforts in this direction.

23. See Sauvant, document IV.D.7.

24. See, Main Documents of the Group of 77 at UNCTAD IV, "Statement Regarding the Forthcoming Conference on Economic Co-operation among the Developing Countries," ibid., document IV.E.I.

25. See, Ministerial Meeting of the Group of 77, "Communiqué," ibid., document X.C.I.a.

26. Ministers for Foreign Affairs of the Group of 77, Fourth Meeting, "Declaration," ibid., document X.B.4.a.

27. Nyerere, infra, pp. 123,122.

28. Producers' associations other than OPEC had not acquired a political significance of their own.

29. In other words, the oil-importing developing countries had nothing to gain from turning against OPEC since this would not have affected the price of oil. Maintaining solidarity, on the other hand, combined with some pressure, could have led to some concessions by the OPEC countries (be it in the form of aid, special price arrangements, or both), and it strengthened the bargaining power of the oil-importing developing countries in their negotiations with the North.

30. Nyerere, infra, p. I 33.

Critical Thinking

1. Why was it necessary for the developing countries to organize the G-77?

2. Why did the drive for a new international economic order fail?

3. What is the current strategy of the G-77 for promoting economic justice in the world economic system?

Create Central

www.mhhe.com/createcentral

Internet References

G-77
   http://www.g/77.org/doc

Post-2015 development Agenda
   http://post2015.org

United Nations Conference on Trade and Development
   http://unctad.org/eu/Pages/home.aspx

KARL P. SAUVANT is Resident Senior Fellow at the Vale Columbia Center on Sustainable International Investment, a joint center of Columbia Law School and the Earth Institute at Columbia University, and the Founding Executive Director of that Center. He retired from UNCTAD in 2005 as the Director of UNCTAD's Investment Division. He is the author of *The Group of 77: Evolution, Structure, Organization* (New York: Oceana (now: Oxford University Press), 1982).

Article Prepared by: Robert Weiner, *University of Massachusetts, Boston*

The Mobile-Finance Revolution: How Cell Phones Can Spur Development

JAKE KENDALL AND RODGER VOORHIES

Learning Outcomes

After reading this article, you will be able to:

- Understand how digital technology can contribute to economic development.

- Understand the relationship between microfinance and business in developing countries.

The roughly 2.5 billion people in the world who live on less than $2 a day are not destined to remain in a state of chronic poverty. Every few years, somewhere between 10 and 30 percent of the world's poorest households manage to escape poverty, typically by finding steady employment or through entrepreneurial activities such as growing a business or improving agricultural harvests. During that same period, however, roughly an equal number of households slip below the poverty line. Health-related emergencies are the most common cause, but there are many more: crop failures, livestock deaths, farming-equipment breakdowns, and even wedding expenses.

In many such situations, the most important buffers against crippling setbacks are financial tools such as personal savings, insurance, credit, or cash transfers from family and friends. Yet these are rarely available because most of the world's poor lack access to even the most basic banking services. Globally, 77 percent of them do not have a savings account; in sub-Saharan Africa, the figure is 85 percent. An even greater number of poor people lack access to formal credit or insurance products. The main problem is not that the poor have nothing to save—studies show that they do—but rather that they are not profitable customers, so banks and other service providers do not try to reach them. As a result, poor people usually struggle to stitch together a patchwork of informal, often precarious arrangements to manage their financial lives.

Over the last few decades, microcredit programs—through which lenders have granted millions of small loans to poor people—have worked to address the problem. Institutions such as the Grameen Bank, which won the Nobel Peace Prize in 2006, have demonstrated impressive results with new financial arrangements, such as group loans that require weekly payments. Today, the microfinance industry provides loans to roughly 200 million borrowers—an impressive number to be sure, but only enough to make a dent in the over two billion people who lack access to formal financial services.

Despite its success, the microfinance industry has faced major hurdles. Due to the high overhead costs of administering so many small loans, the interest rates and fees associated with microcredit can be steep, often reaching 100 percent annually. Moreover, a number of rigorous field studies have shown that even when lending programs successfully reach borrowers, there is only a limited increase in entrepreneurial activity—and no measurable decrease in poverty rates. For years, the development community has promoted a narrative that borrowing and entrepreneurship have lifted large numbers of people out of poverty. But that narrative has not held up.

Despite these challenges, two trends indicate great promise for the next generation of financial-inclusion efforts. First, mobile technology has found its way to the developing world and spread at an astonishing pace. According to the World Bank, mobile signals now cover some 90 percent of the world's poor, and there are, on average, more than 89 cell-phone accounts for every 100 people living in a developing country. That presents an extraordinary opportunity: mobile-based financial tools have the potential to dramatically lower the cost of delivering banking services to the poor.

Second, economists and other researchers have in recent years generated a much richer fact base from rigorous studies to inform future product offerings. Early on, both sides of the debate over the true value of microcredit programs for the poor relied mostly on anecdotal observations and gut instincts. But now, there are hundreds of studies to draw from. The flexible, low-cost models made possible by mobile technology and the evidence base to guide their design have thus created a major opportunity to deliver real value to the poor.

Show Them the Money

Mobile finance offers at least three major advantages over traditional financial models. First, digital transactions are essentially free. In-person services and cash transactions account for the majority of routine banking expenses. But mobile-finance clients keep their money in digital form, and so they can send and receive money often, even with distant counter-parties, without creating significant transaction costs for their banks or mobile service providers. Second, mobile communications generate copious amounts of data, which banks and other providers can use to develop more profitable services and even to substitute for traditional credit scores (which can be hard for those without formal records or financial histories to obtain). Third, mobile platforms link banks to clients in real time. This means that banks can instantly relay account information or send reminders and clients can sign up for services quickly on their own.

The potential, in other words, is enormous. The benefits of credit, savings, and insurance are clear, but for most poor households, the simple ability to transfer money can be equally important. For example, a recent Gallup poll conducted in 11 sub-Saharan African countries found that over 50 percent of adults surveyed had made at least one payment to someone far away within the preceding 30 days. Eighty-three percent of them had used cash. Whether they were paying utility bills or sending money to their families, most had sent the money with bus drivers, had asked friends to carry it, or had delivered the payments themselves. The costs were high; moving physical cash, particularly in sub-Saharan Africa, is risky, unreliable, and slow.

Imagine what would happen if the poor had a better option. A recent study in Kenya found that access to a mobile-money product called M-Pesa, which allows clients to store money on their cell phones and send it at the touch of a button, increased the size and efficiency of the networks within which they moved money. That came in handy when poorer participants endured economic shocks spurred by unexpected events, such as a hospitalization or a house fire. Households with access to M-Pesa received more financial support from larger and more distant networks of friends and family. As a result, they were better able to survive hard times, maintaining their regular diets and keeping their children in school.

To consumers, the benefits of M-Pesa are self-evident. Today, according to a study by Kenya's Financial Sector Deepening Trust, 62 percent of adults in the country have active accounts. And other countries have since launched their own versions of the product. In Tanzania, over 47 percent of households have a family member who has registered. In Uganda, 26 percent of adults are users. The rates of adoption have been extraordinary; by contrast, microlenders rarely get more than 10 percent participation in their program areas.

Mobile money is useful for more than just emergency transfers. Regular remittances from family members working in other parts of the country, for example, make up a large share of the incomes of many poor households. A Gallup study in South Asia recently found that 72 percent of remittance-receiving households indicated that the cash transfers were "very important" to their financial situations. Studies of small-business owners show that they make use of mobile payments to improve their efficiency and expand their customer bases.

These technologies could also transform the way people interact with large formal institutions, especially by improving people's access to government services. A study in Niger by a researcher from Tufts University found that during a drought, allowing people to request emergency government support through their cell phones resulted in better diets for those people, compared with the diets of those who received cash handouts. The researchers concluded that women were more likely than men to control digital transfers (as opposed to cash transfers) and that they were more likely to spend the money on high-quality food.

Governments, meanwhile, stand to gain as much as consumers do. A McKinsey study in India found that the government could save $22 billion each year from digitizing all of its payments. Another study, by the Better Than Cash Alliance, a nonprofit that helps countries adopt electronic payment systems, found that the Mexican government's shift to digital payments (which began in 1997) trimmed its spending on wages, pensions, and social welfare by 3.3 percent annually, or nearly $1.3 billion.

Savings and Phones

In the developed world, bankers have long known that relatively simple nudges can have a big impact on long-term behavior. Banks regularly encourage clients to sign off on automatic contributions to their 401(k) retirement plans, set up automatic deposits into savings accounts from their paychecks, and open special accounts to save for a particular purpose.

Studies in the developing world confirm that, if anything, the poor need such decision aids even more than the rich, owing to the constant pressure they are under to spend their money on immediate needs. And cell phones make nudging easy. For example, a series of studies have shown that when clients receive text messages urging them to make regular savings deposits, they improve their balances over time. More draconian features have also proved effective, such as so-called commitment accounts, which impose financial discipline with large penalty fees.

Many poor people have already demonstrated their interest in financial mechanisms that encourage savings. In Africa, women commonly join groups called rotating savings and credit associations, or ROSCAS, which require them to attend weekly meetings and meet rigid deposit and withdrawal schedules. Studies suggest that in such countries as Cameroon, Gambia, Nigeria, and Togo, roughly half of all adults are members of a ROSCA, and similar group savings schemes are widespread outside Africa, as well. Research shows that members are drawn to the discipline of required regular payments and the social pressure of group meetings.

Mobile-banking applications have the potential to encourage financial discipline in even more effective ways. Seemingly marginal features designed to incentivize financial discipline can do much to set people on the path to financial prosperity. In one experiment, researchers allowed some small-scale farmers in Malawi to have their harvest proceeds directly deposited into commitment accounts. The farmers who were offered this option and chose to participate ended up investing 30 percent more in farm inputs than those who weren't offered the option, leading to a 22 percent increase in revenues and a 17 percent increase in household consumption after the harvest.

Poor households, not unlike rich ones, are not well served by simple loans in isolation; they need a full suite of financial tools that work in concert to mitigate risk, fund investment, grow savings, and move money. Insurance, for example, can significantly affect how borrowers invest in their businesses. A recent field study in Ghana gave different groups of farmers cash grants to fund investments in farm inputs, crop insurance, or both. The farmers with crop insurance invested more in agricultural inputs, particularly in chemicals, land preparation, and hired labor. And they spent, on average, $266 more on cultivation than did the farmers without insurance. It was not the farmers' lack of credit, then, that was the greatest barrier to expanding their businesses; it was risk.

Mobile applications allow banks to offer such services to huge numbers of customers in very short order. In November 2012, the Commercial Bank of Africa and the telecommunications firm Safaricom launched a product called M-Shwari, which enables M-Pesa users to open interest-accruing savings accounts and apply for short-term loans through their cell phones. The demand for the product proved overwhelming. By effectively eliminating the time it would have taken for users to sign up or apply in person, M-Shwari added roughly one million accounts in its first three months.

By attracting so many customers and tracking their behavior in real time, mobile platforms generate reams of useful data. People's calling and transaction patterns can reveal valuable insights about the behavior of certain segments of the client population, demonstrating how variations in income levels, employment status, social connectedness, marital status, creditworthiness, or other attributes shape outcomes. Many studies have already shown how certain product features can affect some groups differently from others. In one Kenyan study, researchers gave clients ATM cards that permitted cash withdrawals at lowered costs and allowed the clients to access their savings accounts after hours and on weekends. The change ended up positively affecting married men and adversely affecting married women, whose husbands could more easily get their hands on the money saved in a joint account. Before the ATM cards, married women could cite the high withdrawal fees or the bank's limited hours to discourage withdrawals. With the cards, moreover, husbands could get cash from an ATM themselves, whereas withdrawals at the branch office had usually required the wives to go in person during the hours their husbands were at work.

Location, Location, Location

The high cost of basic banking infrastructure may be the biggest barrier to providing financial services to the poor. Banks place ATMs and branch offices almost exclusively in the wealthier, denser (and safer) areas of poor countries. The cost of such infrastructure often dwarfs the potential profits to be made in poorer, more rural areas. In contrast, mobile banking allows customers to carry out transactions in existing shops and even market stalls, creating denser networks of transaction points at a much lower cost.

For clients to fully benefit from mobile financial services, however, access to a physical office that deals in cash remains critical. When researchers studying the M-Pesa program in Kenya cross-referenced the locations of M-Pesa agents and the locations of households in the program, they found that the closer a household was to an M-Pesa kiosk, where cash and customer services were available, the more it benefited from the service. Beyond a certain distance, it becomes infeasible for clients to use a given financial service, no matter how much they need it.

Meanwhile, a number of studies have shown that increasing physical access points to the financial system can help lift local economies. Researchers in India have documented the effects of a regulation requiring banks to open rural branches in exchange for licenses to operate in more profitable urban areas.

The data showed significant increases in lending and agricultural output in the areas that received branches due to the program, as well as 4–5 percent reductions in the number of people living in poverty. A similar study in Mexico found that in areas where bank branches were introduced, the number of people who owned informal businesses increased by 7.6 percent. There were also ripple effects: an uptick in employment and a 7 percent increase in incomes.

In the right hands, then, access to financial tools can stimulate underserved economies and, at critical times, determine whether a poor household is able to capture an opportunity to move out of poverty or weather an otherwise debilitating financial shock. Thanks to new research, much more is known about what types of features can do the most to improve consumers' lives. And due to the rapid proliferation of cell phones, it is now possible to deliver such services to more people than ever before. Both of these trends have set the stage for yet further innovations by banks, cell-phone companies, microlenders, and entrepreneurs—all of whom have a role to play in delivering life-changing financial services to those who need them most.

Critical Thinking

1. How does Wizzit bank contribute to microfinance?
2. What is the relationship between cell phones and banking?
3. What is the role of the International Finance Corporation in mobile banking?

Create Central

www.mhhe.com/createcentral

Internet References

International Finance Corporation
  http://www.ifc.org
Wizzit bank
  http://www.wizzit.co.za

JAKE KENDALL is Senior Program Officer for the Financial Services for the Poor program at the Bill & Melinda Gates Foundation. **RODGER VOORHIES** is Director of the Financial Services for the Poor program at the Bill & Melinda Gates Foundation.

Unit 4

UNIT

Prepared by: Robert Weiner, *University of Massachusetts, Boston*

Terrorism

Terrorism usually refers to non-state actors who engage in violent extremist action to draw attention to their goals and ideologies. Terrorists can consist of organizations or single, radicalized individuals, who may be motivated by ideology, religion, or ethno-nationalism. Terrorists believe that it is necessary to use extreme violence to draw attention to their cause. In the age of globalization in the 21st century, terrorism is a transnational phenomenon. Terrorists have taken advantage of modern technology to engage in violent acts, as when the United States found itself attacked by Al-Qaida on 9/11. Terrorism has evolved from bomb throwing and assassinations to the hijacking of aircraft to efforts to kill as many innocent and vulnerable victims as possible. According to experts, terrorist acts have increased fourfold since 9/11, especially marked by an increase in the number of suicide bombers. Even though Osama Bin Laden, the leader of Al -Qaida or the "base" in Arabic, was killed by U.S. Seals in 2011, and many other members of the organization were decimated by drones, franchises of Al-Qaida and other terrorist organizations have morphed and proliferated throughout the Middle East, North Africa, Somalia, and Nigeria. These include organizations such as Al-Shabaab, Boko Haram, and Al-Nusra, among others.

The United States especially focused on the Islamic State in Iraq and the Levant (ISIL) also known as Daesh. By 2015, ISIL, which originally began with Baathist military officers from Saddam Hussein's regime, and was joined by disaffected Sunni tribes who felt repressed by the Shia government in Bagdad, had grown to 25,000–30,000 insurgents, whose recruits were drawn from around the world by ISIL's sophisticated use of the social media. New recruits flocked to ISIL from Europe and the United States. It is estimated that in 2015, about 250 recruits from the United States joined ISIL. Western governments worried about the dangers of blowback as foreign fighters returned to their home states. There was fear of more incidents like the murder of nearly 12 staff members of the French satirical magazine *Charlie Hedro* in Paris by jihadists who had been trained in the Middle East.

ISIL's conduct of the war in Iraq and Syria, which aimed at overthrowing the regimes in both countries, was marked by a great deal of brutality and cruelty, including beheadings, mass executions, persecution of Christians and other non-believers and apostates. ISIL also looted and destroyed Assyrian monuments, sculptures, and treasures in the ancient Assyrian city of Palmyira. ISIL has assumed the attributes of a state, as it attempts to reconstitute an Islamic Caliphate of the Middle Ages.

The United States led a counterterrorist coalition of about 60 countries to degrade and destroy ISIS. However, U.S. efforts to train and use a group of moderate Syrian insurgents to combat ISIL as well as the Assad regime turned out to be a fiasco. In 2015, despite repeated U.S. airstrikes against ISIL, it still remained entrenched in parts of Iraq and Syria. Russia also deployed combat aircraft and other military assets in Syria, to defend President Assad's regime, which reportedly controlled about 25% of the country. This added a new level of complexity to the conflict, as President Putin urged other countries to join Russia in the fight against terrorism. The Russian military intervention was also designed to establish a more permanent presence for Moscow in the Middle East, and restore its credibility as a Great Power.

Article

Prepared by: Robert Weiner, *University of Massachusetts, Boston*

ISIS Is Not a Terrorist Group: Why Counterterrorism Won't Stop the Latest Jihadist Threat

AUDREY KURTH CRONIN

Learning Outcomes

After reading this article, you will be able to:

- Define who and what ISIS is.

- Understand the relationship between ISIS and al Qaeda.

After 9/11, many within the U.S. national security establishment worried that, following decades of preparation for confronting conventional enemies, Washington was unready for the challenge posed by an unconventional adversary such as al Qaeda. So over the next decade, the United States built an elaborate bureaucratic structure to fight the jihadist organization, adapting its military and its intelligence and law enforcement agencies to the tasks of counterterrorism and counterinsurgency.

Now, however, a different group, the Islamic State of Iraq and alSham (ISIS), which also calls itself the Islamic State, has supplanted al Qaeda as the jihadist threat of greatest concern. ISIS' ideology, rhetoric, and long-term goals are similar to al Qaeda's, and the two groups were once formally allied. So many observers assume that the current challenge is simply to refocus Washington's now-formidable counterterrorism apparatus on a new target.

But ISIS is not al Qaeda. It is not an outgrowth or a part of the older radical Islamist organization, nor does it represent the next phase in its evolution. Although al Qaeda remains dangerous—especially its affiliates in North Africa and Yemen—ISIS is its successor. ISIS represents the post-al Qaeda jihadist threat.

In a nationally televised speech last September explaining his plan to "degrade and ultimately destroy" ISIS, U.S. President Barack Obama drew a straight line between the group and al Qaeda and claimed that ISIS is "a terrorist organization, pure and simple." This was mistaken; ISIS hardly fits that description, and indeed, although it uses terrorism as a tactic, it is not really a terrorist organization at all. Terrorist networks, such as al Qaeda, generally have only dozens or hundreds of members, attack civilians, do not hold territory, and cannot directly confront military forces. ISIS, on the other hand, boasts some 30,000 fighters, holds territory in both Iraq and Syria, maintains extensive military capabilities, controls lines of communication, commands infrastructure, funds itself, and engages in sophisticated military operations. If ISIS is purely and simply anything, it is a pseudo-state led by a conventional army. And that is why the counterterrorism and counterinsurgency strategies that greatly diminished the threat from al Qaeda will not work against ISIS.

Washington has been slow to adapt its policies in Iraq and Syria to the true nature of the threat from ISIS. In Syria, U.S. counterterrorism has mostly prioritized the bombing of al Qaeda affiliates, which has given an edge to ISIS and has also provided the Assad regime with the opportunity to crush U.S.-allied moderate Syrian rebels. In Iraq, Washington continues to rely on a form of counterinsurgency, depending on the central government in Baghdad to regain its lost legitimacy, unite the country, and build indigenous forces to defeat ISIS. These approaches were developed to meet a different threat, and they have been overtaken by events. What's needed now is a strategy of "offensive containment": a combination of limited military tactics and a broad diplomatic strategy to halt ISIS' expansion, isolate the group, and degrade its capabilities.

Different Strokes

The differences between al Qaeda and ISIS are partly rooted in their histories. Al Qaeda came into being in the aftermath of the 1979 Soviet invasion of Afghanistan. Its leaders' worldviews and strategic thinking were shaped by the ten-year war against Soviet occupation, when thousands of Muslim militants, including Osama bin Laden, converged on the country. As the organization coalesced, it took the form of a global network focused on carrying out spectacular attacks against Western or Western-allied targets, with the goal of rallying Muslims to join a global confrontation with secular powers near and far.

ISIS came into being thanks to the 2003 U.S. invasion of Iraq. In its earliest incarnation, it was just one of a number of Sunni extremist groups fighting U.S. forces and attacking Shiite civilians in an attempt to foment a sectarian civil war. At that time, it was called al Qaeda in Iraq (aqi), and its leader, Abu Musab al-Zarqawi, had pledged allegiance to bin Laden. Zarqawi was killed by a U.S. air strike in 2006, and soon after, aqi was nearly wiped out when Sunni tribes decided to partner with the Americans to confront the jihadists. But the defeat was temporary; aqi renewed itself inside U.S.-run prisons in Iraq, where insurgents and terrorist operatives connected and formed networks—and where the group's current chief and self-proclaimed caliph, Abu Bakr al-Baghdadi, first distinguished himself as a leader.

In 2011, as a revolt against the Assad regime in Syria expanded into a full-blown civil war, the group took advantage of the chaos, seizing territory in Syria's northeast, establishing a base of operations, and rebranding itself as ISIS. In Iraq, the group continued to capitalize on the weakness of the central state and to exploit the country's sectarian strife, which intensified after U.S. combat forces withdrew. With the Americans gone, Iraqi Prime Minister Nouri al-Maliki pursued a hard-line pro-Shiite agenda, further alienating Sunni Arabs throughout the country. ISIS now counts among its members Iraqi Sunni tribal leaders, former anti-U.S. insurgents, and even secular former Iraqi military officers who seek to regain the power and security they enjoyed during the Saddam Hussein era.

The group's territorial conquest in Iraq came as a shock. When ISIS captured Fallujah and Ramadi in January 2014, most analysts predicted that the U.S.-trained Iraqi security forces would contain the threat. But in June, amid mass desertions from the Iraqi army, ISIS moved toward Baghdad, capturing Mosul, Tikrit, al-Qaim, and numerous other Iraqi towns. By the end of the month, ISIS had renamed itself the Islamic State and had proclaimed the territory under its control to be a new caliphate. Meanwhile, according to U.S. intelligence estimates, some 15,000 foreign fighters from 80 countries flocked to the region to join ISIS, at the rate of around 1,000 per month. Although most of these recruits came from Muslim-majority countries, such as Tunisia and Saudi Arabia, some also hailed from Australia, China, Russia, and western European countries. ISIS has even managed to attract some American teenagers, boys and girls alike, from ordinary middle-class homes in Denver, Minneapolis, and the suburbs of Chicago.

As ISIS has grown, its goals and intentions have become clearer. Al Qaeda conceived of itself as the vanguard of a global insurgency mobilizing Muslim communities against secular rule. ISIS, in contrast, seeks to control territory and create a "pure" Sunni Islamist state governed by a brutal interpretation of sharia; to immediately obliterate the political borders of the Middle East that were created by Western powers in the twentieth century; and to position itself as the sole political, religious, and military authority over all of the world's Muslims.

Not the Usual Suspects

Since ISIS' origins and goals differ markedly from al Qaeda's, the two groups operate in completely different ways. That is why a U.S. counterterrorism strategy custom-made to fight al Qaeda does not fit the struggle against ISIS.

In the post-9/11 era, the United States has built up a trillion-dollar infrastructure of intelligence, law enforcement, and military operations aimed at al Qaeda and its affiliates. According to a 2010 investigation by *The Washington Post,* some 263 U.S. government organizations were created or reorganized in response to the 9/11 attacks, including the Department of Homeland Security, the National Counterterrorism Center, and the Transportation Security Administration. Each year, U.S. intelligence agencies produce some 50,000 reports on terrorism. Fifty-one U.S. federal organizations and military commands track the flow of money to and from terrorist networks. This structure has helped make terrorist attacks on U.S. soil exceedingly rare. In that sense, the system has worked. But it is not well suited for dealing with ISIS, which presents a different sort of challenge. Consider first the tremendous U.S. military and intelligence campaign to capture or kill al Qaeda's core leadership through drone strikes and Special Forces raids. Some 75 percent of the leaders of the core al Qaeda group have been killed by raids and armed drones, a technology well suited to the task of going after targets hiding in rural areas, where the risk of accidentally killing civilians is lower.

Such tactics, however, don't hold much promise for combating ISIS. The group's fighters and leaders cluster in urban areas, where they are well integrated into civilian populations and usually surrounded by buildings, making drone strikes and raids much harder to carry out. And simply killing ISIS' leaders would not cripple the organization. They govern a functioning pseudo-state with a complex administrative structure. At the top of the military command is the emirate, which consists of Baghdadi and two deputies, both of whom formerly served as generals in the Saddam-era Iraqi army: Abu Ali al-Anbari, who controls ISIS' operations in Syria, and Abu Muslim al-Turkmani, who controls operations in Iraq. ISIS' civilian bureaucracy is supervised by 12 administrators who govern territories in Iraq and Syria, overseeing councils that handle matters such as finances, media, and religious affairs. Although it is hardly the model government depicted in ISIS' propaganda videos, this pseudo-state would carry on quite ably without Baghdadi or his closest lieutenants.

ISIS also poses a daunting challenge to traditional U.S. counterterrorism tactics that take aim at jihadist financing, propaganda, and recruitment. Cutting off al Qaeda's funding has been one of U.S. counterterrorism's most impressive success stories. Soon after the 9/11 attacks, the FBI and the CIA began to coordinate closely on financial intelligence, and they were soon joined by the Department of Defense. FBI agents embedded with U.S. military units during the 2003 invasion of Iraq and debriefed suspected terrorists detained at the U.S. facility at Guantánamo Bay, Cuba. In 2004, the U.S. Treasury Department established the Office of Terrorism and Financial Intelligence, which has cut deeply into al Qaeda's ability to profit from money laundering and receive funds under the cover of charitable giving. A global network for countering terrorist financing has also emerged, backed by the UN, the EU, and hundreds of cooperating governments. The result has been a serious squeeze on al Qaeda's financing; by 2011, the Treasury Department reported that al Qaeda was "struggling to secure steady financing to plan and execute terrorist attacks."

But such tools contribute little to the fight against ISIS, because ISIS does not need outside funding. Holding territory has allowed the group to build a self-sustaining financial model unthinkable for most terrorist groups. Beginning in 2012, ISIS gradually took over key oil assets in eastern Syria; it now controls an estimated 60 percent of the country's oil production capacity. Meanwhile, during its push into Iraq last summer, ISIS also seized seven oil-producing operations in that country. The group manages to sell some of this oil on the black market in Iraq and Syria—including, according to some reports, to the Assad regime itself. ISIS also smuggles oil out of Iraq and Syria into Jordan and Turkey, where it finds plenty of buyers happy to pay below-market prices for illicit crude. All told, ISIS' revenue from oil is estimated to be between $1 million and $3 million per day. And oil is only one element in the group's financial portfolio. Last June, when ISIS seized control of the northern Iraqi city of Mosul, it looted the provincial central bank and other smaller banks and plundered antiquities to sell on the black market. It steals jewelry, cars, machinery, and livestock from conquered residents. The group also controls major transportation arteries in western Iraq, allowing it to tax the movement of goods and charge tolls. It even earns revenue from cotton and wheat grown in Raqqa, the breadbasket of Syria.

Of course, like terrorist groups, ISIS also takes hostages, demanding tens of millions of dollars in ransom payments. But more important to the group's finances is a wide-ranging extortion racket that targets owners and producers in ISIS territory, taxing everything from small family farms to large enterprises such as cell-phone service providers, water delivery companies, and electric utilities. The enterprise is so complex that the U.S. Treasury has declined to estimate ISIS' total assets and revenues, but ISIS is clearly a highly diversified enterprise whose wealth dwarfs that of any terrorist organization. And there is little evidence that Washington has succeeded in reducing the group's coffers.

Sex and the Single Jihadist

Another aspect of U.S. counterterrorism that has worked well against al Qaeda is the effort to delegitimize the group by publicizing its targeting errors and violent excesses—or by helping U.S. allies do so. Al Qaeda's attacks frequently kill Muslims, and the group's leaders are highly sensitive to the risk this poses to their image as the vanguard of a mass Muslim movement. Attacks in Morocco, Saudi Arabia, and Turkey in 2003; Spain in 2004; and Jordan and the United Kingdom in 2005 all resulted in Muslim casualties that outraged members of Islamic communities everywhere and reduced support for al Qaeda across the Muslim world. The group has steadily lost popular support since around 2007; today, al Qaeda is widely reviled in the Muslim world. The Pew Research Center surveyed nearly 9,000 Muslims in 11 countries in 2013 and found a high median level of disapproval of al Qaeda: 57 percent. In many countries, the number was far higher: 96 percent of Muslims polled in Lebanon, 81 percent in Jordan, 73 percent in Turkey, and 69 percent in Egypt held an unfavorable view of al Qaeda.

ISIS, however, seems impervious to the risk of a backlash. In proclaiming himself the caliph, Baghdadi made a bold (if absurd) claim to religious authority. But ISIS' core message is about raw power and revenge, not legitimacy. Its brutality—videotaped beheadings, mass executions—is designed to intimidate foes and suppress dissent. Revulsion among Muslims at such cruelty might eventually undermine ISIS. But for the time being, Washington's focus on ISIS' savagery only helps the group augment its aura of strength.

For similar reasons, it has proved difficult for the United States and its partners to combat the recruitment efforts that have attracted so many young Muslims to ISIS' ranks. The core al Qaeda group attracted followers with religious arguments and a pseudo-scholarly message of altruism for the sake of the ummah, the global Muslim community. Bin Laden and his longtime second-in-command and successor, Ayman alZawahiri, carefully constructed an image of religious legitimacy and piety. In their propaganda videos, the men appeared as ascetic warriors, sitting on the ground in caves, studying in libraries, or taking refuge in remote camps. Although some of al Qaedas affiliates have better recruiting pitches, the core group cast the establishment of a caliphate as a long-term, almost utopian goal: educating and mobilizing the ummah came first. In al Qaeda, there is no place for alcohol or women. In this sense, al Qaeda's image is deeply unsexy; indeed, for the young al Qaeda recruit, sex itself comes only after marriage—or martyrdom.

Even for the angriest young Muslim man, this might be a bit of a hard sell. Al Qaeda's leaders' attempts to depict themselves as moral—even moralistic—figures have limited their appeal. Successful deradicalization programs in places such as Indonesia and Singapore have zeroed in on the mismatch between what al Qaeda offers and what most young people are really interested in, encouraging militants to reintegrate into society, where their more prosaic hopes and desires might be fulfilled more readily.

ISIS, in contrast, offers a very different message for young men, and sometimes women. The group attracts followers yearning for not only religious righteousness but also adventure, personal power, and a sense of self and community. And, of course, some people just want to kill—and ISIS welcomes them, too. The group's brutal violence attracts attention, demonstrates dominance, and draws people to the action.

ISIS operates in urban settings and offers recruits immediate opportunities to fight. It advertises by distributing exhilarating podcasts produced by individual fighters on the frontlines. The group also procures sexual partners for its male recruits; some of these women volunteer for this role, but most of them are coerced or even enslaved. The group barely bothers to justify this behavior in religious terms; its sales pitch is conquest in all its forms, including the sexual kind. And it has already established a self-styled caliphate, with Baghdadi as the caliph, thus making present (if only in a limited way, for now) what al Qaeda generally held out as something more akin to a utopian future.

In short, ISIS offers short-term, primitive gratification. It does not radicalize people in ways that can be countered by appeals to logic. Teenagers are attracted to the group without even understanding what it is, and older fighters just want to be associated with ISIS' success. Compared with fighting al Qaeda's relatively austere message, Washington has found it much harder to counter ISIS' more visceral appeal, perhaps for a very simple reason: a desire for power, agency, and instant results also pervades American culture.

2015 ≠ 2006

Counterterrorism wasn't the only element of national security practice that Washington rediscovered and reinvigorated after 9/11; counterinsurgency also enjoyed a renaissance. As chaos erupted in Iraq in the aftermath of the U.S. invasion and occupation of 2003, the U.S. military grudgingly started thinking about counterinsurgency, a subject that had fallen out of favor in the national security establishment after the Vietnam War. The most successful application of U.S. counterinsurgency doctrine was the 2007 "surge" in Iraq, overseen by General David Petraeus. In 2006, as violence peaked in Sunni-dominated Anbar Province, U.S. officials concluded that the United States was losing the war. In response, President George W. Bush decided to send an additional 20,000 U.S. troops to Iraq. General John Allen, then serving as deputy commander of the multinational forces in Anbar, cultivated relationships with local Sunni tribes and nurtured the so-called Sunni Awakening, in which some 40 Sunni tribes or subtribes essentially switched sides and decided to fight with the newly augmented U.S. forces against aqi. By the summer of 2008, the number of insurgent attacks had fallen by more than 80 percent.

Looking at the extent of ISIS' recent gains in Sunni areas of Iraq, which have undone much of the progress made in the surge, some have argued that Washington should respond with a second application of the Iraq war's counterinsurgency strategy. And the White House seems at least partly persuaded by this line of thinking: last year, Obama asked Allen to act as a special envoy for building an anti-ISIS coalition in the region. There is a certain logic to this approach, since ISIS draws support from many of the same insurgent groups that the surge and the Sunni Awakening neutralized—groups that have reemerged as threats thanks to the vacuum created by the withdrawal of U.S. forces in 2011 and Maliki's sectarian rule in Baghdad.

But vast differences exist between the situation today and the one that Washington faced in 2006, and the logic of U.S. counterinsurgency does not suit the struggle against ISIS. The United States cannot win the hearts and minds of Iraq's Sunni Arabs, because the Maliki government has already lost them. The Shiite-dominated Iraqi government has so badly undercut its own political legitimacy that it might be impossible to restore it. Moreover, the United States no longer occupies Iraq. Washington can send in more troops, but it cannot lend legitimacy to a government it no longer controls. ISIS is less an insurgent group fighting against an established government than one party in a conventional civil war between a breakaway territory and a weak central state.

Divide and Conquer?

The United States has relied on counterinsurgency strategy not only to reverse Iraq's slide into state failure but also to serve as a model for how to combat the wider jihadist movement. Al Qaeda expanded by persuading Muslim militant groups all over the world to turn their more narrowly targeted nationalist campaigns into nodes in al Qaeda's global jihad—and, sometimes, to convert themselves into al Qaeda affiliates. But there was little commonality in the visions pursued by Chechen, Filipino, Indonesian, Kashmiri, Palestinian, and Uighur militants, all of whom bin Laden tried to draw into al Qaeda's tent, and al Qaeda often had trouble fully reconciling its own goals with the interests of its far-flung affiliates.

That created a vulnerability, and the United States and its allies sought to exploit it. Governments in Indonesia and the Philippines won dramatic victories against al Qaeda affiliates in their countries by combining counterterrorism operations with relationship building in local communities, instituting deradicalization programs, providing religious training in prisons, using rehabilitated former terrorist operatives as government spokespeople, and sometimes negotiating over local grievances.

Some observers have called for Washington to apply the same strategy to ISIS by attempting to expose the fault lines between the group's secular former Iraqi army officers, Sunni tribal leaders, and Sunni resistance fighters, on the one hand, and its veteran jihadists, on the other. But it's too late for that approach to work. ISIS is now led by well-trained, capable former Iraqi military leaders who know U.S. techniques and habits because Washington helped train them. And after routing Iraqi army units and taking their U.S.-supplied equipment, ISIS is now armed with American tanks, artillery, armored Humvees, and mine-resistant vehicles.

Perhaps ISIS' harsh religious fanaticism will eventually prove too much for their secular former Baathist allies. But for now, the Saddam-era officers are far from reluctant warriors for ISIS: rather, they are leading the charge. In their hands, ISIS has developed a sophisticated light infantry army, brandishing American weapons.

Of course, this opens up a third possible approach to ISIS, besides counterterrorism and counterinsurgency: a full-on conventional war against the group, waged with the goal of completely destroying it. Such a war would be folly. After experiencing more than a decade of continuous war, the American public simply would not support the long-term occupation and intense fighting that would be required to obliterate ISIS. The pursuit of a full-fledged military campaign would exhaust U.S. resources and offer little hope of obtaining the objective. Wars pursued at odds with political reality cannot be won.

Containing the Threat

The sobering fact is that the United States has no good military options in its fight against ISIS. Neither counterterrorism, nor counterinsurgency, nor conventional warfare is likely to afford Washington a clear-cut victory against the group. For the time being, at least, the policy that best matches ends and means and that has the best chance of securing U.S. interests is one of offensive containment: combining a limited military campaign with a major diplomatic and economic effort to weaken ISIS and align the interests of the many countries that are threatened by the group's advance.

ISIS is not merely an American problem. The wars in Iraq and Syria involve not only regional players but also major global actors, such as Russia, Turkey, Iran, Saudi Arabia, and other Gulf states. Washington must stop behaving as if it can fix the region's problems with military force and instead resurrect its role as a diplomatic superpower.

Of course, U.S. military force would be an important part of an offensive containment policy. Air strikes can pin ISIS down, and cutting off its supply of technology, weapons, and ammunition by choking off smuggling routes would further weaken the group. Meanwhile, the United States should continue to advise and support the Iraqi military, assist regional forces such as the Kurdish Pesh Merga, and provide humanitarian assistance to civilians fleeing ISIS' territory. Washington should also expand its assistance to neighboring countries such as Jordan and Lebanon, which are struggling to contend with the massive flow of refugees from Syria. But putting more U.S. troops on the ground would be counterproductive, entangling the United States in an unwinnable war that could go on for decades. The United States cannot rebuild the Iraqi state or determine the outcome of the Syrian civil war. Frustrating as it might be to some, when it comes to military action, Washington should stick to a realistic course that recognizes the limitations of U.S. military force as a long-term solution.

The Obama administration's recently convened "summit on countering violent extremism"—which brought world leaders to Washington to discuss how to combat radical jihadism—was a valuable exercise. But although it highlighted the existing threat posed by al Qaeda's regional affiliates, it also reinforced the idea that ISIS is primarily a counterterrorism challenge. In fact, ISIS poses a much greater risk: it seeks to challenge the current international order, and, unlike the greatly diminished core al Qaeda organization, it is coming closer to actually achieving that goal. The United States cannot single-handedly defend the region and the world from an aggressive revisionist theocratic state—nor should it. The major powers must develop a common diplomatic, economic, and military approach to ensure that this pseudo-state is tightly contained and treated as a global pariah. The good news is that no government supports ISIS; the group has managed to make itself an enemy of every state in the region—and, indeed, the world. To exploit that fact, Washington should pursue a more aggressive, top-level diplomatic agenda with major powers and regional players, including Iran, Saudi Arabia, France, Germany, the United Kingdom, Russia, and even China, as well as Iraq's and Syria's neighbors, to design a unified response to ISIS.

That response must go beyond making a mutual commitment to prevent the radicalization and recruitment of would-be jihadists and beyond the regional military coalition that the United States has built. The major powers and regional players must agree to stiffen the international arms embargo currently imposed on ISIS, enact more vigorous sanctions against the group, conduct joint border patrols, provide more aid for displaced persons and refugees, and strengthen UN peacekeeping missions in countries that border Iraq and Syria. Although some of these tools overlap with counterterrorism, they should be put in the service of a strategy for fighting an enemy more akin to a state actor: ISIS is not a nuclear power, but the group represents a threat to international stability equivalent to that posed by North Korea. It should be treated no less seriously.

Given that political posturing over U.S. foreign policy will only intensify as the 2016 U.S. presidential election approaches, the White House would likely face numerous attacks on a containment approach that would satisfy neither the hawkish nor the anti-interventionist camp within the U.S. national security establishment. In the face of such criticism, the United States must stay committed to fighting ISIS over the long term in a manner that matches ends with means, calibrating and improving U.S. efforts to contain the group by moving past outmoded forms of counterterrorism and counterinsurgency while also resisting pressure to cross the threshold into full-fledged war.

Over time, the successful containment of ISIS might open up better policy options. But for the foreseeable future, containment is the best policy that the United States can pursue.

Critical Thinking

1. What is the difference between ISIS and Al Qaeda?
2. What is the best strategy for the United States to use to deal with ISIS and why?
3. What explains the success of ISIS so far?

Internet References

Defense Intelligence Agency
www.dni.gov/index.php/intelligence-community/members-of-the ic#dia

Department of Homeland Security
www.dhs.gov/

SITE Intelligence Group
https//:ent.siteintelgroup.com/

U.S. Department of State.Bureau of Counter-Terrorism
www.state.gov/j/ct/

AUDREY KURTH CRONIN is Distinguished Professor and Director of the International Security Program at George Mason University and the author of *How Terrorism Ends: Understanding the Decline and Demise of Terrorist Campaigns.* Follow her on Twitter @akcronin.

Article   Prepared by: Robert Weiner, *University of Massachusetts, Boston*

ISIS and the Third Wave of Jihadism

"There is no simple or quick solution to rid the Middle East of ISIS because it is a manifestation of the breakdown of state institutions and the spread of sectarian fires in the region."

FAWAZ A. GERGES

Learning Outcomes

After reading this article, you will be able to:

- Discuss what factors contributed to the emergence of ISIS.

- Explain the differences between ISIS and al-Qaeda.

In order to make sense of the so-called Islamic State (known as ISIS or ISIL, or by its Arabic acronym, Daesh) and its sudden territorial conquests in Iraq and Syria, it is important to place the organization within the broader global jihadist movement. By tracing ISIS's social origins and comparing it with the first two jihadist waves of the 1980s and 1990s, we can gauge the extent of continuity and change, and account for the group's notorious savagery.

Although ISIS is an extension of the global jihadist movement in its ideology and worldview, its social origins are rooted in a specific Iraqi context, and, to a lesser extent, in the Syrian war that has raged for almost four years. While al-Qaeda's central organization emerged from an alliance between ultraconservative Saudi Salafism and radical Egyptian Islamism, ISIS was born of an unholy union between an Iraq-based al-Qaeda offshoot and the defeated Iraqi Baathist regime of Saddam Hussein, which has proved a lethal combination.

Bitter Inheritance

The causes of ISIS's unrestrained extremism lie in its origins in al-Qaeda in Iraq (AQI), founded by Abu Musab al-Zarqawi, who was killed by the Americans in 2006. The US-led invasion and occupation of Iraq caused a rupture in an Iraqi society already fractured and bled by decades of war and economic sanctions. America's destruction of Iraqi institutions, particularly its dismantling of the Baath Party and the army, created a vacuum that unleashed a fierce power struggle and allowed non-state actors, including al-Qaeda, to infiltrate the fragile body politic.

ISIS's viciousness reflects the bitter inheritance of decades of Baathist rule that tore apart Iraq's social fabric and left deep wounds that are still festering. America's bloody vanquishing of Baathism and the invasion's aftermath of sectarian civil war plunged Iraq into a sustained crisis, inflaming Sunnis' grievances over their disempowerment under the new Shia ascendancy and preponderant Iranian influence.

Iraqi Sunnis have been protesting the marginalization and discrimination they face for some time, but their complaints fell on deaf ears in Baghdad and Washington. This created an opening for ISIS to step in and instrumentalize their grievances. A similar story of Sunni resentment unfolded in Syria, where the minority Alawite sect dominates the regime of President Bashar al-Assad. Thousands of embittered Iraqi and Syrian Sunnis fight under ISIS's banner, even though many do not subscribe to its extremist Islamist ideology. While its chief, Abu Bakr al-Baghdadi, has anointed himself as the new caliph, on a more practical level he blended his group with local armed insurgencies in Syria and Iraq, building a base of support among rebellious Sunnis.

ISIS is a symptom of the broken politics of the Middle East and the fraying and delegitimation of state institutions, as well as the spreading of civil wars in Syria and Iraq. The group has filled the resulting vacuum of legitimate authority. For almost two decades, "al-Qaeda Central" leaders Osama bin Laden and Ayman al-Zawahiri were unable to establish the kind of social movement that Baghdadi has created in less than five years.

Unlike its transnational, borderless parent organization, ISIS has found a haven in the heart of the Levant. It has done so by exploiting the chaos in war-torn Syria and the sectarian, exclusionary policies of former Iraqi Prime Minister Nuri Kamal al-Maliki. More like the Taliban in Afghanistan in the 1990s than al-Qaeda Central, ISIS is developing a rudimentary infrastructure of administration and governance in captured territories in Syria and Iraq. It now controls a landmass as large as the United Kingdom. ISIS's swift military expansion stems from its ability not only to terrorize enemies but also to co-opt local Sunni communities, using networks of patronage and privilege. It offers economic incentives such as protection of contraband trafficking activity and a share of the oil trade and smuggling in eastern Syria.

Sectarian War

Building a social base from scratch in Iraq, AQI exploited the Sunni-Shia divide that opened after the United States toppled Hussein's Sunni-dominated regime. The group carried out wave after wave of suicide bombings against the Shia. Zarqawi's goal was to trigger all-out sectarian war and to position AQI as the champion of the embattled Sunnis. He ignored repeated pleas from his mentors, bin Laden and Zawahiri, to stop the indiscriminate killing of Shia and to focus instead on attacking Western troops and citizens.

Although Salafi jihadists are nourished on an anti-Shia propaganda diet, al-Qaeda Central prioritized the fight against the "far enemy"—America and its European allies. In contrast, AQI and its successor, ISIS, have so far consistently focused on the Shia and the "near enemy" (the Iraqi and Syrian regimes, as well as all secular, pro-Western regimes in the Muslim world). Baghdadi, like Zarqawi before him, has a genocidal worldview, according to which Shias are infidels—a fifth column in the heart of Islam that must either convert or be exterminated. The struggle against America and Europe is a distant, secondary goal that must be deferred until liberation at home is achieved. At the height of the Israeli assault on Gaza during the summer of 2014, militants criticized ISIS on social media for killing Muslims while failing to help the Palestinians. ISIS retorted that the struggle against the Shia comes first.

Baghdadi has exploited the deepening Sunni-Shia rift across the Middle East, intensified by a new regional cold war between Sunni-dominated Saudi Arabia and Shia-dominated Iran. He depicts his group as the vanguard of persecuted Sunni Arabs in a revolt against sectarian-based regimes in Baghdad, Damascus, and beyond. He has amassed a Sunni army of more than 30,000 fighters (including some 18,000 core members, plus affiliated groups). By contrast, at the height of its power in the late 1990s, al-Qaeda Central mustered only 1,000 to 3,000 fighters, a fact that shows the limits of transnational jihadism and its small constituency compared with the "near enemy" or local jihadism of the ISIS variety.

The weakest link of ISIS as a social movement is its poverty of ideas.

Numbers alone do not explain ISIS's rapid military advances in Syria and Iraq. After Baghdadi took charge of AQI in 2010, when it was in precipitous decline, he restructured its military network and recruited experienced officers from Hussein's disbanded army, particularly the Republican Guards, who turned ISIS into a professional fighting force. It has been toughened by fighting in neighboring Syria since the civil war there began in 2011. According to knowledgeable Iraqi sources, Baghdadi relies on a military council made up of 8 to 13 officers who all served in Saddam Hussein's army.

Rational Savagery

In a formal sense, ISIS is an effective fighting force. But it has become synonymous with viciousness, carrying out massacres, beheadings, and other atrocities. It has engaged in religious and ethnic cleansing against Yazidis and Kurds as well as Shia. Such savagery might seem senseless, but for ISIS it appears to be a rational choice, intended to terrorize its enemies and to impress potential recruits. ISIS's brutality also stems from the ruralization of this third wave of jihadism. Whereas the two previous waves had leaders from the social elite and a rank and file mainly composed of lower-middle-class university graduates, ISIS's cadre is rural and lacking in both theological and intellectual accomplishment. This social profile helps ISIS thrive among poor, disenfranchised Sunni communities in Iraq, Syria, Lebanon, and elsewhere.

ISIS adheres to a doctrine of total war, with no constraints. It disdains arbitration or compromise, even with Sunni Islamist rivals. Unlike al-Qaeda Central, it does not rely on theology to justify its actions. "The only law I subscribe to is the law of the jungle," retorted Baghdadi's second-in-command and right-hand man, Abu Muhammed al-Adnani, to a request more than a year ago by rival militant Islamists in Syria who called for ISIS to submit to a Sharia court so that a dispute with other factions could be properly adjudicated. For the top ideologues of Salafi jihadism, such statements and actions are sacrilegious, "smearing the reputation" of the global jihadist movement, in the words of Abu Mohammed al-Maqdisi, a Jordan-based mentor to Zarqawi and many jihadists worldwide.

New Wave

The scale and intensity of ISIS's brutality, stemming from Iraq's blood-soaked modern history, far exceed either of the first two jihadist waves of recent decades. Disciples of Sayyid Qutb—a radical Egyptian Islamist known as the master theoretician of modern jihadism—led the first wave. Pro-Western, secular Arab regimes, which they called the "near enemy," would be the main targets. Their first major act was the assassination of Egyptian President Anwar Sadat in 1981.

This first wave included militant religious activists of Zawahiri's generation. They wrote manifestos in an effort to obtain theological legitimation for their attacks on "renegade" and "apostate" rulers, such as Sadat, and their security services. On balance, though, they showed restraint in the use of political violence. Conscious of the importance of Egyptian and wider Arab opinion, Zawahiri spent considerable energy over the years trying to explain the circumstances that led to the killing of two children in Egypt and Sudan, and repeatedly insisted that his group, Egyptian Islamic Jihad, did not target civilians.

The first wave had subsided by the end of the 1990s. During the 1980s, many militants had traveled to Afghanistan to fight the Soviet occupation, a cause that launched the second jihadist wave. After the withdrawal of the Soviets from Afghanistan, bin Laden emerged as the leader of the new wave. The focus shifted to the "far enemy" in the West—the United States and, to a lesser degree, Europe.

To win support, bin Laden justified his actions as a form of self-defense. He portrayed al-Qaeda's September 11, 2001, attack on the United States as an act of "defensive jihad," or a just retaliation for American domination of Muslim countries. Baghdadi, by contrast, cares little for world opinion. Indeed, ISIS makes a point of displaying its barbarity in its internet videos. Stressing violent action rather than theology, it has offered no ideas to sustain its followers. Baghdadi has not fleshed out his vision of a caliphate but merely declared it by fiat, which contradicts Islamic law and tradition.

Ironically, Baghdadi—who has a doctorate from the Islamic University of Baghdad, with a focus on Islamic culture, history, sharia, and jurisprudence—is more steeped in religious education than al-Qaeda's past and current leaders, bin Laden (an engineer) and Zawahiri (a medical doctor), who had no such credentials. Yet he surrounds himself with former Baathist army officers, rather than ideologues, and has not issued a single manifesto laying out his claim to either the caliphate or the leadership of the global jihadist movement. ISIS's brutality has alienated senior radical preachers who have publicly disowned it, though some have softened their criticism in the wake of

US-led airstrikes against the group in Iraq and Syria, which one ideologue described as "the aggression of crusaders."

Bin Laden said, "When people see a strong horse and a weak horse, by nature they will like the strong horse." Baghdadi's slogan of "victory through fear and terrorism" signals to friends and foes alike that ISIS is a winning horse. Increasing evidence shows that over the past few months, hundreds, if not thousands, of die-hard former Islamist enemies of ISIS, including members of groups such as the Nusra Front and the Islamic Front, have declared allegiance to Baghdadi.

For now, ISIS has taken operational leadership of the global jihadist movement by default, eclipsing its parent organization, al-Qaeda Central. Baghdadi has won the first round against his former mentor, Zawahiri, who triggered an intra-jihadist civil war by unsuccessfully trying to elevate his own man, Abu Mohammed al-Golani, head of the Nusra Front, over Baghdadi in Syria.

Recruiting Tactics

However, the so-called Islamic State is much more fragile than Baghdadi would like us to believe. His call to arms has not found any takers among either top jihadist preachers or leaders of mainstream Islamist organizations, while Islamic scholars, including the most notable Salafi clerics, have dismissed his declaration of a caliphate as null and void. In fact, many of these same renowned Salafi scholars have equated ISIS with the extremist Kharijites of the Prophet's time. ISIS also threatens the vital interests of regional and international powers, a fact that explains the large coalition organized by the United States to combat the group.

Nevertheless, ISIS's sophisticated outreach campaign appeals to disaffected Sunni youth around the world by presenting the group as a powerful vanguard movement capable of delivering victory and salvation. It provides them with both a utopian worldview and a political project. Young recruits do not abhor its brutality; on the contrary, its shock-and-awe methods against the enemies of Islam are what attract them.

ISIS adheres to a doctrine of total war, with no constraints.

ISIS's exploits on the battlefield, its conquest of vast swaths of territory in Syria and Iraq, and its declaration of a caliphate have resonated widely, facilitating recruitment. Increasing evidence shows that the US-led airstrikes have not slowed down the flow of foreign recruits to Syria—far from it. The *Washington Post* reported that more than 1,000 foreign fighters are streaming into Syria each month. Efforts by other countries, especially Turkey, to stem the flow of recruits (many of them from European countries) have proved largely ineffective, according to US intelligence officials. ISIS fighters have also highlighted the important role of Chechen trainers in developing the group's military capabilities. Some reportedly have set up a Russian school in Raqqa for their children, to prepare them for jihad back home.

Muslims living in Western countries join ISIS and other extremist groups because they want to be part of a tight-knit community with a potent identity. ISIS's vision of resurrecting an idealized caliphate gives them the sense of serving a sacred mission. Corrupt Arab rulers and the crushing of the Arab Spring uprisings have provided further motivation for recruits. Many young men from Western Europe and elsewhere migrate to the lands of jihad because they feel a duty to defend persecuted coreligionists. Yet many of those who join the ranks of ISIS find themselves persecuting innocent civilians of other faiths and committing atrocities.

Hearts and Minds

Now that the United States and Europe have joined the fight against ISIS, the group might garner backing from quarters of the Middle Eastern public sphere that oppose Western intervention in internal Arab affairs, though there has been no such blowback so far. More than bin Laden and Zawahiri, Baghdadi has mastered the art of making enemies. He has failed to nourish a broad constituency beyond a narrow, radical sectarian base.

There is no simple or quick solution to rid the Middle East of ISIS because it is a manifestation of the breakdown of state institutions and the spread of sectarian fires in the region. ISIS is a creature of accumulated grievances, of ideological and social polarization and mobilization a decade in the making. As a non-state actor, it represents a transformative movement in the politics of the Middle East, one that is qualitatively different from al-Qaeda Central's.

The key to weakening ISIS lies in working closely with local Sunni communities that it has co-opted, a bottom-up approach that requires considerable material and ideological investment. The most effective means to degrade ISIS is to dismantle its social base by winning over the hearts and minds of local communities. This is easier said than done, given the gravity of the crisis in the heart of the Arab world. The jury is still out on whether the new Iraqi prime minister, Haider al-Abadi, will be able to appeal to mistrustful Sunnis and reconcile warring communities. Rebuilding trust takes hard work and time, both of which play to ISIS's advantage.

Equally important, there is an urgent need to find a diplomatic solution to the civil war in Syria, which has empowered ISIS, fueling its surge after its predecessor, AQI, was vanquished in Iraq. Syria is the nerve center of ISIS—the location of its de facto capital, the northern city of Raqqa, and of its major sources of income, including the oil trade, taxation, and criminal activities. More than two-thirds of its fighters are deployed in Syria, according to US intelligence officials.

In the short- to medium-term, it would take a political miracle to engineer a settlement in Syria, given the disintegration of the country and the fragmentation of power among rival warlords and fiefdoms, not to mention the regional and great power proxy wars playing out there. Until there is a regional and international agreement to end the Syrian civil war, ISIS will continue to entrench itself in the country's provinces and cities.

Yet even ISIS's dark cloud has a silver lining. Once Baghdadi's killing machine is dismantled, he will leave behind no ideas, no theories, and no intellectual legacy. The weakest link of ISIS as a social movement is its poverty of ideas. It can thrive and sustain itself only in an environment of despair, state breakdown, and war. If these social conditions can be reversed, its appeal and potency will wither away, though its bloodletting will likely leave deep scars on the consciousness of Arab and Muslim youth.

Critical Thinking

1. Why has ISIS been able to attract so many young recruits?
2. What is the difference between the concepts of the "Far Enemy" and the "Near Enemy"?
3. Is ISIS both a non-state actor and a state actor? why?

Internet References

Bureau of Counter-Terrorism, U.S. Department of State
www.state.gov/j/ct/

Department of Homeland Security
www.dhs.gov/

SITE Intelligence Group
https://ent.siteintelgroup.com

FAWAZ A. GERGES is a professor of international relations and Middle Eastern politics at the London School of Economics and Political Science. His books include The Far Enemy: Why Jihad Went Global (Cambridge University Press, 2005) and, most recently, The New Middle East: Protest and Revolution in the Arab World (Cambridge, 2014).

Article   Prepared by: Robert Weiner, *University of Massachusetts, Boston*

Fixing Fragile States

Dennis Blair, Ronald Neumann, and Eric Olson

Learning Outcomes

After reading this article, you will be able to:

- Talk about the operations of Al Qaeda in Yemen.

- Discuss how terrorist organizations can establish themselves in fragile states.

- Understand the role of bureaucratic politics in implementing U.S. foreign policy.

Since the 9/11 attacks, the United States has waged major postwar reconstruction campaigns in Iraq and Afghanistan and similar but smaller programs in other countries that harbor Al Qaeda affiliates. Continued complex political, economic and military operations will be needed for many years to deal with the continuing threat from Al Qaeda and its associated organizations, much of it stemming from fragile states with weak institutions, high rates of poverty and deep ethnic, religious or tribal divisions. Despite 13 years of experience—and innumerable opportunities to learn lessons from both successes and mistakes—there have been few significant changes in our cumbersome, inefficient and ineffective approach to interagency operations in the field.

We believe the time has come to look to a new, more effective operational model. For fragile states in which Al Qaeda is present, the United States should develop, select and support with strong staff a new type of ambassador with more authority to plan and direct complex operations across department and agency lines, and who will be accountable for their success or failure. We need to develop the plans to protect American interests and strengthen these countries out in the field, where local realities are understood, before Washington agencies bring their inside-the-Beltway perspectives to bear. Congress and the executive branch need to authorize field leaders to shift resources across agency lines to meet new threats. It is, in short,

a time for change—change that upends our complacent and antiquated approach toward foreign societies and cultures.

The 9/11 attacks offered us a painful reminder of an old verity, which is that fragile states unable to enforce their laws and control their territory are the progenitors of potent threats that can be carried out simply and effectively. Such states provide safe havens from which Al Qaeda and its affiliates plan and launch terror attacks against the United States and other countries. Al Qaeda in the Arabian Peninsula (AQAP) operates in Yemen; Al Qaeda in the Islamic Maghreb operates across Algeria, Mali and other neighboring countries; and Al Shabab operates in Somalia. Civil war in Syria, spreading violence in Iraq and continued turmoil in Africa will most likely open new havens for similar groups.

Until now, the American response to the threat from fragile states has had three major components. First, we have greatly strengthened the control of our own borders. Second, American intelligence and military forces, particularly the CIA and the U.S. Special Operations Command, have taken the fight to Al Qaeda. Third, the United States, along with other countries and international organizations, has increased economic and civil assistance to many fragile states using existing programs and authorities.

How much have these approaches achieved? The American-led reconstruction efforts in Iraq and Afghanistan have been prolonged and massive, but cannot be considered successful.

A dysfunctional system of authorities and procedures hampered effectiveness. Plans were made in Washington by committees of the representatives of different departments and agencies; individual departments and agencies sent instructions to their representatives in the field; and the allocations of resources to country programs were based in large part on individual departmental and agency priorities and available funding, not on overall national priorities. Each of the departments maintained direct authority over its field personnel and resources. Short-term staffing was endemic and cooperation in the field was voluntary, with neither the ambassador nor any official in Washington below the president authorized to resolve disputes or set overall priorities. Budget resources for a particular program could not be shifted smoothly to others when local conditions changed, and congressional oversight was split among committees that oversaw only individual aspects of the overall program in a country.

Even when the president, the National Security Council and an energetic interagency process in Washington were fully engaged—as they were in later years in Iraq and Afghanistan—the results have not matched the commitment of resources. Numerous accounts by journalists and memoirs of participants have documented the interdepartmental frictions, inefficient bureaucratic compromises and delayed decisions that have hampered progress. The authors of this article know personally most of those involved in leading the long wars in Iraq and Afghanistan. They are to a person—whether military officers or civilian officials—diligent and dedicated patriots. They have often worked across departmental lines to integrate security, governance and economic-assistance programs to achieve real successes. However, when officials and officers in the field did not get along, the deficiencies of the system allowed their disputes to bring in-country progress to a halt. What is needed is an overall system that will make cooperation and integration the norm, not the exception.

Yemen and Libya provide smaller-scale but more contemporary illustrations of the shortcomings of today's approaches. Although American officials have gained more experience, the authorities and procedures have not changed.

Yemen is the home base for AQAP, generally considered the most dangerous franchise of Al Qaeda. The speeches of American officials paint a picture of a comprehensive, balanced set of U.S. government programs not only to attack AQAP, but also to assist the current Yemeni government with both political and economic development. In congressional testimony in November 2013, for example, Deputy Assistant Secretary of State for the Arabian Peninsula Barbara Leaf emphasized American "support for Yemen's historic transition and continued bilateral security cooperation." She mentioned the $39 million that the United States had provided to support the national reconciliation process, U.S. encouragement of economic reform, its support for restructuring the Yemeni armed forces, and its participation in a weekly meeting among outside countries and international organizations to "compare notes, compare approaches, and coordinate tightly." She said nothing of the American military and intelligence attacks on AQAP fighters, yet these actions are the most costly U.S. programs dealing with Yemen, and they feature prominently in Yemeni popular opinion.

Even in this friendly hearing, however, the shortcomings of American and international programs were made clear.

Congressman Ted Deutch noted, "U.S. assistance to Yemen totaled $256 million for Fiscal Year 2013, but these funds come from 17 different accounts, all with very different objectives." He asked a fundamental question that went unanswered: "What exactly is our long-term strategy for Yemen?"

The view on the ground in Yemen is considerably darker. Two weeks before Leaf's testimony, an op-ed in the Yemen Post under the headline "Law of the Jungle in Yemen" stated:

> People have lost hope in the National Dialogue. . . . Billions of US dollars are still looted in the poverty stricken Yemen with not one corrupt senior official prosecuted. . . . Laws are only practiced against the weak and helpless. . . . An internal war is ongoing in the north of Yemen. . . . Al-Qaeda is regrouping and seeking to become a power once again . . . Safety and security in Yemen is nowhere to be seen. Government authority and presence over many parts of the country is limited, and where they are present they are almost useless.

U.S. policy in Yemen has been cobbled together in Washington through the typical interagency process. Because congressional funding for counterterrorist programs, both military and intelligence, is still flowing relatively freely, they are the largest American programs in Yemen. According to press reporting, there are two independent task forces—one military, one CIA—operating drones over Yemen, and U.S. security assistance to the Yemeni armed forces is focused on the creation of small, well-trained counterterrorist forces. The Saudis and other Gulf Cooperation Council (GCC) states have promised over $3 billion in economic support to Yemen. American economic assistance to Yemen is a small fraction of this amount; thus, the American plan must leverage the greater GCC contribution. The overall picture in Yemen, then, is one of unbalanced, uncoordinated and suboptimal U.S. and international programs based on no coherent plan.

Libya is another excellent illustration of an American assistance program that is not meeting the needs of the country. American commitment of financial resources to assist Libya has been modest: the State Department estimates about $240 million since the beginning of operations to oust Muammar el-Qaddafi in 2011. Far more money was spent by the United States on the NATO air operations that pushed Qaddafi out of power. American assistance to Libya has been spread across different government programs, depending on the other bills for those programs in the rest of the world. With few resources at their command, the country team needed an integrated plan to make the actions they could take effective, to set priorities and to leverage the actions of other countries. Yet the various American agencies working in Libya, as usual, cooperated as best they could, under no integrated plan, with little experienced leadership either in

Tripoli or Washington. As crises occurred and conditions deteriorated, responses were improvised.

Like Iraq in 2003, Libya was coming out of a long and brutal dictatorship. Rebuilding the country would require extraordinary actions by the Libyans themselves and by outside countries like the United States that had helped bring down Qaddafi and had a stake in a favorable outcome. International security-support programs, including those by the United States, have been notably weak. Any doubt about the conditions on the ground ended when Ambassador J. Christopher Stevens was murdered in Benghazi in September 2012, the first U.S. ambassador to die in the line of duty since 1979. Yet it was over another year later that the United States and several other European countries began belatedly to take actions to strengthen the army and police. NATO began a program to train about twenty thousand Libyan soldiers. The program is not scheduled to be completed for many more months, and the result will be trained soldiers who perform their duties with mostly inadequate medical, communications and logistical support.

Yet this belated program to strengthen basic security in Libya still is not part of an overall plan to help Libya become a competent, functioning state. According to two experts at the Atlantic Council and the European Council on Foreign Relations, respectively:

The current western agenda for Libya lacks a political strategy and is focused almost exclusively on the training of the Libyan army. If experience elsewhere is an indication, it will take between 5 and 13 years for that to conclude. The same experience tells us that "strengthening the central government" is an insufficient goal if the country is to become stable and under the rule of law.

Meanwhile, the official U.S. activities in Libya, as described by the current U.S. ambassador, Deborah K. Jones, are directed toward "a broad process encompassing a National Dialogue, constitutional development, and governance capacity-building to increase public confidence." In American pronouncements, there is little sense of priorities, combined programs, milestones or urgency.

The embassies of the United States and its international partners in these fragile countries must do more than just be supportive of individual areas needing improvement. Their approach has to be selective, hands-on, tailored, flexible and integrated.

Selective: Resources are limited, and the approach needs to be sustained over an extended period of time. It should be applied only to the handful of countries in which the threat is high and host government capacity is low.

Hands-on: The United States and other international partners cannot simply transfer money to government departments in fragile states, as it will likely be stolen or misused. Instead, they must take an active role in building competent local government organizations that can use increased resources effectively. American and other international operators cannot train local organizations to be replicas of their Western equivalents, or models of counterparts in other countries; conditions are too different. Likewise, they cannot simply fly in for a two-week stint and then head home; there will be no follow-through. Experienced, carefully selected and trained officials who can influence host officials and build local capacity without causing resentment are essential.

Tailored: Sometimes existing security or law-enforcement organizations or judicial systems can be strengthened; other times they must be created. Sometimes putting the national finances of a country in order will unleash economic growth; other times training and economic support in a particular region of the country are vital. Sometimes training and assisting central government officials is important; other times it is competent provincial officials that are essential for success. The key to a tailored approach is for the American representatives in a country to have the authority and responsibility both for planning and for carrying out the plan.

Flexible: Requirements are always dynamic in fragile states. Plans need to be revised quickly in response to events on the ground; money and personnel need to be shifted quickly to meet new problems and to take advantage of new opportunities.

Integrated: Integration depends on setting a common set of priorities across all programs. Once security forces stabilize a city or region, improved governance and economic opportunity must follow immediately, or the security gains will be wasted. Integration depends on realistic sequencing of different programs. Unless the judiciary and prison systems are improved along with police forces, criminals will be released or tortured after their arrest. Policemen can be trained or retrained in weeks and prison systems can be improved quickly, yet training a core of judges and lawyers takes years. There must be practical interim plans that will ensure progress.

Finally, current operations to capture or kill hardcore Al Qaeda members need to continue, without stirring up local resentment that will make it more difficult to make the necessary longer-term improvements. However, these operations need to be consolidated and integrated into an overall plan in each country.

There is duplication, overlap and sometimes competition between the traditional military operations of the Department of Defense and the covert paramilitary operations of the CIA against Al Qaeda. To fully understand the issue, it's important to be clear about the significant difference between clandestine and covert operations. A "clandestine" operation is one that is secret, and no government official is to talk about it. Clandestine operations are routinely conducted by the Department of Defense, and on occasion by other U.S. government agencies. A "covert" operation is one in which the involvement of the

U.S. government is to be kept secret, to the point of official denial. The CIA has generally conducted covert operations, and an executive order gives this preference, but the basic legislation authorizes them to be conducted by other departments or agencies as directed by the president.

Although geopolitical conditions have changed fundamentally since the Cold War, when covert operations were originally authorized, there has been no serious consideration of updating the authorities for covert action. The 9/11 Commission's recommendation to assign paramilitary operations to the Department of Defense was not adopted either by Congress or by two successive administrations. The result has been continued complicated, duplicative and costly operations against Al Qaeda. It has only been experienced, dedicated and mission-focused operators in the field that have permitted the current system to work, and their successes have obscured the need for clarity and simplification. This recommendation should be seriously revisited based on our experiences of the last decade.

Two types of armed operations against Al Qaeda are the most important: raids and armed drone strikes. For raids—the helicopter raid that killed Osama bin Laden in Abbottabad is the best known—all the operational skills are in the Department of Defense, mostly within components of the Special Operations Command. Yet, there are often questions and disputes about whether they should be conducted as clandestine traditional military operations commanded by the secretary of defense under Title 10, or covert intelligence operations controlled by the director of the CIA under Title 50.

In reality, the president has the legal authority to order these operations under either title, using either organization. In 2011, the president decided to authorize the Abbottabad raid, entirely conducted by Department of Defense personnel, as a Title 50 covert action, under the control of the director of the CIA. It was a "clandestine military operation" conducted under authorities that were designated for "covert action." There was never any reason or intention to deny the role of the U.S. government—the primary rationale for covert action—once the operation commenced and inevitably became public. To the contrary, government officials were running for the microphones as soon as the helicopters returned from Pakistani airspace. Fortunately, experienced military commanders made all the tactical decisions, and the raid was a success. Had anything gone wrong—the loss of a helicopter and the capture of its crew by Pakistan, or a dispute between CIA officers and special-operations officers during the raid, with each group appealing to its own chain of command—the results could have been quite different.

For armed attacks by drones, the CIA and the Department of Defense have set up duplicate organizations, each authorized by separate legislation. There are reasons for the current arrangements. The bottom line, however, is that it is the Department of Defense that is established, trained and authorized to kill enemy combatants. For reasons of competence, accountability and effectiveness, the armed drone campaign should be assigned to the secretary of defense, with the entire intelligence community, including the CIA, playing an essential role in identifying, prioritizing and tracking the targets.

A new model for interagency operations in fragile states would be strongest and longest lasting if it were established by legislation. However, much can be done by executive order, policy and practice.

The foundational process change should be to assign the task of developing a comprehensive plan for a fragile state to the team on the ground in that state, rather than to an interagency group in Washington. It is axiomatic in both business and military planning that a plan ought to be drafted by those responsible for carrying it out. Only in American interagency planning is it done by a committee at headquarters, then passed to the field for implementation. Washington's plans are subject to pressures that often make them unrealistic and unsuitable for conditions in the field. An in-country planning team is much more likely to deliver a plan that is balanced between the short and the long term, that includes the most effective applications of the capabilities of the different departments and that realistically matches the needs on the ground. During interagency review in Washington, there will be plenty of opportunity for adding other considerations and good ideas.

However, for an embassy to submit a good plan takes a uniquely qualified and experienced ambassador with a dedicated, competent supporting interagency staff, in addition to the usual country team, comprising the representatives from the various departments and agencies.

Foreign Service officers spend most of their careers in staff positions, responsible for observing, reporting, negotiating, and making policy recommendations that are heavily weighted toward the short term and tactical. Their career pattern develops a high level of expertise, observational and writing skills, and diplomatic abilities. The leader of American in-country operations in a fragile state needs high-order managerial and leadership skills for complex program execution as well as a deep knowledge of the capabilities and limitations of other American organizations, especially military and intelligence. Some Foreign Service officers who became ambassadors have developed these skills. James Jeffrey, Ryan Crocker and Anne Patterson are among several in the recent past. However, although such training has been recommended, the Foreign Service is not geared toward producing such skills broadly. A qualification-and-selection process is needed for ambassadors to places like Yemen, Libya, Pakistan, Mali, Somalia, Afghanistan and Iraq to identify candidates with the experience, knowledge and stature to direct an integrated, multiagency task force.

The current manning of embassies does not include a central staff to support an ambassador in designing and implementing a country plan. What is needed is a small, separate staff of perhaps a dozen experienced officers, drawn from different agencies, to help the ambassador formulate the plan, and then to monitor its execution to determine if it is achieving its objectives and recommend adjustments as circumstances on the ground change. While maintaining strong links back to their parent organizations for advice, support and guidance, these staff officers would primarily serve the ambassador in developing and coordinating his or her plans. Such help is beginning to be available from the State Department's Bureau of Conflict and Stabilization Operations, but this falls short of an integrated interagency effort.

Within the overall integrated plan in a fragile state, the ambassador should recommend the military and intelligence actions to be taken directly against Al Qaeda personnel and units. The country plan must establish the priority and scope of these activities within the overall mission of strengthening security. It needs to define the areas in which the raids and drone strikes will be conducted and the intensity of the campaign. The overall objective is to capture or kill more enemies than are created. The ambassador should recommend whether these actions be taken as military activities under Title 10 or intelligence activities under Title 50. With special-operations forces and CIA planners as part of his team, the ambassador is in the best position to recommend both the actions themselves and the most appropriate authorities under which to conduct them. During the course of the campaign, the ambassador needs the authority to approve direct actions—drone strikes as well as raids and conventional military strikes—to ensure that they are integrated into the overall plan.

When the ambassador has formulated an integrated plan for the country, incorporating diplomatic, economic, intelligence, military and other aspects, including milestones that the plan will achieve on specified dates, it should be sent to Washington for interagency comment and for the allocation of resources—people and money. The Office of Management and Budget should participate at all levels of interagency review to ensure that budget plans are realistic. Ultimately, a resourced, comprehensive plan for a fragile state should be approved by the president.

Virtually every fragile state both affects and is affected by its neighbors. Tribal, ethnic and religious influences cross national boundaries; borders are often porous; pressuring groups in one country pushes them into others. The cooperation of neighboring states is thus essential to success within fragile states. To ensure that these factors are considered to obtain regional buy-in, each country plan should be sent to neighboring embassies (in the case of the State Department and other agencies without regional organizations) and to regional and global combatant commands (in the case of the Department of Defense) for review and comment.

No plan survives first contact with the enemy; success in the implementation of a plan depends on flexibility and adaptation. Yet currently those carrying out military, economic, diplomatic and other programs in fragile states have very little authority or capability to react. A change in one aspect of a plan will always cause changes in other aspects, yet because authorities in the American national-security system pass directly from departments and agencies to representatives in the field, it is very difficult to gain approval for necessary adjustments. Congressional oversight, based on jurisdiction over appropriated budgets, further hinders flexibility. Economic-development programs depend on successful security operations, yet there is no authority in a country that can direct adjustments when setbacks in one area require changes in another. No financial or personnel reserves are available to cover unexpected problems or to take advantage of surprise opportunities—the budget incentive is "use it or lose it," whether or not a program is effective, or whether or not the money could be used more effectively elsewhere. Again, dedicated, hardworking officers and officials cooperate with each other as best they can, but the current system does not support flexibility.

The solution is to give the ambassador both the responsibility for overall progress on the plan and directive authority over the programs in country within the limits of the plan that was approved. Once budgets have been allocated to U.S. programs to strengthen a fragile state, an ambassador should be able to shift them, within realistic thresholds, as needs and opportunities develop.

These reforms will go a long way toward improving American support for fragile states and dealing with Al Qaeda groups that find refuge in them. However, additional improvements are needed.

It is only the Department of Defense that has either the authority or the tradition of assigning personnel to difficult overseas postings, whether they volunteer or not. All other agencies rely on volunteers. The result has been chronic short-changing of the nonmilitary billets in fragile states—short assignments for officers, or the use of contractors. Authority must be granted to department and agency heads to assign their personnel as needed to support the national interest. Without this change, American campaigns in these countries will be unbalanced and heavily influenced by military considerations, since it is the military personnel who show up.

Although the Department of Defense has the authority to send personnel overseas as needed, some key skills for assisting fragile states have deteriorated within the military services in recent years. In the past, there were experienced civil engineers, utility company officials, local government administrators and transportation officials in the Army and Marine Corps Reserve. They had the skills and experience to help establish competent organizations to provide basic infrastructure in fragile states. The civil-affairs personnel in military reserve units are more junior and much less experienced now. The contractors and individually mobilized reservists that are now used to assist struggling government organizations in fragile states are inadequate for the importance and difficulty of the need. The Reserve Components of the Army and Marine Corps must reestablish strong civilian-affairs components.

Successive secretaries of state in recent administrations have made strong attempts to improve the numbers and qualifications of civilian officials sent to fragile states. Continued emphasis is needed, as the State Department still has difficulty filling even established billets in Afghanistan, language skills do not meet existing requirements in fragile states, and there are too many short-term assignments of personnel to jobs that require sustained interaction with local officials to build trust. The State Department must continue to develop a cadre of officers who can be effective in the tough tasks of strengthening fragile states.

In virtually every fragile state, some of the weakest institutions are the police, courts and prisons. The U.S. government has very little capacity to help strengthen them. The Department of Justice has only a limited training-and-advisory capacity similar to that in the Department of Defense or the Department of State, and generally requires outside funding to mount training programs. Other countries have some capacity, but retired state and local police officers, private contractors, or volunteer judges and prison officials man the American assistance programs. In Afghanistan, the mission of police training was assigned to the Department of Defense, despite the fundamental differences between the military and law-enforcement missions. We need to develop a cadre of advisers and trainers for police, courts and prisons, and a means to supplement the cadre with qualified and supervised private volunteers.

Finally, Congress will need to establish new oversight procedures for an integrated country strategy, rather than the disjointed current system in which generals testify in front of one committee, ambassadors in front of another, and no executive branch official below the president has the responsibility for overall success or failure to strengthen a fragile state in which the United States has important interests.

Countries with weak governments, high levels of poverty, and internal ethnic, religious and tribal tensions that provide sanctuaries for Al Qaeda or its affiliates will remain a perennial source of instability and threats for America and its allies. Most of the discussion of the challenges of dealing with fragile states

has been dominated by abstract debates over vital American interests, fears of long-term commitments, sterile arguments over military versus civil components, disagreements over deadlines and often ill-informed applications of the perceived lessons of the U.S. experience in one country to another. This is unfortunate. What has been missing from the discussion is an understanding of the very segmented, rigid and inefficient system under which the United States attempts to help these countries stabilize their governments and societies and control outside terrorist groups. The current system guarantees that the resources—people and dollars—that are allocated to these countries do not produce the results they could and should. The United States should likely provide more funding for its programs in countries like Yemen, Libya and Mali. However, what is even more important is for the Obama administration and Congress to improve the basic system for organizing and conducting these programs. The improvements in authorities and procedures that we recommend here will go a long way toward making America safer with a very small expenditure of additional resources. It's time to replace decades of failure with a new approach that protects American security by transforming fragile states into genuinely secure ones.

Critical Thinking

1. Why has the United States failed to bring peace and stability to Yemen?
2. How can U.S. diplomacy stabilize Libya?
3. Should drones be used in armed attacks against Al Qaeda?

Internet References

Fragile States Index
fsi.fundforpeace.org

Libya
https://www.cia.gov/library/publications/the-world-factbook/geos/ly.html

Yemen
https://www.cia.gov/library/publications/the-world-factbook/geos/ym.html

---

DENNIS BLAIR is the former Director of National Intelligence and former Commander in Chief of U.S. Pacific Command. Ronald Neumann is president of the American Academy of Diplomacy and former U.S. Ambassador to Algeria, Bahrain and Afghanistan. Eric Olson is the former Commander of U.S. Special Operations Command.

Blair, Dennis and Neumann, Ronald and Olson, Eric. "Fixing Fragile States," *The National Interest,* October 2014. Copyright © 2014 by National Interest. Used with permission.

Unit 5

UNIT

Prepared by: Robert Weiner, *University of Massachusetts, Boston*

Conflict and Peace

There is no single cause of war, as wars in any event have multiple causes. For example, wars can be about scarce resources, such as water, especially because the world's water supply is not endless, and a small number of countries actually possess the bulk of the world's water supply. Political scientists argue that the causes of war can be found at different levels of analysis, ranging from the individual level, to the domestic level (regime type and system), to the international level (interstate relations) and to the global level (international communications such as cyberwar) and international terrorism. Recent empirical studies, conducted by credible research institutions, have concluded that there has been a decline in the amount of interstate warfare in the international system, but an increase in internal or civil conflicts (sometimes with significant external intervention) since the end of the Cold War.

In 2014, the 100th anniversary of World War I was observed, as the legacy of the "Great War" is still being felt around the globe. Historians and political scientists are still debating the factors that led to World War I. At the level of the international system, the emergence of two rival alliance systems—the Entente and the Central Powers—may have been an important factor that contributed to the eruption of the war. The outbreak of World War I shattered the "Long Peace" (1815–1914) that had prevailed in Europe since the end of the Napoleonic Wars. This century of peace has been seen as a period of "golden diplomacy" in which the maintenance of a finely calibrated balance of power preserved stability in Europe.

World War I was called the "Great War" because no one could ever imagine that another bout of such atavistic bloodletting would happen until the occurrence of World War II. Some historians have viewed the Second World War as a continuation of World War I, based on a 20-year interlude between the two wars. Moreover, some respected experts on international relations believe that another Great War will never take place. On the other hand, Norman Friedman and John Mearsheimer both argue that there is the possibility of a Great War taking place between China and the United States, especially since the problem of Taiwan has yet to be resolved. Like Britain before it, the United States is a great trading state, which finds its sea power position in Asia challenged by China, which is a rising power. Beijing seems to be pursuing a Grand Strategy that is based on establishing itself as a regional hegemon in the South China and East China Seas, as well as projecting its power into Eurasia. Beijing also sees

a strategic opportunity to project its air and naval power as the United States declines. The U.S. response to China's rise has been to pursue a "pivot" or rebalancing to Asia by moving some of its military assets from Europe and the Middle East to the Pacific region, thereby raising the level of tension with China. Japan plays an important strategic role in the rebalancing, as Washington has supported a revision of Japan's self-defense forces to develop a more offensive capability. As part of the overall strategy of rebalancing, Japan has signed on to the U.S. free trade arrangement known as the Trans-Pacific Partnership (TPP). The TPP consists of 12 Pacific nations, including Canada and Mexico.

Ethnic and religious sectarian differences are also major causes of civil conflicts. For example, separatists in Eastern Ukraine consolidated their position. The Ukrainian situation settled into the pattern of a "frozen conflict," somewhat similar to other "frozen conflicts" that had prevailed in other parts of the former Soviet Union, such as Transnistria. The inability of the West to persuade the Russians to withdraw their support from the Ukrainian separatists also resulted in a rise in the level of tension between the Baltic States and Russia. The Baltic States also contain significant Russian minorities, viewed by Moscow as kith and kin. NATO's reaction was to reconfigure its strategy, to reassure the security concerns of the Baltics, Poland, as well as Finland and Sweden.

President Obama had campaigned on the theme of ending the U.S. involvement in the wars in the Middle East and Afghanistan. The United States had claimed to withdraw its combat forces from Iraq in 2014, but re-engaged in the conflict in Iraq and found itself drawn into the civil conflict in Syria, which had begun in 2011. According to media reports, the conflict in Syria had cost about 250,000 lives by 2015. The U.S. strategy in Iraq and Syria was complicated by the victories that were scored by what the United states viewed as an extremist group of Sunni jihadists, who were responsible for the videotaped executions of American and British citizens. The group was known in Arabic as Daesh, and in English as the Thalamic State in Iraq and the Levant (ISIL). ISIS had been able to gain control of a significant amount of territory in Iraq and Syria, including the Iraqi city of Mosul, which was an important oil center. The Islamic State had been able to take advantage of the power vacuum that was created in Syria by the civil war, and used the country as a sanctuary and base from which to expand its control of Iraqi territory, thereby putting Iraq once again on the brink of disaster.

The Obama administration made the decision to launch airstrikes against the Islamic State, even though the airstrikes had the effect of helping the Assad regime maintain its power. However, the policy of the Obama administration was Iraq-centric, designed to degrade and destroy the Islamic State. The Islamic State received some support from dissatisfied Sunni tribes which had been excluded from key power-sharing arrangements by the Maliki administration. In 2015, Moscow also decided to engage in a military intervention in Syria to support its client, President Assad. With both Russian and U.S. aircraft operating in Syria, Washington expressed its concern about the Russian move.

The Obama administration also pursued a policy of trying to extricate the United States from the long-running war in Afghanistan, but with a commitment to remain into 2017 to aid the new Afghan government that followed the Karzai regime. U.S. efforts to withdraw from Afghanistan illustrated the difficulties associated with terminating a war in which there was no clear-cut winner. President Obama was not able to achieve his objective of terminating what seemed to be an endless war in Afghanistan. Opportunities for peace talks with the Taliban had been missed, as another complicating factor was added with the involvement of the Islamic State in the war in Afghanistan.

Philosophers and political scientists have worked on the problem of creating a system of universal peace as the underpinning of world order for centuries. The idea is to realize Kant's age-old dream of instituting a global system of perpetual peace as opposed to the Hobbesian system of a cruel world order based on perpetual warfare, where life is mean, nasty, short, and brutish, and consists of the war of each against all. Advocates of the possibility of a world order based on peace and justice have a much more optimistic view of human nature than classical realists, who seem to believe that human beings are inherently evil. Liberal internationalists especially believe that human beings are rational creatures, who find it in their interest to cooperate with each other. Human beings also have the capacity to work with each other, and construct a peaceful world society. Liberals believe that international institutions like the League of Nations and the United Nations can make a difference in preventing conflicts from occurring in the first place. The central problem of international relations is the reconciliation of order with justice, and liberals argue that international law and morality can contribute significantly to the creation of a peaceful world order as well.

Another central tenet of liberalism is that the domestic political system of a state has an effect on its foreign policy in the international arena. Liberal democratic states, according to democratic peace theory, which some political scientists view as the closest thing to an iron law of political science, have less of a tendency to go to war with other liberal democratic states. The reasons for this may range from the system of checks and balances that function to mitigate the decision to go to war in a democratic state, the values and morality that are associated with democracy, and the fact that liberal democratic states may be connected by a set of economic and trade linkages that enmesh them in a web of cooperation. Liberals also stress that economic ties between states in general may reduce the likelihood of a war taking place between them, because the economic costs of a war may jeopardize the benefits of a peaceful relationship.

Finally, arms control and disarmament have been viewed as a means of reducing the possibility of conflict and war between states in the international system. The age-old dream has been for the creation of a system of general and complete disarmament where the biblical injunction of beating swords into plowshares will mean that humans will never make war on each other again. However, military technology and the trade in conventional weapons have made the task of establishing a system of general and complete disarmament extremely difficult. The international community, however, has made progress in dealing with the conventional weapons that are supplied by both governmental and private "merchants of death" with the conclusion of an international treaty regulating the global arms trade.

Furthermore, a network of treaties has been negotiated since the end of World War II to deal with reducing, and hopefully eliminating, weapons of mass destruction (WMDs), as the technology associated with these weapons has spread. The focus of WMD treaties has been to prevent the spread of nuclear weapons. Perhaps with lessons learned from the Cuban missile crisis, the United States recently focused its efforts on preventing Iran from developing the bomb. The United States and Iran reached an agreement in the summer of 2015, known as the Joint Comprehensive Plan of Action. The agreement had been approved by the Iranian Parliament, but had yet to be approved by the U.S. Congress. The nuclear deal between Iran and the United States, which was the result of months of intensive negotiations, represented an attempt to contain Iran's nuclear weapons program for 15 years. The Iranians agreed in essence to freeze their weapons program, by placing thousands of their centrifuges, which produced the enriched uranium needed to make the bomb, in storage. In return, U.S. and international sanctions that Iran had been subjected to would be lifted, and the Iranians would also be able to export their oil and natural gas, as well as gain access to at least $100 billion in frozen assets.

Teheran also agreed to open up its nuclear program to inspection by the International Atomic Energy Agency (IAEA). However, this portion of the agreement was open to criticism by the opponents of the deal, because Iran would have a 24 day notice before the IAEA inspectors arrived at the sites. It was argued that the Iranians would have enough time to clean up any evidence that they were secretly working on the development of nuclear weapons in violation of the agreement. However, the agreement was seen by President Obama as a way to engage in a deeper dialogue with the Iranians. (It is worth noting that the Obama administration established diplomatic relations with Cuba in 2015.) It was hoped that Iran would moderate its behavior in the region and restore a sense of equilibrium as the United States attempted to withdraw from the area.

Article | Prepared by: Robert Weiner, *University of Massachusetts, Boston*

The Growing Threat of Maritime Conflict

"What makes these disputes so dangerous . . . is the apparent willingness of many claimants to employ military means in demarking their offshore territories and demonstrating their resolve to keep them."

Michael T. Klare

Learning Outcomes

After reading this article, you will be able to:

- Describe why these conflicts are increasingly dangerous.
- Identify the role of oil and natural gas in these disputes.
- Identify specific zones of conflict.

For centuries, nations and empires have gone to war over disputed colonies, territories, and border regions. Although usually justified by dynastic, religious, or nationalistic claims, such contests have largely been driven by the pursuit of valuable resources and the taxes or other income derived from the inhabitants of the disputed lands. Many of the great international conflicts of recent centuries—the Seven Years War, the Franco-German War, and World Wars I and II, for example—were sparked in large part by territorial disputes of this type. By the end of the twentieth century, however, most international boundary disputes had been resolved, and few states possessed the will or the capacity to alter existing territorial arrangements through military force.

Yet, even as the prospects for conflict over disputed land boundaries seem to have dwindled, the risk of conflict over contested maritime boundaries is growing. From the East China Sea to the Eastern Mediterranean, from the South China Sea to the South Atlantic, littoral powers are displaying fresh resolve to retain control over contested offshore territories.

The most recent expression of this phenomenon, and one of the most dangerous, is the clash between China and Japan over a group of uninhabited islands in the East China Sea that are claimed by both. Friction over the islands—known as the Diaoyu in China and the Senkaku in Japan—has persisted for years, but it reached an especially high level of intensity in the summer of 2012 after Japanese authorities arrested 14 Chinese citizens who attempted to land on one of the islands to press China's claims, provoking widespread anti-Japanese protests across China and a series of naval show-of-force operations in nearby waters.

Senior Chinese and Japanese officials have met privately in an attempt to reduce tensions, but no solution to the dispute has yet been announced, and both sides continue to deploy armed vessels in the area—often in close proximity to one another. Although the Barack Obama administration would like to see a negotiated outcome to the dispute. China views Washington as too close to Japan, so Beijing has rebuffed US mediation efforts.

Risk of conflict has also arisen in another disputed maritime area, the South China Sea, where China is again one of the major offshore claimants. As in the East China Sea, the dispute centers on a collection of (largely) uninhabited islands: the Paracels in the northwest, the Spratlys in the southeast, and Macclesfield Bank in the northeast (known in China as the Xisha, Nansha, and Zhongsha islands, respectively). China and Taiwan claim all of the islands, while Brunei, Malaysia, the Philippines, and Vietnam claim some among them, notably those lying closest to their shorelines.

Friction over these contested claims led to a series of nasty naval encounters in 2012, some involving China and Vietnam, and some China and the Philippines. In one such incident, armed

Chinese marine surveillance ships blocked efforts by a Philippine Navy warship to inspect Chinese fishing boats believed to be engaged in illegal fishing activities, leading to a tense stand-off that lasted weeks. Chinese officials announced recently that, beginning January 1, their patrol ships will be empowered to stop, search, and repel foreign ships that enter the 12-nautical-mile zone surrounding the South China Sea islands claimed by Beijing, setting the stage for further confrontations.

Maritime disputes of this sort, also involving the use or threatened use of military force, have surfaced in other parts of the world, including the Sea of Japan, the Celebes Sea, the South Atlantic, and the Eastern Mediterranean. In these and other such cases, adjacent states have announced claims to large swaths of ocean (and the seabed below) that are also claimed in whole or in part by other nearby countries. The countries involved cite various provisions of the United Nations Convention on the Law of the Sea (UNCLOS) to justify their claims—provisions that in some cases seem to contradict one another.

Because the legal machinery for adjudicating offshore boundary disputes remains underdeveloped, and because many states are reluctant to cede authority over these matters to as-yet untested international courts and agencies, most disputants have refused to abandon any of their claims. This makes resolution of the quarrels especially difficult.

What makes these disputes so dangerous, however, is the apparent willingness of many claimants to employ military means in demarking their offshore teritories and demonstrating their resolve to keep them. This is evident, for example, in both the East and South China Seas, where China has repeatedly deployed its naval vessels in an aggressive fashion to assert its claims to the contested islands and chase off ships from all the other claimants. In response, Japan, Vietnam, and the Philippines have also employed their navies in a muscular manner, clearly aiming to show that they will not be intimidated by Beijing. Although shots have rarely been fired in these encounters, the ships often sail very close to each other and engage in menacing maneuvers of one sort or another, compounding the risk of accidental escalation.

What accounts for this growing emphasis on offshore disputes at a time when few states appear willing to fight over more traditional causes of war?

For some governments, offshore disputes may be seen as a sort of release valve for nationalistic impulses that might prove more dangerous if applied to other issues, or as a distraction from domestic woes. China's conflict with Japan over the Diaoyu/Senkaku Islands, for example, has provoked strong nationalistic passions in both countries—passions that leaders on each side no doubt would prefer to keep separate from the more important realm of economic relations. Likewise, Argentina's renewed focus on the Falklands/Malvinas is widely considered to be a deliberate response to political and economic difficulties at home. But these considerations are only part of the picture; far more important, in most cases, is a desire to exploit the oil and natural gas potential of the disputed areas.

The Lure of Oil and Gas

The world needs more oil and gas than ever before, and an ever-increasing share of this energy is likely to be derived from offshore reservoirs. According to the Energy Information Administration (EIA) of the US Department of Energy, global petroleum use will rise by 31 percent over the next quarter-century, climbing from 85 million to 115 million barrels per day. Consumption of natural gas will grow by an even faster rate, jumping from 111 trillion to 169 trillion cubic feet per year. Older industrialized nations, led by the United States and European countries, are expected to generate some of this growth in consumption. But most of it is projected to come from the newer industrial powers, including China, India, Brazil, and South Korea. These four nations alone, predicts the EIA, will account for 57 percent of the total global increase in energy demand between now and 2035.

Until now, the world's ever-increasing thirst for oil has been satisfied with supplies obtained from fields on land or shallow coastal areas that can be exploited without specialized drilling rigs. But many of the world's major onshore fields have been producing oil for a long time, and are now yielding diminishing levels of output: likewise, production in shallow areas of the Gulf of Mexico and the North Sea has long since fallen from peak levels. Some of the loss from existing reservoirs will be offset through the accelerated extraction of petroleum from shale rock, made possible by new technologies like hydraulic fracturing. But any significant increase in global oil production will require the accelerated exploitation of offshore—especially deep-offshore—reserves.

According to analysts at Douglas-Westwood, a United Kingdom–based energy consultancy, the share of world oil production supplied by offshore fields will rise from 25 percent in 1990 to 34 percent in 2020. More important, the share of world oil provided by deep wells (over 1,000 feet in depth) and ultra-deep wells (over one mile) will grow from zero in 1990 to a projected 13 percent in 2020. Douglas-Westwood further projects that onshore and shallow-water fields will yield no additional production increases after 2015, so all additional growth subsequently will have to come from deep and ultra-deep reserves. Meanwhile, the world's reliance on natural gas is likely to exhibit a similar trajectory: Whereas in 2000 approximately 27 percent of the world's gas supply came from offshore fields, by the year 2020 that share is projected to reach 41 percent.

Driving this shift toward greater reliance on offshore oil and gas is not only the depletion of onshore fields but also advances in drilling technology. Until recently, it was considered impossible to extract oil or gas from reserves located in waters over a mile deep. Now drilling at such depths is becoming almost routine, and extraction at even greater depths—up to two miles—is about to commence. Specialized rigs have also been developed for operations in the Arctic Ocean, and in areas that pose unusual climatic and environmental challenges, such as the Caspian Sea and the Sea of Okhotsk, off Russia's Sakhalin Island. In the future, technology may allow the extraction of natural gas from so-called methane hydrates—dense nodules of frozen gas that are trapped in ice crystals lying at the bottom of some northerly oceans.

It follows from all this that the world's major energy consumers—led by China, the United States, Japan, and the European Union countries—will become increasingly reliant on oil and gas supplies derived offshore. Some of this energy can be acquired from fields in areas with no outstanding territorial disputes, such as the North Sea and the Gulf of Mexico. Other large reservoirs, such as Brazil's "pre-salt" fields in the deep Atlantic, lie far enough from other coastal states to eliminate the potential for boundary conflict. But many promising fields are located in bodies of water where maritime boundaries remain undefined. And, as the perceived value of these resources grows, the potential for discord to take a military form will increase as well. This risk is greatest in areas thought to harbor large reserves of oil and gas, where the contending parties have repeatedly rebuffed efforts to adopt precise, mutually acceptable offshore boundaries, and where one or more of the claimants have employed (or threatened the use of) military means.

Contested Seas

The risk is especially great in the East and South China Seas. Both regions are thought to sit atop substantial reserves of oil and gas, both lack mutually accepted offshore boundaries, and both have witnessed repeated military encounters. The East China Sea, bounded by China to the west, Taiwan to the south, Japan to the east, and Korea to the north, harbors several large natural gas fields in areas claimed by China, Japan, and Taiwan. The South China Sea, bounded by China and Taiwan to the north. Vietnam to the west, the island of Borneo (divided among Brunei, Indonesia, and Malaysia) to the south, and the Philippines to the east, is believed to possess both oil and gas deposits: China and Taiwan claim the entire region, while Brunei, Malaysia, Vietnam, and the Philippines claim large portions of it. All of these countries have engaged in negotiations aimed at resolving the various overlapping claims—without

achieving notable success—and all have taken military steps of one sort or another to defend their offshore interests.

Considerable debate persists among industry professionals as to exactly how much oil and gas is buried beneath the East and South China Seas. Because limited drilling has been conducted in these areas (except on the margins), analysts possess little detailed information from which to derive estimates of recoverable reserves. Nevertheless, Chinese experts regularly offer highly optimistic assessments of the seas' potential. The East China Sea, they claim, contains between 175 trillion and 210 trillion cubic feet of natural gas—approximately equivalent to the proven reserves of Venezuela, the world's seventh largest gas power. Chinese estimates of the oil and gas lying beneath the South China Sea are even more exalted: These place the region's ultimate oil potential at over 213 billion barrels (an amount exceeded only by the proven reserves of Saudi Arabia and Venezuela), and that of gas at 900 trillion cubic feet (exceeded only by Russia and Iran). Western analysts, such as those employed by the EIA, are reluctant to embrace such lofty estimates in the absence of actual drilling results, but acknowledge the two areas' great potential.

Whatever the precise scale of the East and South China Seas' hydrocarbon reserves, the various littoral states clearly see them as promising sources of energy. China, Japan, Malaysia, Vietnam, and the Philippines have awarded contracts to different combinations of private and state-owned firms to exploit oil and gas reserves in the areas they claim, and more such awards are being announced all the time. The Chinese have been particularly active, drilling for natural gas in the East China Sea and for oil in the South China Sea. Their efforts took a big step forward in May 2012, when the China National Offshore Oil Corporation (CNOOC) deployed the country's first Chinese-made deep-sea drilling platform in the South China Sea, at a point some 200 miles southeast of Hong Kong.

The Vietnamese have long extracted oil and gas from their coastal waters, and are now seeking to operate in deeper waters of the South China Sea. Across the sea to the east, the Philippines' Philex Petroleum Corporation has been exploring a major natural gas find off Reed Bank—another uninhabited islet claimed by China as well as the Philippines and a site of recent clashes between Chinese and Filipino vessels. Although Chinese leaders say they want to promote cooperative development of the East and South China Seas, Beijing has often taken steps to deter efforts by its neighbors to explore for oil and gas in these areas. In May 2011, for example, Chinese patrol boats repeatedly harassed exploration ships operated by state-owned PetroVietnam in the South China Sea, in two instances slicing cables attached to underwater survey equipment.

Despite the high expectations for oil and gas extraction in the two seas, therefore, any significant progress will have to await the resolution of outstanding territorial disputes or some agreement allowing drilling to proceed without risk of interference. Yet none of the parties to these disputes appears willing to retreat from long-established positions or eschew the use of force. Efforts to seek negotiated outcomes have been frustrated, moreover, by contending historical narratives and a lack of clarity in international law regarding the demarcation of offshore boundaries.

Legal Confusion

In the East China Sea, both China and Japan draw on competing provisions of UNCLOS (which both have signed) to justify their maritime claims. Each set of provisions defines a state's outer maritime boundary in a different way: One set allows coastal states to establish an exclusive economic zone (EEZ) extending up to 200 nautical miles offshore, in which they possess the sole right to exploit marine life and undersea resources, such as oil and gas; the other allows coastal states to exert such control over the "natural prolongation" of their outer continental shelf, even if it exceeds 200 nautical miles.

China, citing the latter provision, says that its maritime boundary in the East China Sea is defined by its continental shelf, an underwater feature that extends nearly to the Japanese islands. Japan, citing the former provision, insists that the boundary should be drawn along a median line equidistant between the two countries, since the distance separating them is less than 400 nautical miles.

Lying between these two hypothetical boundary lines is a contested area of approximately 81,000 square miles (nearly the size of Kansas) that is thought to harbor large volumes of natural gas—a resource that each side claims is its alone to exploit. The contested Diaoyu/Senkaku Islands lie at the southern edge of this area, and so neither side is willing to relinquish control over them, each fearing that doing so would jeopardize its claim to the adjacent seabed. Negotiations to resolve the impasse have produced talk of joint development efforts in the contested area, but no willingness to compromise on the basic issues.

The dispute in the South China Sea is even more complex. Drawing on ancient maps and historical accounts, the Chinese and Taiwanese insist that the sea's two island chains, the Spratlys and the Paracels, were long occupied by Chinese fisherfolk, and so the entire region belongs to them. The Vietnamese also assert historical ties to the two chains based on long-term fishing activities, while the other littoral states each claim a 200-nautical mile EEZ stretching into the heart of the sea. When combined, these various claims produce multiple overlaps, in some instances with three or more states involved—but always including China and Taiwan as claimants. Efforts to devise a formula to resolve the disputes through negotiations sponsored by the Association of Southeast Asian Nations (ASEAN) have so far met with failure: While China has offered to negotiate one-on-one with individual states but not in a roundtable with all claimants, the other countries—mindful of China's greater wealth and power—prefer to negotiate en masse.

Again, the various claimants in these conflicts have, on a regular basis, employed military force to demonstrate their determination to retain control over the territories they have claimed and to deter economic activities in these areas by competing countries. Few such actions have resulted in bloodshed—one major exception was a 1988 clash between Chinese and Vietnamese warships near Johnson Reef in the Spratly Islands that resulted in the loss of more than 70 lives—but many have prompted countermoves by other countries, posing a significant risk of escalation. In September 2005, for example, Chinese warships patrolling along the median line claimed by Japan in the East China Sea aimed their guns at a Japanese Navy surveillance plane, nearly leading to a serious incident.

More such engagements have occurred in the South China Sea, where there are a larger number of claimants and greater uncertainty over the location of boundaries. In one such incident Vietnamese troops fired on a Philippine air force plane on a reconnaissance mission in the Spratlys; in another, Malaysian and Filipino aircraft came close to firing on each other while flying over a Malaysian-occupied reef in the Spratlys.

Recognizing the potential for escalation, leaders of the countries involved in such encounters have taken some steps to avert a serious clash. Chinese and Japanese officials have met on several occasions to discuss the boundary dispute in the East China Sea, pledging to avoid the use of force. Likewise, China and the 10 members of ASEAN signed a Joint Declaration on the Conduct of Parties in 2002, pledging to resolve their territorial disputes in the South China Sea by peaceful means. However, these measures have not prevented the major parties from continuing to employ military means to reinforce their bargaining positions. Worried that such activities could lead to more serious conflict, endangering vital US interests, the Obama administration has offered to act as a mediator—only to provoke a hostile response from Beijing, which sees this as an unwelcome form of American meddling in its backyard.

Girding for Conflict

If the East and South China Seas represent the most conspicuous cases of offshore territorial conflicts driven in large part by the competitive pursuit of energy resources, they are by no means the only ones with a potential to spark violence. Others that exhibit many of the same characteristics include quarrels over the Falklands/Malvinas, the eastern Mediterranean, and the Caribbean near Nicaragua and Colombia.

The dispute over the Falklands/Malvinas Islands and their surrounding waters, claimed both by Britain and Argentina, is well known from the 1982 war over the islands, in which the British defeated an invasion by Argentina. At the time, the primary impulses for conflict were thought to be national pride and the political fortunes of the key leaders involved: Margaret Thatcher in Britain, and an unpopular military junta in Argentina. Now, however, a new factor has emerged: competing claims to undersea energy reserves. Large reservoirs of oil are thought to lie beneath areas of the South Atlantic to the north and south of the islands, and both Argentina and Britain say the reserves belong exclusively to them. A number of companies have obtained permits from British and Falkland Islands authorities to sink test wells within a 200 mile EEZ surrounding the islands claimed by London after it ratified UNCLOS in 1997.

Until now, neither side has engaged in provocative military action of the sort seen in the other offshore disputes, but both sides appear to be girding for the possibility. The British have replaced older ships and aircraft in the Falklands with more modern equipment, including Typhoon combat aircraft of the type used during the 2011 Libyan campaign. The Argentines have responded by blocking access to Argentine ports for British cruise ships that first dock in the Falklands—a largely symbolic act, to be sure, but one that hints of stronger actions to come. How this will play out remains to be seen, but neither side has budged on any of the fundamental issues, and the prospect of significant oil production by British firms on what the Argentines consider to be their sovereign territory is bound to increase resentment in the years ahead.

The Eastern Mediterranean, like the Falklands/Malvinas, is also a site of earlier conflict. In addition to the recurring Arab-Israeli wars, there are the ongoing Greek-Turkish dispute over governance of Cyprus—the backdrop for a war in 1974—and a growing schism between Israel and Turkey. But now, again, the discovery of potentially vast energy reserves is aggravating traditional rivalries. The offshore Levant Basin, stretching from Cyprus in the north to Egypt in the south and bounded by Israel, Lebanon, and the Gaza Strip on the east, is thought to hold 120 trillion cubic feet of natural gas, and perhaps much more. Production of this gas could prove a boon to the nations involved—few of which have experienced any benefit from the oil boom in neighboring countries.

At this point, the most advanced projects are under way in Israeli-claimed territory. Noble Energy, a Houston-based firm, is developing a number of giant gas fields in waters off the northern port of Haifa. The largest of these, named Leviathan, lies astride the EEZ claimed by the Republic of Cyprus, where Noble has also found substantial gas reservoirs. Although significant hurdles remain, both Israel and Cyprus hope to extract natural gas from these fields by the middle of the decade and

to ship considerable volumes to Europe via new pipelines to be installed on the Mediterranean sea-bed, or in the form of liquefied natural gas.

Seeing the potential for cooperation in exporting gas, Israel and Cyprus have discussed common transportation options and signed a maritime border agreement in December 2011. But both countries face significant challenges from other nations in the region. The Leviathan field and other gas reservoirs being developed by Noble are located at the northern edge of the EEZ staked out by Israel, in waters also claimed by Lebanon. Lebanese authorities, who refuse to negotiate with Israel, have urged the UN to pressure Israel to recognize Lebanon's sovereignty over the area, but to no avail. Far more worrisome are threats by Hezbollah, the Iranian-backed Shiite militia based in Lebanon, to attack Israeli drilling rigs in waters claimed by the Lebanese. These threats have prompted Israel's air force to deploy drones over the facilities, allowing for a prompt response to any potential terrorist attack. Meanwhile, Noble's operations in Cypriot-claimed waters have been challenged by Turkey, which does not recognize the Republic of Cyprus or its claim to an EEZ. The Turks have deployed air and naval craft off the Turkish Republic of Northern Cyprus, an ethnic separatist entity that only they recognize, in what is viewed as an implied threat to Noble and other companies operating in the Cypriot EEZ.

In the Western Hemisphere, a dispute has arisen between Colombia and Nicaragua over a swath of the Caribbean claimed by both of them. On November 19, 2012, the International Court of Justice in The Hague awarded control of some 35,000 square miles of the Caribbean—believed to harbor valuable undersea reserves of oil and gas—to Nicaragua. The decision infuriated the Colombians, who rejected the ruling and withdrew from a pact recognizing the court's jurisdiction over its territorial disputes. Leaders of both countries have pledged to seek a peaceful resolution, but the situation remains tense. "Of course no one wants a war," said Colombian President Juan Manuel Santos. "That is a last resort."

Options for Resolution

As should now be evident, the accelerated pursuit of oil and gas reserves in disputed offshore territories entails significant potential for international friction, crisis, and conflict. This is so because such efforts combine unusually high economic stakes with intense nationalism and the absence of clearly defined boundaries. Add to this the lack of clearly defined mechanisms for resolving boundary disputes of this sort, and the magnitude of the problem becomes apparent. Unless a concerted effort is made to resolve these and other such disputes, what is now latent or low-level conflict could erupt into full-scale violence.

The problem is not a lack of viable solutions. In several contested maritime regions, countries that were unable to agree on their offshore boundaries have been able to establish joint development areas (JDAS) in which drilling has proceeded while negotiations continue regarding the demarcation of final borders. The first of these special zones, the Malaysia-Thailand Joint Development Area, was created in 1979 and has been producing gas since 2005: Vietnam has also become a party to an additional slice of the JDA. A similar formula has been adopted by Nigeria and the island state of São Tomé and Principe to develop offshore fields in a contested stretch of the Gulf of Guinea. China and Japan once agreed to employ a solution of this sort to develop the contested area claimed by both in the East China Sea, but so far little has come of the effort.

Meanwhile, UNCLOS, as amended, incorporates various measures for resolving disputes over the location of offshore territories. Essentially, it mandates that such disputes be resolved peacefully, through negotiations among the affected parties. UNCLOS also includes provisions for arbitration by third parties and referral of disputes to the International Court of Justice, or to the newly established International Tribunal for the Law of the Sea (based in Hamburg). Also, to help determine the validity of a state's claim to offshore territories based on the natural prolongation of its continental shelf, the UN has established a Commission on the Limits of the Continental Shelf.

However, all of these measures have limitations. For one thing, they do not apply to countries that have failed to ratify UNCLOS, such as Turkey and the United States. They have little effect, moreover, when contending states refuse to negotiate, as is the case with Israel and Lebanon; or eschew arbitration and outside involvement, as China has done in the East and South China Seas. Clearly, something more is needed.

What appears most lacking in all of these situations is a perception by the larger world community that disputes like these pose a significant threat to international peace and stability. Were these disputes occurring on land, one suspects, world leaders would pay much closer attention to the risks involved and take urgent steps to avoid military action and escalation. But because they are taking place at sea, away from population centers and the media, they seem to have attracted less concern.

This is a dangerous misreading of the perils involved: Because the parties to these disputes appear more inclined to employ military force than they might elsewhere, and boundaries are harder to define, the risk of miscalculation is greater, and so is the potential for violent confrontation. The risks can only grow as the world becomes more reliant on offshore energy and coastal states become less willing to surrender maritime claims.

To prevent the outbreak of serious conflict, the international community must acknowledge the seriousness of these disputes and call on all parties involved to solve them through peaceful means, as quickly as possible. This could occur through

resolutions by the UN Security Council, or statements by leaders meeting in such forums as the Group of 20 governments. Such declarations need not specify the precise nature of any particular outcome, but rather must articulate a consensus view that a resolution of some sort is essential for the common good. Arbitration by neutral, internationally respected "elders" can be provided as necessary. To facilitate this process, ambiguities in UNCLOS should be resolved and holdouts from the treaty—including the United States—should be encouraged to sign.

After Consensus

Assuming such a consensus can be forged, solutions to the various maritime disputes should be within reach. China and Japan should jointly develop the gas field in the disputed area of the East China Sea until a final boundary is adopted—an option already embraced in principle by the two countries. In the South China Sea, a JDA should be established on the model of the Malaysia-Thailand Joint Authority, consisting of representatives of all littoral states and empowered to award exploration contracts (and allocate revenues) on an equitable basis. A similar authority should oversee drilling in the waters surrounding the Falklands, the Israel-Lebanon offshore area, and the waters around Cyprus. At the same time, negotiations leading to a permanent border settlement in these areas should be undertaken under international auspices.

If the countries involved cannot agree to such measures, they should be pressured to submit their competing claims to an international tribunal with the authority to determine the final demarcation of boundaries, while international energy companies should be required to abide by the outcome of such decisions or face legal action and the possible loss of revenues.

Such measures are important for another reason: to help reduce the risk of environmental damage. As demonstrated by the Deepwater Horizon disaster of April 2010 in the Gulf of Mexico and more recent oil leakages from Brazil's pre-salt fields, deep offshore drilling poses a significant threat to the environment if not conducted under the most scrupulous production methods. Clearly, maritime areas that lack an accepted regulatory and jurisdictional regime, such as the South China Sea, are more likely to experience spills and other disasters than areas with well-established boundaries and effective supervision.

The establishment of clear maritime boundaries and the promotion of collaborative offshore enterprises rank among the most important tasks facing the international community as the global competition for resources moves from traditional areas of struggle, such as the Middle East, to seas where the rules of engagement are less defined. The exploitation of offshore oil and gas could help compensate for the decline of existing reserves on land, but will result in increased levels of friction and conflict unless accompanied by efforts to resolve maritime boundary disputes. Defining borders at sea may not be as easy as it is on land, where natural features provide obvious reference points, but it will become increasingly critical as more of the world's vital resources are extracted from the deep oceans.

Critical Thinking

1. How are these conflicts changing Japanese military policy?
2. How do these conflicts reflect the emergence of China as a military power?
3. What role is the United States likely to play in these maritime conflicts?

Create Central

www.mhhe.com/createcentral

Internet References

Japanese Ministry of Defense
www.mod.go.jp/e

Institute for Defence Studies and Analysis
www.idsa.in

Chinese Ministry of National Defense
http://eng.mod.gov.cn

Association of Southeast Asian Nations
www.aseansec.org

U.S. Department of Defense: Quadrennial Defense Review
www.defense.gov/qdr

MICHAEL T. KLARE, a *Current History* contributing editor, is a professor at Hampshire College and the author, most recently, of *The Race for What's Left: The Global Scramble for the World's Last Resources* (Metropolitan Books, 2012)

Article Prepared by: Robert Weiner, *University of Massachusetts, Boston*

Time to Negotiate in Afghanistan

JAMES DOBBINS AND CARTER MALKASIAN

Learning Outcomes

After reading this article, you will be able to:

- Identify missed opportunities for peacemaking in Afghanistan.
- Discuss U.S. opportunities for moving forward with peacemaking in Afghanistan.

How to Talk to the Taliban

Peace talks, if not peace itself, may be close at hand in Afghanistan. Over the past few months, Afghanistan, Pakistan, and the Afghan Taliban have made unexpected strides toward talks. In early May, members of the Taliban and the Afghan government even met in Qatar and expressed real interest in starting official negotiations—a heartening step.

Since 2001, opportunities for peace talks have come and gone. Sometimes, the process has stalled for political reasons, such as the United States' reticence to engage with the Taliban. Other times, discussions have broken down due to miscommunications or a lack of political consensus. It was not until 2010 that the United States fully embraced peace talks as the best way to end the violence in Afghanistan, and even then, progress was slow and halting.

But this time may be different. Ashraf Ghani, Afghanistan's new president, has placed peace talks at the center of his agenda. Pakistan and China both appear willing to help jumpstart the process. And the Taliban themselves have hinted that they may be willing to support an end to violence.

The United States must seize the moment, doing what it can to move the peace process forward. Washington will need to employ a mix of carrots and sticks while remaining committed to Afghanistan's security. It should help Afghan forces hold the line on the battlefield, pressure Pakistan to keep the Taliban at the table, and accept that in the end some concessions will be necessary. Most important, it will need to stay flexible on the withdrawal timeline and dedicated to supporting Afghanistan into 2017 and beyond.

Of course, peace talks may not yield a lasting peace. In 2007, the political scientist James Fearon noted that just 16 percent of civil wars and insurgencies end through a negotiated peace settlement. But even if negotiations are a long shot, they are the best option for Afghanistan and the United States. To stick with the status quo would be to consign Afghanistan to a long war of attrition that would ravage the country, upend regional stability, and strain the budgets of the United States and its allies.

One Step Forward, Two Steps Back

In December 2001, a group of high-ranking Taliban officials met with Hamid Karzai, the soon-to-be Afghan president, whose own anti-Taliban fighters were then advancing on Kandahar, the Taliban's southern capital. According to the journalists Anand Gopal and Bette Dam, the members of the delegation were willing to lay down their arms in return for immunity. They gave Karzai a letter—possibly signed by the Taliban's supreme commander, Mullah Omar—detailing how the Taliban might step down peacefully. The opportunity never came to anything. U.S. officials denied immunity to Mullah Omar, and U.S. and Afghan forces advanced precipitously on Kandahar City. Whether for these or other reasons, Mullah Omar and the bulk of the Taliban's leadership fled to fight another day. Angered by 9/11 and buoyed by its battlefield victories, the United States did not involve the Taliban in a postinvasion settlement.

In 2002, senior Taliban delegations reached out to Karzai once again. Karzai mentioned the contacts to U.S. officials, only to have the United States strongly discourage his government from negotiating with the Taliban. That same year, U.S. troops even imprisoned the former Taliban foreign minister, Wakil Ahmad Muttawakil, when he arrived in Kabul to meet with the Afghan government. By 2003, the Taliban had shifted their focus to taking territory, and once the Taliban offensives began in 2006, peace feelers fell away.

It was not until the last months of the Bush administration that peace talks regained momentum. Within the Taliban, a moderate faction had retained an interest in negotiations, and in 2008, Mullah Abdul Ghani Baradar, Mullah Omar's deputy, allowed subordinates to meet with Afghan government officials under Saudi auspices. He also began communicating directly with members of the Karzai family, who happen to be his fellow tribesmen. Around the same time, a Taliban delegation began meeting with Kai Eide, then the UN envoy to Afghanistan, in Dubai. But all conversations came to a halt in February 2010, when Pakistani officials detained Mullah Baradar in Karachi, a move widely interpreted as a Pakistani veto on direct negotiations between Kabul and the Taliban. As a Pakistani security official admitted to *The New York Times* in 2010: "We picked up Baradar . . . because [the Taliban] were trying to make a deal without us. We protect the Taliban. They are dependent on us. We are not going to allow them to make a deal with Karzai and the Indians."

Meanwhile, the idea of a negotiated peace, first championed within the administration of U.S. President Barack Obama by Richard Holbrooke, then Obama's special representative to Afghanistan and Pakistan, and Barnett Rubin, one of Holbrooke's top advisers, was gaining traction in the United States. In May 2010, Karzai visited Washington, and Obama lifted the Bush-era ban on talking to the Taliban leadership. As a result, a month later, Karzai held a loya jirga, or grand assembly, to discuss the possibility of peace negotiations. And in September, he created the High Peace Council, which would be the public face of his peace effort, a 70-member body led by former Afghan President Burhanuddin Rabbani and filled with Afghan mujahideen commanders and former Taliban members.

Around the same time, the White House encouraged Lakhdar Brahimi, the UN's former top official in Kabul, and Thomas Pickering, a former U.S. Undersecretary of State for Political Affairs, to examine the possibility for peace talks in Afghanistan. They led an international group of diplomats that traveled to Afghanistan and Pakistan and met with former and active Taliban representatives. They reported back to Washington that the Taliban were interested in the possibility of talks with the United States.

The ball was rolling. In November 2010, U.S. diplomats and Taliban representatives met for the first time, in Germany. In February 2011, U.S. Secretary of State Hillary Clinton announced that the United States was officially ready to begin peace negotiations, although she cautioned that any settlement would have to require the Taliban to lay down their arms, accept the Afghan constitution, and sever ties with al Qaeda. After some delay, talks between U.S. and Taliban representatives proceeded in late 2011 and continued into the early months of 2012, at which point the

Taliban broke off contact, rejecting a request from Washington that they begin negotiating with Kabul.

It was a particularly substantial missed opportunity: a failure to initiate a peace process at the peak of U.S. leverage, as NATO troops were retaking large swaths of the Taliban's heartland in Kandahar, Helmand, and nearby provinces. All parties were to blame. On the Afghan side, Karzai did his best to obstruct a process he feared would marginalize him and demanded that the Taliban speak to his government directly. The Taliban refused to negotiate with Kabul unless they first secured the release of several of their former leaders from the U.S. detention facility in Guantánamo Bay, Cuba. The United States, for its part, followed up on Clinton's initial offer cautiously, hindered by lengthy interagency wrangling and indecision. The Defense Department could not agree with the State Department on a variety of issues relating to the negotiations. General David Petraeus, for example, who commanded the NATO-led security mission in Afghanistan from 2010 to 2011, preferred to hold off on peace talks until the surge produced greater military success. Other Pentagon officials balked at the suggestion that the United States should release prisoners from Guantánamo in exchange for Bowe Bergdahl, a U.S. Army sergeant being held by the Taliban. The White House was slow to forge agreement on a way forward, and so the opportunity slipped away.

The "will they, won't they" saga continued into 2013, when the Taliban sent signals to Washington that they were willing to reopen peace talks and also to meet with the Afghan government. Through intermediaries in Qatar, the Taliban planned to open a political office in Doha dedicated to the negotiations. The initiative foundered at the last moment, however, due to a miscommunication. Taliban leaders knew that U.S. and Afghan officials refused to address them as representatives of the Islamic Emirate of Afghanistan, the name adopted by their former government. But they believed, based partly on discussions with Qatari officials, that they could use the title to describe themselves to the outside world. When it opened, the office displayed the flag of the Islamic Emirate and a sign with the name. The United States, having been assured by the Qatari government that the office would not describe itself as part of the Islamic Emirate, demanded that Qatari officials remove the flag and the sign. In response, the Taliban closed the office and cut off all contact with Washington and Kabul.

The experience taught both sides to be more careful when communicating through third parties. In 2014, working again through Qatari intermediaries, the United States and the Taliban were able to arrange the release of Bergdahl in return for the transfer of five former Taliban officials from Guantánamo to Doha, where they would remain for a year. The agreement was not perfect: it sparked a lively controversy in the United States over the legitimacy of the five-for-one exchange rate and whether Congress should have been notified in advance of the deal. But it did demonstrate to each side that the other could deliver on an agreement once reached. Neither side made any attempt to follow up on this success, however, and the momentum for peace talks stalled once again.

A Golden Opportunity

After a period of radio silence, the opportunity for peace talks reemerged suddenly in February of this year—and this time, the prospects of success may be better. That month, Pakistan's army chief, General Raheel Sharif, went to Kabul and told the newly elected Afghan president that the Taliban would be willing to begin official meetings with the Afghan government as early as the next month and that the Taliban were being told by Pakistani officials that it was no longer acceptable to carry on the war. Although months passed as Taliban moderates and hard-liners worked out what to do, in early May, ranking members of the Taliban met openly and unofficially with members of the Afghan High Peace Council in Qatar. During the meeting, the Taliban participants stressed their interest in peace talks and in reopening their Doha office.

A variety of factors make this particular opportunity more promising than the ones before. The first is new leadership in Kabul. Karzai had an embittered relationship with the United States. He was nearly a decade ahead of Washington in seeking to reach out to the Taliban, but by the time U.S. officials came around to his view, he no longer trusted them. Convinced that the United States wanted to cut a separate deal with the Taliban that would divide Afghanistan, Karzai sought to monopolize any talks with the group. He began to believe that the United States was deliberately sabotaging negotiations in an attempt to prolong the war and keep a U.S. military presence in the region. Other governments, such as France and Japan, tried to foster intraAfghan dialogue, but Karzai objected to these forums, which he felt reduced his government to simply another Afghan faction.

Ghani, who succeeded Karzai as president in late 2014, promises to be a different sort of leader. Both he and Abdullah Abdullah, the country's chief executive officer, campaigned on their support for a negotiated peace, and unlike Karzai, they appear willing to make concessions and work with other governments to get there. During a trip to Beijing last October, Ghani encouraged other governments to support his country's reconciliation process, implicitly endorsing China's desire to help launch peace talks. Ghani went on to discuss the peace process with representatives from China, Pakistan, and the United States.

The second promising development is Pakistan's positive attitude toward negotiations. Since 2002, Pakistan has offered the Taliban sanctuary, a place to rest, regroup, and hide. Pervez Musharraf, who served as Pakistan's president from 2001 to 2008, has admitted that his government purposely helped the Taliban in order to secure his country's interests in Afghanistan

and counter Indian influence in the region. In recent years, Pakistan's civilian and military leaders have pledged to end the practice, but little has changed. And although Pakistan has occasionally played a positive role in the reconciliation process—releasing Mullah Baradar, for example—it has never brought key Taliban leaders to the table.

That seems to be changing. True to Sharif's word, since February, Pakistani officials have been meeting with Taliban leaders and encouraging negotiations. Although Pakistan's leadership is divided over how hard to pressure the Taliban to seek peace, Islamabad appears to feel that it has more of a stake in a peaceful Afghanistan than originally thought. Without a plan for a negotiated peace, the departure of U.S. troops cannot end well for Pakistan. The drawdown might give the Taliban the opportunity to seize more ground, which would increase Pakistan's influence in Afghanistan. But the Afghan government would then almost certainly turn to India for money and arms, leaving Pakistan to fight a long-term proxy war against its rival—or, worse, accede to an Indian protectorate over northern Afghanistan. For Pakistan, this is debatably a worse outcome than a neutral Afghanistan committed to staying out of the Indian-Pakistani rivalry.

Taliban battlefield successes might have other drawbacks as well. The extremist threat to Pakistan could grow. Emboldened by such successes, the Afghan Taliban and the Pakistani Taliban might start collaborating more, and safe havens for Pakistani terrorists could emerge on the Afghan side of the border, a long-standing fear of the Pakistani government. That risk was underscored on December 16, 2014, when the Pakistani Taliban attacked the Army Public School in the northwestern Pakistani city of Peshawar, killing 132 schoolchildren.

If Pakistan is beginning to realize that it has more to gain from an Afghanistan led by Ghani than one led by the Taliban, the new Afghan government deserves part of the credit. Whereas Karzai let the Afghan-Pakistani relationship sour—in 2011, he even signed a strategic partnership agreement with India— Ghani has made an effort to reassure Islamabad, going so far as to take military action against the Pakistani Taliban and cancel a weapons deal with India. Still, it is too early to tell if Pakistan will stand fully behind peace. Not all Pakistani officials and military officers agree that rapprochement with Afghanistan is the best way to secure their country against India.

China has also played a role in galvanizing Pakistani support for peace talks. After Ghani's visit to Beijing, the Chinese government hosted Taliban delegations and offered Pakistan additional aid to encourage the Taliban to join the peace process. China's requests carry weight in Pakistan. The two countries have enjoyed a long and close bilateral relationship. China, for its part, has a strong interest in a stable Afghanistan, since it wants to prevent extremism from spreading to its western region of Xinjiang, which contains a large Muslim population. China also has mineral and energy investments in Afghanistan, and so it would lose out if the country were torn apart by a civil war. More broadly, as China grows into its status as a global superpower, it has been willing to play a greater role in promoting regional stability, especially as the United States steps back.

What the Taliban Want

Of the various players, the Taliban themselves may be the most reluctant to negotiate. A moderate faction, including members of the Quetta Shura (the movement's central organization) and influential religious leaders, wants to put an end to years of bloodshed. But other Taliban leaders, such as Mullah Omar's current deputy, Mullah Akhtar Muhammad Mansour, have taken a harder line. Having observed the Taliban's post-2001 comeback, Mansour believes the movement has a chance of outright victory in a protracted war. News reports suggest that it is this internal divide that has slowed the Taliban's coming to the table.

Whether moderate or hard-line, the Taliban have not stopped fighting, nor are they likely to do so before any negotiations are concluded. In 2014, the Quetta Shura launched its biggest offensive in years, pushing back Afghan forces in the southern province of Helmand and striking the provinces of Kandahar, Kunduz, and Nangarhar. Our contacts in the Taliban say they expect to take more ground this year and next, including provincial capitals. If outright victory on the battlefield seems feasible, Taliban leaders will be unlikely to negotiate. Pakistan and China may have leverage over the Taliban, but the Quetta Shura will be sure to resist foreign pressure that it sees as outside its interests.

If the Taliban do decide to participate in peace talks, the next question will be how much they will concede. According to some Afghanistan experts, such as Thomas Ruttig, Michael Semple, and Theo Farrell, the Taliban may be willing to meet the most important of the three U.S. conditions for peace: the renunciation of al Qaeda. Plenty of Taliban leaders have denied any desire to wage international jihad, and in 2009, the Quetta Shura announced that if foreign forces left Afghanistan, the Taliban would not seek to attack other countries, nor would they let outside terrorist groups use Afghanistan as a base of operations. The Taliban have also made clear, however, that they will officially renounce al Qaeda only once they have gotten what they want out of a peace deal.

A bigger sticking point involves the Afghan constitution. For many in the Taliban, the demand that they accept it is untenable, since doing so would force them to cede the legitimacy of what they see as a puppet regime. The Taliban will also want to elect a new government, in which they will expect to participate. In this sense, a peace agreement would mean not merely a cease-fire but also a reconceptualization of the Afghan state.

The Taliban's other major demand is likely to be the removal of all U.S. forces from Afghanistan. Foreign occupation is a major reason the Taliban's rank and file fight. At the May meeting in Qatar, Taliban participants allegedly said that they would accept a cease-fire only after the withdrawal of all foreign forces. Given the salience of this issue, there can be little doubt that the initial Taliban position in any negotiations will be that all U.S. troops must leave.

Of course, hard-liners within the Taliban—or even within outside groups, such as the self-proclaimed Islamic State, or ISIS—could always take matters into their own hands. If extremists assassinated Mullah Omar, for example, negotiations would collapse. Although the Islamic State currently has little influence in Afghanistan, the death of a leader such as Mullah Omar could allow the group to gain a foothold, win over extremists, and carry on an even more violent and vicious war.

War and Peace

A tiny window of opportunity for a negotiated settlement has opened up, and the United States should take advantage of it while it can. Although all sides agree that the talks should be led by Afghanistan, at least three outside powers—China, Pakistan, and the United States—will be directly or indirectly involved. The United States, for its part, can take five concrete steps to keep the negotiations moving forward.

First, it must do its best to prevent large-scale Taliban military victories. Peace begins on the battlefield: if the Taliban capture more ground, particularly provincial capitals, the Quetta Shura will see little reason to bargain, believing that an Afghan government defeat is imminent. The summer fighting season will be particularly critical to Taliban decision-making, as the leadership will take note of successes and failures on the battlefield to decide whether war will be more profitable than peace. A strong performance by the Afghan army could therefore deal a serious blow to the Taliban's confidence, pushing the peace process forward.

To beef up Afghan military capabilities, the United States and its allies should continue to provide financial and material support until a settlement is reached, and possibly beyond. Obama made the right decision in March, when he granted Afghan requests to slow the drawdown of U.S. troops from the country, promising to maintain a force of 9,800 through the end of 2015. He should be just as flexible when it comes to drawdowns in 2016 and 2017. Obama should also continue to grant U.S. forces the authority to carry out limited special operations and air strikes, both of which give the Afghan army and police a strategic edge. Strikes against Quetta Shura members in Afghanistan and Pakistan should not be ruled out, especially so that additional pressure can be brought to bear in the course of the negotiations, if needed.

Second, the United States should weigh in behind the scenes to help Ghani and Abdullah form a disciplined government, capable of the executive action necessary to wage war and broker peace. So far, the Afghan government has been a model of indecision. It took Ghani and Abdullah seven months just to choose their cabinet. Such gridlock, whether over cabinet posts or military policy, emboldens the Quetta Shura. A weak, disjointed government will undermine peace talks. The United States, along with the rest of the international community, should continue to press both camps to work together more effectively.

The third area in which the United States can help involves Pakistan. Washington should do what it can to ensure that Islamabad keeps the Taliban at the bargaining table. The United States has many interests in Pakistan—including securing Pakistan's nuclear weapons and working with Islamabad to weed out al Qaeda—that have distracted it from focusing on ending Pakistan's support for the Taliban. Luckily, the drawdown in U.S. forces will largely eliminate one of these interests: the U.S. military's dependence on Pakistani ports and roads to support its presence in Afghanistan. Washington should condition its substantial military and civilian assistance on Pakistan's agreeing to support the peace process and deny a safe haven to the Taliban.

Fourth, the United States must accept that a workable peace settlement will have to include a new Afghan constitution or institutional arrangements that allow the Taliban to become a legitimate part of the Afghan government. In fact, Washington should assume that a settlement will provide for a loya jirga in which representatives of the Taliban, the Afghan government, and civil society come together to amend the current constitution or write an entirely new one. In such a restructuring, certain civil freedoms, particularly women's rights, would be endangered. The Taliban hold deeply conservative views on women, to put it mildly. Prior to any cease-fire, therefore, the United States should seek to secure from all parties a commitment to leave current civil rights protections unchanged in a new constitution.

The fifth step will come if and when a settlement is reached. At that point, the United States may need to keep troops on the ground only until the constitutional debate is over and any subsequent election has taken place. But even when its troops have departed, the United States should remain committed to a strategic partnership with Afghanistan and continue to provide a base level of military aid. Otherwise, the balance of power may shift to the Taliban, undoing the peace.

Most Afghanistan experts believe that the war will continue for years to come. They generally agree that the Afghan government will stay in power only with continued U.S. economic and military assistance, without which violent militant groups will reign freely. The peace process offers an alternative future, one that the United States should pursue with determination and patience. Success is far from guaranteed—in fact, it's a long shot-but the attempt is worth the effort.

The alternatives would be costly. One is to keep paying for the Afghan security forces, at between $2 billion and $5 billion a year, and let the war go on. In this scenario, an outright government victory would be unlikely, even if the Obama administration left military forces in Afghanistan past 2016. Another option is for the United States to get out of Afghanistan, cut off funding, and accept the attendant Taliban resurgence in Kabul. In either case, the United States might be tempted to bet that the mutual interest of the Afghan government, Pakistan, and China in avoiding regional instability will ultimately bring peace. That would be quite a gamble. Without U.S. pressure on all players, negotiations may never happen, and a full-blown civil war may become inevitable. In that event, extremism would grow: there is little evidence that the Taliban would unilaterally break from al Qaeda or be able to stop al Qaeda or the Islamic State from operating in Afghanistan. And if Iraq is any lesson, even total withdrawal may not prevent the United States from being sucked right back into the morass.

Critical Thinking

1. Why do the authors think that China and Pakistan now support peace negotiations in Afghanistan?

2. What internal differences affect the Taliban's position on peace negotiations?

3. Why did President Karzai distrust the U.S.?

4. What is the role of India in the conflict?

Internet References

Afghanistan Research and Evaluation Unit
http://www.areu.org/af/default.aspx?Lang=en-us

The Afghanistan Analysts Network
www.afghanistan-analysts.org/publications

The Taliban in Afghanistan
http://www.cfr.org/afghanistan/taliban-afghanistan/p10551

JAMES DOBBINS is a Senior Fellow and Distinguished Chair in Diplomacy and Security at the RAND Corporation. He represented the United States at the 2001 Bonn conference and from 2013 to 2014 served as the Obama administration's Special Representative for Afghanistan and Pakistan. **CARTER MALKASIAN** works at the Center for Naval Analyses and is the author of *War Comes to Garmser: Thirty Years of Conflict on the Afghan Frontier.* He was Senior Political Adviser to General Joseph Dunford, commander of U.S. forces in Afghanistan from 2013 to 2014.

Article

Prepared by: Robert Weiner, *University of Massachusetts, Boston*

Water Wars

A Surprisingly Rare Source of Conflict

GREGORY DUNN

Learning Outcomes

After reading this article, you will be able to:

- Understand the problems caused by the growing scarcity of freshwater.

- Understand what has contributed to the growing scarcity of freshwater.

Water seems an unlikely cause of war, but many commentators believe it could define 21st century conflict. A February 2013 article in U.S. News and World Report warns that "the water-war surprises will come," and laments that "traditional statesmanship will only take us so far in heading off water wars." A 2012 article in Al Jazeera notes that "strategists from Israel to Central Asia" are preparing for strife caused by water conflict. Even the United States National Intelligence Estimate predicts wars over water within 10 years. Their concern is understandable—humanity needs fresh water to live, but a rise in population coupled with a fall in available resources would seem to be a perfect catalyst for conflict. This thinking, although intuitively appealing, has little basis in reality—humans have contested water supplies for ages, but disputes over water tend to be resolved via cooperation, rather than conflict. Water conflict, rather than being a disturbing future source of conflict, is instead a study in the prevention of conflict through negotiation and agreement.

To understand the problems with arguments about the importance of water wars, it is first important to understand the arguments themselves. Drinking water is fundamentally necessary for humans to survive, and thus every human needs a reliable source of water to survive. If people are denied access to water they face death, and thus are more likely to go to war—even

a war with only a small chance of resulting in access to water is preferable to certain death through dehydration. In ancient times, this sort of calculus was not necessary, since migration allowed humans to travel to areas that had water if water supplies were exhausted or inaccessible. However, the development of nations, cities, and governments has restricted the extent to which humans can migrate in pursuit of clean water. Additionally, in some areas—notably, the deserts of the Middle East and Africa—water may be so scarce that migration is futile. Additionally, industrial growth has exacerbated water scarcity in some areas. Dammed rivers, water diversion for irrigation, the extraction of water from underground aquifers, and the pollution of water supplies has made water even scarcer for some, and, critically, climate change threatens to dry up many people's sources of water. As water becomes scarcer, people without access to water resources face the choice of fighting or dying of dehydration, and water wars erupt. These wars are not necessarily world-encompassing conflagrations, but they are deadly conflicts between armed parties spurred by water scarcity. This logic of calamity driven by resource scarcity is in many ways simply an updated version of resource scarcity-based apocalypse that have been around since Malthus.

However, a casual look at dryer areas of the world suggests that Malthusian resource scarcity might finally be occurring. In East Africa, diplomatic rows between nations along the Nile grow increasingly heated, and lack of access to water fuels Somalia's conflict and division. Many of the governments in this region have been or are currently being threatened by insurgencies, waging war against the government and thus the current system of resource allocation. Southern Asia, Pakistan, India, and Bangladesh all face issues with regards to water, and the Southern Asian region remains a source of conflict and instability. Even in the developed United States of America, drug wars rage in the Southwest of the country, a desert region supplied by rivers whose water is increasingly diverted for agricultural purposes.

Given these seemingly disturbing conditions, it is not surprising that the United States National Intelligence Estimate on Water, one of the most useful documents for understanding how nations think about water issues, predicts that beyond the year 2022 upstream nations are likely to use their ability to control water supplies coercively, and water scarcity "will likely increase the risk of instability and state failure, exacerbate regional tensions and distract countries from working with the United States on important policy objectives." There is little doubt that climate change will deny people access to the water they need to survive, which seems a convincing argument that future conflict will occur to secure this valuable resource.

A Familiar Concern

However, this analysis does not take into account the economic, geopolitical, and governmental contexts that such changes will occur in. Economic growth, international organizations, and political leaders are powerful forces that dampen the tendency for water scarcity to cause conflict. The most powerful reason why the future does not hold water wars is the reason typically used to refute Malthusian arguments—technological and economic growth. Malthus correctly predicted the explosion in human population, and the amount of humans on earth would increase by five billion by the year 2000. However, the collapse of society Malthus envisioned failed to occur. The failure of human society to collapse was largely due to the economic and technological developments that occurred around the world. Economic growth allowed more access to resources, thus enabling people to invest in technology to increase their productivity. This investment in technology enabled incredible leaps in the productivity of farmers, thanks to devices like tractors, new practices in irrigation and crop rotation, and improvements in crops due to breeding and genetic modification. Although the data is somewhat inconclusive, estimates in literature reviewing the increase in farming productivity agree that farm productivity has increased many times over since the publication of the Essay on the Principle of Population, thus averting the collapse Malthus predicted.

A similar line of thought can he applied to water. Currently, many people access water from wells or rivers, sources that are susceptible to environmental changes. However, technological and economic growth allows for the development of aqueducts to service areas with little water, and the adoption of more efficient methods of using water (notably, watering plants with drip irrigation results in substantially less water loss), resulting in greater water availability. As evidenced by the development of the arid West of the United States, a lack of water does

not necessarily mean that humans cannot survive, it merely means that technology and capital is required for survival. As nations continue to grow economically, they can acquire more resources and develop new technologies, such as water sanitation and treatment or desalination, to give their people better access to water, thus decreasing water scarcity over time. In fact, the University of California, San Diego's Erik Gartzke notes that global warming is associated with a reduction, rather than increase, in interstate conflict. He goes on to note that while resource depletion associated with global warming may contribute to instability, the economic growth that is associated with it results in an overall reduction of crime. Gartzke concludes that the only way climate-induced conflict might come about is if efforts to stem global warming at the expense of economic growth lead to a loss of wealth, and thus conflict. Although water scarcity may be a factor that can cause conflict, the economic development associated with modem water scarcity results in more peace, not more war. As nations develop, they gain the technology by which they can mitigate the effects of climate change, and the capital with which to implement these technological advances.

Modern times are associated with increasing rates of water depletion, but also with a rise of international institutions, diplomacy, and conflict mediation. History has shown that these forces are not always powerful enough to overcome wars fought for political or strategic reasons (notably, the Iraq war was launched to destroy the military threat of Weapons of Mass Destruction). However, water scarcity is a problem related to economic development. Thus, wars associated with water scarcity are not based in the wishes of leaders, but rather a failure of environment or leadership. International organizations are able to respond to a nation's failures, and leaders are generally willing to receive aid to complete tasks they have been unable to accomplish. Failures in water supply and distribution can be remedied with aid, which can install wells, aqueducts, and water purification facilities to improve access to clean water. Additionally, educational aid can help develop better practices for water use and conservation in an area of water scarcity. A large proportion of drinkable water is wasted or contaminated before it is available to those who need it to survive, a problem that can be solved through proper education and infrastructure development.

Examples of the power of aid to solve water issues are plentiful. In the United States, the state of California used federal assistance to construct an aqueduct from the wet North of the state to the arid South, allowing the city of Los Angeles to prosper as well as providing water to farmers along the fertile Central Valley of California. The international Non-Governmental Organization WaterAid approached the city of Takkas, in Nigeria. They installed wells, latrines, and instructed locals in best practices with regards to sanitation, resulting in a decrease in waterborne disease and an increase in water availability and thus quality of life. However, doubts about the long-term sustainability of water development projects remain since many nations do not have the capability to perform maintenance on the facilities provided to them. Thus, in terms of development, aid serves as a stopgap measure, providing critical water resources until economic growth allows nations to develop the infrastructure to indigenously refine and maintain water infrastructure. However, with regards to war, water aid is extremely effective, since temporary aid can be used to reduce tempers in the short term. Although a series of stop-gap measures is not substitute for indigenous production and maintenance of water supplies, stop-gap measures can prevent the humanitarian issue of water scarcity from causing international conflicts.

Although international aid and involvement are effective tools in development assistance, international aid is perhaps even more effective in aiding negotiations regarding the provision of water. Conflict over water is relatively easy to detect, since water scarcity builds over time. International tensions regarding water trigger a series of escalating diplomatic incidents and concerns that are easy to identify and thus resolve. Since the potential conflict is over a future where one or more parties lack access to water, rather than a nation's immediate needs, international organizations can foster negotiations to solve the problem before it gets out of hand.

Not Water Wars, Water Deals

Perhaps the best example of international organization facilitating water resource allocation is the Indus Waters Treaty. The Indus River, a key source of water for Pakistan, has headwaters and tributaries in both Pakistan and India. When the partition between India and Pakistan occurred, there was great animosity between the two nations, which eventually led to a series of wars. One future source of conflict was the Indus River, a river whose resources were contested by two bitter rivals. While in the late 1950s Pakistan and India were not at war, there was great potential for water to play a role in future hostilities between the nations, perhaps exacerbating conflict. At the time, the World Bank was playing an active role in the region, seeking to aid the development of the new countries. They held substantial sway in the region thanks to their ability to provide loans to the new nations, and were therefore able to bring both India and Pakistan to the negotiating table to determine use of the river. Pakistan was concerned that India could use water as a weapon in future conflict, while India was concerned that Indians (especially those in the north of the country) would be unable to access water resources that had historically been theirs. Over a period of 6 years from 1954

to 1960, the World Bank helped orchestrate talks which determined which river systems were under control of India, which systems were under control of Pakistan, and how infrastructure necessary for the control of water in the river system was to be developed and funded. In 1960, thanks in part to development assistance provided by the United States and the United Kingdom, an agreement was found and the treaty was signed. After the signing of the treaty, three wars occurred, but the treaty was not broken, a testament to the power of the international agreement. Water allocation difficulties are a problem of developing nations, since developed nations can make up for scarcity with infrastructure. Thus, developing nations are most prone to water conflict, but they are also in the most need of staying in the good graces of the international community. Therefore, these countries are quick to negotiate with international organizations, making treaties and negotiation a powerful tool in addressing water conflict.

> "...the government has little incentive to start a war over water shortages impacting those the state is already failing—their protests are inevitable, and the shortages do not impact the government."

Furthermore, the involvement of international organizations can redirect anger, turning potential conflicts into political matters. In 2000, the World Bank compelled Bolivia to privatize the water provider in Cochabamba, a large Bolivian city, to fund the construction of a dam. This move proved massively unpopular, sparking widespread riots. This anger over the provision of water was not directed at the Bolivian government, but instead the anger was directed at the World Bank, an international organization that mainly interacted with Bolivia through financial, rather than physical means. The World Bank and the privatized companies it endorsed became the targets, and thus rage was harmlessly fired at an international organization, rather than targeted upon the Bolivian government. In this way, international organizations served as a scapegoat, absorbing criticism in the place of the government, which was left alone to maintain the peace.

The government of Bolivia, like many governments in region susceptible to water conflict, was not itself affected by the water scarcity. Governments have the power, resources, and authority to find and secure water in their country, and a water shortage is generally unlikely to severely affect those within a government. Rather, a water shortage is felt most acutely by those with almost no power, little money, and few resources. Water shortages hit the poorest hard, and the government is slow to respond since governmental officials are generally not impacted by such shortages. While this might seem at first consideration like a factor that is more likely to exacerbate water conflicts by allowing scarcity to rise undetected, it is ultimately a major dampener on the chances of water war. While individual citizens may protest their condition, and in extreme cases mount ail insurgency, these actions are unlikely to have a substantial effect on the country. The most powerless in a country already have much to protest about, and the addition of water scarcity is unlikely to dramatically alter the frequency or fervor of protest. The government of a nation must expect that some citizens cannot be fully provided for, and therefore protests are inevitable. The propensity of water shortages to impact this segment of the population means that the net effect of water shortage will be relatively small, reducing the necessity of the government to respond to the crisis. Even an insurgency will be mounted by those with many grievances and few resources, which makes the insurgency comparatively simple to combat. Critically, the government has little incentive to start a war over water shortages impacting those the state is already failing—their protests are inevitable, and the shortages do not impact the government. While water shortages will of course trigger mass protest if enough of the population is impacted, the tendency of water shortages to prey upon the most vulnerable makes the onset of such mass protest less likely.

The idea of water wars fits many contemporary narratives well. In an era where we are forced to face the consequences of economic growth—pollution, climate change, and unrest—water wars seems a convenient instance of our failure to properly safeguard our natural resources. While it is easy to think of local consequences of the corruption of natural resources (for example, lung cancer resulting from air pollution), it is more difficult to give examples of widespread social change spurred by pollution. Despite a litany of international conferences issuing increasingly urgent manifestos demanding dramatic change, society has changed its patterns of consumption comparatively little, with seemingly few more widespread societal (rather than local) consequences. Although global warming threatens to destroy our way of life, society has not responded to the impacts of a warmer climate. Water wars seem to make up for this lack of action, since they are a powerful social problem easily attributable to the degradation of national resources. However, they have so far failed to meaningfully transpire, thanks to the very forces—the international geopolitical order and economic growth—that would presumably cause water wars in the first place. While the degradation of natural resources is a serious problem with modern society, the lack of water wars serves as a reminder of the power of the forces of peace and prosperity that are an inherent part of the modern world.

Critical Thinking

1. Why was it expected that the growing scarcity of water in the 21st century would lead to growing conflict?

2. What factors explain the low incidence of water wars?

3. Provide an example of a water war.

Create Central

www.mhhe.com/createcentral

Internet References

International Freshwater Treaties
http://ocid.macse.org/4fdd/treaties.php

The International Water Events Data Base
http://www.transboundarywaters.orst.edu/database/event_bar_scale.html

Article Prepared by: Robert Weiner, *University of Massachusetts, Boston*

Taiwan's Dire Straits

JOHN J. MEARSHEIMER

Learning Outcomes

After reading this article, you will be able to:

- Understand how to apply the theory of offensive realism to the rise of China.

- Understand the Chinese approach to world order.

W hat are the implications for Taiwan of China's continued rise? Not today. Not next year. No, the real dilemma Taiwan will confront looms in the decades ahead, when China, whose continued economic growth seems likely although not a sure thing, is far more powerful than it is today.

Contemporary China does not possess significant military power; its military forces are inferior, and not by a small margin, to those of the United States. Beijing would be making a huge mistake to pick a fight with the American military nowadays. China, in other words, is constrained by the present global balance of power, which is clearly stacked in America's favor.

But power is rarely static. The real question that is often overlooked is what happens in a future world in which the balance of power has shifted sharply against Taiwan and the United States, in which China controls much more relative power than it does today, and in which China is in roughly the same economic and military league as the United States. In essence: a world in which China is much less constrained than it is today. That world may seem forbidding, even ominous, but it is one that may be coming.

It is my firm conviction that the continuing rise of China will have huge consequences for Taiwan, almost all of which will be bad. Not only will China be much more powerful than it is today, but it will also remain deeply committed to making Taiwan part of China. Moreover, China will try to dominate Asia the way the United States dominates the Western Hemisphere, which means it will seek to reduce, if not eliminate, the American military presence in Asia. The United States, of course, will resist mightily, and go to great lengths to contain China's growing power. The ensuing security competition will not be good for Taiwan, no matter how it turns out in the end. Time is not on Taiwan's side. Herewith, a guide to what is likely to ensue between the United States, China, and Taiwan.

I n an ideal world, most Taiwanese would like their country to gain de jure independence and become a legitimate sovereign state in the international system. This outcome is especially attractive because a strong Taiwanese identity—separate from a Chinese identity—has blossomed in Taiwan over the past 65 years. Many of those people who identify themselves as Taiwanese would like their own nation-state, and they have little interest in being a province of mainland China.

According to National Chengchi University's Election Study Center, in 1992, 17.6 percent of the people living in Taiwan identified as Taiwanese only. By June 2013, that number was 57.5 percent, a clear majority. Only 3.6 percent of those surveyed identified as Chinese only. Furthermore, the 2011 Taiwan National Security Survey found that if one assumes China would not attack Taiwan if it declared its independence, 80.2 percent of Taiwanese would in fact opt for independence. Another recent poll found that about 80 percent of Taiwanese view Taiwan and China as different countries.

However, Taiwan is not going to gain formal independence in the foreseeable future, mainly because China would not tolerate that outcome. In fact, China has made it clear that it would go to war against Taiwan if the island declares its independence. The antisecession law, which China passed in 2005, says explicitly that "the state shall employ nonpeaceful means and other necessary measures" if Taiwan moves toward de jure independence. It is also worth noting that the United States does not recognize Taiwan as a sovereign country, and according to President Obama, Washington "fully supports a one-China policy."

Thus, the best situation Taiwan can hope for in the foreseeable future is maintenance of the status quo, which means de facto independence. In fact, over 90 percent of the Taiwanese surveyed this past June by the Election Study Center favored maintaining the status quo indefinitely or until some later date.

The worst possible outcome is unification with China under terms dictated by Beijing. Of course, unification could happen in a variety of ways, some of which are better than others. Probably the least bad outcome would be one in which Taiwan ended up with considerable autonomy, much like Hong Kong enjoys today. Chinese leaders refer to this solution as "one country, two systems." Still, it has little appeal to most Taiwanese. As Yuan-kang Wang reports: "An overwhelming majority of Taiwan's public opposes unification, even under favorable circumstances. If anything, longitudinal data reveal a decline in public support of unification."

In short, for Taiwan, de facto independence is much preferable to becoming part of China, regardless of what the final political arrangements look like. The critical question for Taiwan, however, is whether it can avoid unification and maintain de facto independence in the face of a rising China.

What about China? How does it think about Taiwan? Two different logics, one revolving around nationalism and the other around security, shape its views concerning Taiwan. Both logics, however, lead to the same endgame: the unification of China and Taiwan.

The nationalism story is straightforward and uncontroversial. China is deeply committed to making Taiwan part of China. For China's elites, as well as its public, Taiwan can never become a sovereign state. It is sacred territory that has been part of China since ancient times, but was taken away by the hated Japanese in 1895—when China was weak and vulnerable. It must once again become an integral part of China. As Hu Jintao said in 2007 at the Seventeenth Party Congress: "The two sides of the Straits are bound to be reunified in the course of the great rejuvenation of the Chinese nation."

The unification of China and Taiwan is one of the core elements of Chinese national identity. There is simply no compromising on this issue. Indeed, the legitimacy of the Chinese regime is bound up with making sure Taiwan does not become a sovereign state and that it eventually becomes an integral part of China.

The continuing rise of China will have huge consequences for Taiwan, almost all of which will be bad.

Chinese leaders insist that Taiwan must be brought back into the fold sooner rather than later and that hopefully it can be done peacefully. At the same time, they have made it clear that force is an option if they have no other recourse.

The security story is a different one, and it is inextricably bound up with the rise of China. Specifically, it revolves around a straightforward but profound question: How is China likely to behave in Asia over time, as it grows increasingly powerful? The answer to this question obviously has huge consequences for Taiwan.

The only way to predict how a rising China is likely to behave toward its neighbors as well as the United States is with a theory of great-power politics. The main reason for relying on theory is that we have no facts about the future, because it has not happened yet. Thomas Hobbes put the point well: "The present only has a being in nature; things past have a being in the memory only; but things to come have no being at all." Thus, we have no choice but to rely on theories to determine what is likely to transpire in world politics.

My own realist theory of international relations says that the structure of the international system forces countries concerned about their security to compete with each other for power. The ultimate goal of every major state is to maximize its share of world power and eventually dominate the system. In practical terms, this means that the most powerful states seek to establish hegemony in their region of the world, while making sure that no rival great power dominates another region.

To be more specific, the international system has three defining characteristics. First, the main actors are states that operate in anarchy, which simply means that there is no higher authority above them. Second, all great powers have some offensive military capability, which means they have the wherewithal to hurt each other. Third, no state can know the intentions of other states with certainty, especially their future intentions. It is simply impossible, for example, to know what Germany's or Japan's intentions will be toward their neighbors in 2025.

In a world where other states might have malign intentions as well as significant offensive capabilities, states tend to fear each other. That fear is compounded by the fact that in an anarchic system there is no night watchman for states to call if trouble comes knocking at their door. Therefore, states recognize that the best way to survive in such a system is to be as powerful as possible relative to potential rivals. The mightier a state is, the less likely it is that another state will attack it. No Americans, for example, worry that Canada or Mexico will attack the United States, because neither of those countries is strong enough to contemplate a fight with Uncle Sam.

But great powers do not merely strive to be the strongest great power, although that is a welcome outcome. Their ultimate aim is to be the hegemon—which means being the only great power in the system.

What exactly does it mean to be a hegemon in the modern world? It is almost impossible for any state to achieve global hegemony, because it is too hard to sustain power around the globe and project it onto the territory of distant great powers. The best outcome a state can hope for is to be a regional hegemon, to dominate one's own geographical area. The United States has been a regional hegemon in the Western Hemisphere since about 1900. Although the United States is clearly the most powerful state on the planet today, it is not a global hegemon.

States that gain regional hegemony have a further aim: they seek to prevent great powers in other regions from duplicating their feat. Regional hegemons, in other words, do not want peer competitors. Instead, they want to keep other regions divided among several great powers, so that those states will compete with each other and be unable to focus their attention and resources on them. In sum, the ideal situation for any great power is to be the only regional hegemon in the world. The United States enjoys that exalted position today.

What does this theory say about how China is likely to behave as it rises in the years ahead? Put simply, China will try to dominate Asia the way the United States dominates the Western Hemisphere. It will try to become a regional hegemon. In particular, China will seek to maximize the power gap between itself and its neighbors, especially India, Japan, and Russia. China will want to make sure it is so powerful that no state in Asia has the wherewithal to threaten it.

It is unlikely that China will pursue military superiority so it can go on a rampage and conquer other Asian countries, although that is always possible. Instead, it is more likely that it will want to dictate the boundaries of acceptable behavior to neighboring countries, much the way the United States lets other states in the Americas know that it is the boss.

An increasingly powerful China is also likely to attempt to push the United States out of Asia, much the way the United States pushed the European great powers out of the Western Hemisphere in the 19 century. We should expect China to come up with its own version of the Monroe Doctrine, as Japan did in the 1930s.

These policy goals make good strategic sense for China. Beijing should want a militarily weak Japan and Russia as its neighbors, just as the United States prefers a militarily weak Canada and Mexico on its borders. What state in its right mind would want other powerful states located in its region? All Chinese surely remember what happened in the previous two centuries when Japan was powerful and China was weak.

Furthermore, why would a powerful China accept U.S. military forces operating in its backyard? American policy makers, after all, go ballistic when other great powers send military forces into the Western Hemisphere. Those foreign forces are invariably seen as a potential threat to American security. The same logic should apply to China. Why would China feel safe with U.S. forces deployed on its doorstep? Following the logic of the Monroe Doctrine, would China's security not be better served by pushing the American military out of Asia?

Why should we expect China to act any differently than the United States did? Are Chinese leaders more principled than American leaders? More ethical? Are they less nationalistic? Less concerned about their survival? They are none of these things, of course, which is why China is likely to imitate the United States and try to become a regional hegemon.

What are the implications of this security story for Taiwan? The answer is that there is a powerful strategic rationale for China—at the very least—to try to sever Taiwan's close ties with the United States and neutralize Taiwan. However, the best possible outcome for China, which it will surely pursue with increasing vigor over time, would be to make Taiwan part of China.

Unification would work to China's strategic advantage in two important ways. First, Beijing would absorb Taiwan's economic and military resources, thus shifting the balance of power in Asia even further in Chinas direction. Second, Taiwan is effectively a giant aircraft carrier sitting off China's coast; acquiring that aircraft carrier would enhance China's ability to project military power into the western Pacific Ocean.

In short, we see that nationalism as well as realist logic give China powerful incentives to put an end to Taiwan's de facto independence and make it part of a unified China. This is clearly bad news for Taiwan, especially since the balance of power in Asia is shifting in China's favor, and it will not be long before Taiwan cannot defend itself against China. Thus, the obvious question is whether the United States can provide security for Taiwan in the face of a rising China. In other words, can Taiwan depend on the United States for its security?

Let us now consider America's goals in Asia and how they relate to Taiwan. Regional hegemons go to great lengths to stop other great powers from becoming hegemons in their region of the world. The best outcome for any great power is to be the sole regional hegemon in the system. It is apparent from the historical record that the United States operates according to this logic. It does not tolerate peer competitors.

During the 20th century, there were four great powers that had the capability to make a run at regional hegemony:

Imperial Germany from 1900 to 1918, Imperial Japan between 1931 and 1945, Nazi Germany from 1933 to 1945 and the Soviet Union during the Cold War. Not surprisingly, each tried to match what the United States had achieved in the Western Hemisphere.

How did the United States react? In each case, it played a key role in defeating and dismantling those aspiring hegemons.

An increasingly powerful China is likely to attempt to push the United States out of Asia, much the way the United States pushed the European great powers out of the Western Hemisphere.

The United States entered World War I in April 1917 when Imperial Germany looked like it might win the war and rule Europe. American troops played a critical role in tipping the balance against the Kaiserreich, which collapsed in November 1918. In the early 1940s, President Franklin Roosevelt went to great lengths to maneuver the United States into World War II to thwart Japan's ambitions in Asia and Germany's ambitions in Europe. The United States came into the war in December 1941, and helped destroy both Axis powers. Since 1945, American policy makers have gone to considerable lengths to put limits on German and Japanese military power. Finally, during the Cold War, the United States steadfastly worked to prevent the Soviet Union from dominating Eurasia and then helped relegate it to the scrap heap of history in the late 1980s and early 1990s.

Shortly after the Cold War ended, the George H. W. Bush administration's controversial "Defense Planning Guidance" of 1992 was leaked to the press. It boldly stated that the United States was now the most powerful state in the world by far and it planned to remain in that exalted position. In other words, the United States would not tolerate a peer competitor.

That same message was repeated in the famous 2002 National Security Strategy issued by the George W. Bush administration. There was much criticism of that document, especially its claims about "preemptive" war. But hardly a word of protest was raised about the assertion that the United States should check rising powers and maintain its commanding position in the global balance of power.

The bottom line is that the United States—for sound strategic reasons—worked hard for more than a century to gain hegemony in the Western Hemisphere. Since achieving regional dominance, it has gone to great lengths to prevent other great powers from controlling either Asia or Europe.

Thus, there is little doubt as to how American policy makers will react if China attempts to dominate Asia. The United States can be expected to go to great lengths to contain China and ultimately weaken it to the point where it is no longer capable of ruling the roost in Asia. In essence, the United States is likely to behave toward China much the way it acted toward the Soviet Union during the Cold War.

China's neighbors are certain to fear its rise as well, and they too will do whatever they can to prevent it from achieving regional hegemony. Indeed, there is already substantial evidence that countries like India, Japan, and Russia as well as smaller powers like Singapore, South Korea, and Vietnam are worried about China's ascendancy and are looking for ways to contain it. In the end, they will join an American-led balancing coalition to check China's rise, much the way Britain, France, Germany, Italy, Japan, and even China joined forces with the United States to contain the Soviet Union during the Cold War.

How does Taiwan fit into this story? The United States has a rich history of close relations with Taiwan since the early days of the Cold War, when the Nationalist forces under Chiang Kai-shek retreated to the island from the Chinese mainland. However, Washington is not obliged by treaty to come to the defense of Taiwan if it is attacked by China or anyone else.

Regardless, the United States will have powerful incentives to make Taiwan an important player in its anti-China balancing coalition. First, as noted, Taiwan has significant economic and military resources and it is effectively a giant aircraft carrier that can be used to help control the waters close to China's all-important eastern coast. The United States will surely want Taiwan's assets on its side of the strategic balance, not on China's side.

Second, America's commitment to Taiwan is inextricably bound up with U.S. credibility in the region, which matters greatly to policy makers in Washington. Because the United States is located roughly 6,000 miles from East Asia, it has to work hard to convince its Asian allies—especially Japan and South Korea—that it will back them up in the event they are threatened by China or North Korea. Importantly, it has to convince Seoul and Tokyo that they can rely on the American nuclear umbrella to protect them. This is the thorny problem of extended deterrence, which the United States and its allies wrestled with throughout the Cold War.

If the United States were to sever its military ties with Taiwan or fail to defend it in a crisis with China, that would surely send a strong signal to America's other allies in the region that they cannot rely on the United States for protection. Policy makers in Washington will go to great lengths to avoid that outcome and instead maintain America's reputation as a reliable partner. This means they will be inclined to back Taiwan no matter what.

While the United States has good reasons to want Taiwan as part of the balancing coalition it will build against China, there are also reasons to think this relationship is not sustainable over the long term. For starters, at some point in the next decade or so it will become impossible for the United States to help Taiwan defend itself against a Chinese attack. Remember that we are talking about a China with much more military capability than it has today.

In addition, geography works in China's favor in a major way, simply because Taiwan is so close to the Chinese mainland and so far away from the United States. When it comes to a competition between China and the United States over projecting military power into Taiwan, China wins hands down. Furthermore, in a fight over Taiwan, American policy makers would surely be reluctant to launch major attacks against Chinese forces on the mainland, for fear they might precipitate nuclear escalation. This reticence would also work to China's advantage.

One might argue that there is a simple way to deal with the fact that Taiwan will not have an effective conventional deterrent against China in the not-too-distant future: put America's nuclear umbrella over Taiwan. This approach will not solve the problem, however, because the United States is not going to escalate to the nuclear level if Taiwan is being overrun by China. The stakes are not high enough to risk a general thermonuclear war. Taiwan is not Japan or even South Korea. Thus, the smart strategy for America is to not even try to extend its nuclear deterrent over Taiwan.

There is a second reason the United States might eventually forsake Taiwan: it is an especially dangerous flashpoint, which could easily precipitate a Sino-American war that is not in America's interest. U.S. policy makers understand that the fate of Taiwan is a matter of great concern to Chinese of all persuasions and that they will be extremely angry if it looks like the United States is preventing unification. But that is exactly what Washington will be doing if it forms a close military alliance with Taiwan, and that point will not be lost on the Chinese people.

It is important to note in this regard that Chinese nationalism, which is a potent force, emphasizes how great powers like the United States humiliated China in the past when it was weak and appropriated Chinese territory like Hong Kong and Taiwan. Thus, it is not difficult to imagine crises breaking out over Taiwan or scenarios in which a crisis escalates into a shooting war. After all, Chinese nationalism will surely be a force for trouble in those crises, and China will at some point have the military wherewithal to conquer Taiwan, which will make war even more likely.

There was no flashpoint between the superpowers during the Cold War that was as dangerous as Taiwan will be in a Sino-American security competition. Some commentators liken Berlin in the Cold War to Taiwan, but Berlin was not sacred territory for the Soviet Union and it was actually of little strategic importance for either side. Taiwan is different. Given how dangerous it is for precipitating a war and given the fact that the United States will eventually reach the point where it cannot defend Taiwan, there is a reasonable chance that American policy makers will eventually conclude that it makes good strategic sense to abandon Taiwan and allow China to coerce it into accepting unification.

All of this is to say that the United States is likely to be somewhat schizophrenic about Taiwan in the decades ahead. On one hand, it has powerful incentives to make it part of a balancing coalition aimed at containing China. On the other hand, there are good reasons to think that with the passage of time the benefits of maintaining close ties with Taiwan will be outweighed by the potential costs, which are likely to be huge. Of course, in the near term, the United States will protect Taiwan and treat it as a strategic asset. But how long that relationship lasts is an open question.

So far, the discussion about Taiwan's future has focused almost exclusively on how the United States is likely to act toward Taiwan. However, what happens to Taiwan in the face of Chinas rise also depends greatly on what policies Taiwan's leaders and its people choose to pursue over time. There is little doubt that Taiwan's overriding goal in the years ahead will be to preserve its independence from China. That aim should not be too difficult to achieve for the next decade, mainly because Taiwan is almost certain to maintain close relations with the United States, which will have powerful incentives as well as the capability to protect Taiwan. But after that point Taiwan's strategic situation is likely to deteriorate in significant ways, mainly because China will be rapidly approaching the point where it can conquer Taiwan even if the American military helps defend the island. And, as noted, it is not clear that the United States will be there for Taiwan over the long term.

In the face of this grim future, Taiwan has three options. First, it can develop its own nuclear deterrent. Nuclear weapons are the ultimate deterrent, and there is no question that a Taiwanese nuclear arsenal would markedly reduce the likelihood of a Chinese attack against Taiwan.

Taiwan pursued this option in the 1970s, when it feared American abandonment in the wake of the Vietnam War. The United States, however, stopped Taiwan's nuclear-weapons program in its tracks. And then Taiwan tried to develop a bomb secretly in the 1980s, but again the United States found out and forced Taipei to shut the program down. It is unfortunate for

Taiwan that it failed to build a bomb, because its prospects for maintaining its independence would be much improved if it had its own nuclear arsenal.

No doubt Taiwan still has time to acquire a nuclear deterrent before the balance of power in Asia shifts decisively against it. But the problem with this suggestion is that both Beijing and Washington are sure to oppose Taiwan going nuclear. The United States would oppose Taiwanese nuclear weapons, not only because they would encourage Japan and South Korea to follow suit, but also because American policy makers abhor the idea of an ally being in a position to start a nuclear war that might ultimately involve the United States. To put it bluntly, no American wants to be in a situation where Taiwan can precipitate a conflict that might result in a massive nuclear attack on the United States.

China will adamantly oppose Taiwan obtaining a nuclear deterrent, in large part because Beijing surely understands that it would make it difficult—maybe even impossible—to conquer Taiwan. What's more, China will recognize that Taiwanese nuclear weapons would facilitate nuclear proliferation in East Asia, which would not only limit China's ability to throw its weight around in that region, but also would increase the likelihood that any conventional war that breaks out would escalate to the nuclear level. For these reasons, China is likely to make it manifestly clear that if Taiwan decides to pursue nuclear weapons, it will strike its nuclear facilities, and maybe even launch a war to conquer the island. In short, it appears that it is too late for Taiwan to pursue the nuclear option.

There was no flashpoint between the superpowers during the Cold War that was as dangerous as Taiwan will be in a Sino-American security competition.

Taiwan's second option is conventional deterrence. How could Taiwan make deterrence work without nuclear weapons in a world where China has clear-cut military superiority over the combined forces of Taiwan and the United States? The key to success is not to be able to defeat the Chinese military— that is impossible—but instead to make China pay a huge price to achieve victory. In other words, the aim is to make China fight a protracted and bloody war to conquer Taiwan. Yes, Beijing would prevail in the end, but it would be a Pyrrhic victory. This strategy would be even more effective if Taiwan could promise China that the resistance would continue even after its forces were defeated on the battlefield. The threat that Taiwan might turn into another Sinkiang or Tibet would foster deterrence for sure.

This option is akin to Admiral Alfred von Tirpitz's famous "risk strategy," which Imperial Germany adopted in the decade before World War I. Tirpitz accepted the fact that Germany could not build a navy powerful enough to defeat the mighty Royal Navy in battle. He reasoned, however, that Berlin could build a navy that was strong enough to inflict so much damage on the Royal Navy that it would cause London to fear a fight with Germany and thus be deterred. Moreover, Tirpitz reasoned that this "risk fleet" might even give Germany diplomatic leverage it could use against Britain.

There are a number of problems with this form of conventional deterrence, which raise serious doubts about whether it can work for Taiwan over the long haul. For starters, the strategy depends on the United States fighting side by side with Taiwan. But it is difficult to imagine American policy makers purposely choosing to fight a war in which the U.S. military is not only going to lose, but is also going to pay a huge price in the process. It is not even clear that Taiwan would want to fight such a war, because it would be fought mainly on Taiwanese territory—not Chinese territory—and there would be death and destruction everywhere. And Taiwan would lose in the end anyway.

Furthermore, pursuing this option would mean that Taiwan would be constantly in an arms race with China, which would help fuel an intense and dangerous security competition between them. The sword of Damocles, in other words, would always be hanging over Taiwan.

Finally, although it is difficult to predict just how dominant China will become in the distant future, it is possible that it will eventually become so powerful that Taiwan will be unable to put up major resistance against a Chinese onslaught. This would certainly be true if America's commitment to defend Taiwan weakens as China morphs into a superpower.

Taiwan's third option is to pursue what I will call the "Hong Kong strategy." In this case, Taiwan accepts the fact that it is doomed to lose its independence and become part of China. It then works hard to make sure that the transition is peaceful and that it gains as much autonomy as possible from Beijing. This option is unpalatable today and will remain so for at least the next decade. But it is likely to become more attractive in the distant future if China becomes so powerful that it can conquer Taiwan with relative ease.

So where does this leave Taiwan? The nuclear option is not feasible, as neither China nor the United States would accept a nuclear-armed Taiwan. Conventional deterrence in the form of a "risk strategy" is far from ideal, but it makes sense as long as China is not so dominant that it can subordinate Taiwan without difficulty. Of course, for that strategy to work, the United States must remain committed to the defense of Taiwan, which is not guaranteed over the long term.

Once China becomes a superpower, it probably makes the most sense for Taiwan to give up hope of maintaining its de facto independence and instead pursue the "Hong Kong strategy." This is definitely not an attractive option, but as Thucydides argued long ago, in international politics "the strong do what they can and the weak suffer what they must."

By now, it should be glaringly apparent that whether Taiwan is forced to give up its independence largely depends on how formidable China's military becomes in the decades ahead. Taiwan will surely do everything it can to buy time and maintain the political status quo. But if China continues its impressive rise, Taiwan appears destined to become part of China.

There is one set of circumstances under which Taiwan can avoid this scenario. Specifically, all Taiwanese should hope there is a drastic slowdown in Chinese economic growth in the years ahead and that Beijing also has serious political problems on the home front that work to keep it focused inward. If that happens, China will not be in a position to pursue regional hegemony and the United States will be able to protect Taiwan from China, as it does now. In essence, the best way for Taiwan to maintain de facto independence is for China to be economically and militarily weak. Unfortunately for Taiwan, it has no way of influencing events so that this outcome actually becomes reality.

When China started its impressive growth in the 1980s, most Americans and Asians thought this was wonderful news, because all of the ensuing trade and other forms of economic intercourse would make everyone richer and happier. China, according to the reigning wisdom, would become a responsible stakeholder in the international community, and its neighbors would have little to worry about. Many Taiwanese shared this optimistic outlook, and some still do.

They are wrong. By trading with China and helping it grow into an economic powerhouse, Taiwan has helped create a burgeoning Goliath with revisionist goals that include ending Taiwan's independence and making it an integral part of China. In sum, a powerful China isn't just a problem for Taiwan. It is a nightmare.

Critical Thinking

1. Why doesn't the United States recognize Taiwan as a sovereign state?

2. Will the United States go to war with China to defend Taiwan? Why or why not?

3. Do you think that China's rise as a regional hegemon in Asia is peaceful? Why or why not?

Create Central

www.mhhe.com/createcentral

Internet References

Kissinger Institute on China and the United States
http://www.wilsoncenter.org/program/kissinger-institute-china-and-the-UnitedStates

Minister of Foreign Affairs, Republic of China
http://www.frnprc.gov.cn/eng

Minister of Foreign Affairs, Republic of China (Taiwan)
http://www.mofa.gov.tw/en

JOHN J. MEARSHEIMER is the R. Wendell Harrison Distinguished Service Professor of Political Science at the University of Chicago. He serves on the Advisory Council of *The National Interest*. This article is adapted from a speech he gave in Taipei on December 7, 2013, to the Taiwanese Association of International Relations. An updated edition of his book *The Tragedy of Great Power Politics* will be published in April by W. W. Norton.

John J. Mearsheimer, "Taiwan's Dire Straits," *The National Interest*, March/April 2014, pp. 29–38.

Article   Prepared by: Robert Weiner, *University of Massachusetts, Bosto*

Why 1914 Still Matters

NORMAN FRIEDMAN

Learning Outcomes

After reading this article, you will be able to:

- Understand the causes of World War I.

- Understand the legacy of the war.

T oday, as a century ago, the fact that war between trading nations would be ruinous does not necessarily mean that its outbreak is impossible.

Imagine that your closest trading partner is also your most threatening potential enemy. Imagine, too, that this partner is building a large navy specifically targeted at yours, hence at the overseas trade vital to you. Does that sound like the current U.S. situation with respect to China? It was certainly the British situation relative to Germany a century ago, on the eve of World War I. History never repeats, but it is often instructive to look at the mistakes of the past. The worse the mistakes, the more instructive. No one looking at the outbreak and then the course of World War I can see it as anything but a huge mistake. Hopefully we can do better.

The worst mistake, from a British point of view, was to forget that this was a maritime war. Had the British not entered the war at all, it would have been a European land war. Once Britain entered, the character of the war changed, not only because Britain was the world's dominant sea power, but also because the British Empire—including vital informal elements—was a seaborne entity, drawing much of its strength from overseas. As an island, Britain was almost impossible to invade. Centuries earlier, Sir Francis Bacon had written that he who controls the sea can take as much or as little of the war as he likes. The sea power did not have to place a mass army ashore. That was not necessarily its appropriate contribution to a coalition effort.

Our memory of World War I overwhelmingly emphasizes the blood and horror of the Western Front, to which the U.S. Army and Marines were assigned when entering the conflict

in 1917. The war at sea is usually dismissed as at best an enabler for the more important action view obscures the reality that the war was shape considerations, and, at least as importantly, the seaborne mobility offered the British and the instance of a strategic attack from the sea, Ga danelles campaign), is usually dismissed as an Lord of the Admiralty Winston Churchill to the Royal Navy. In fact, it was a high-risk, h tion supported by the British cabinet for ver That it failed does not make it a foolish bit only proves that planning and execution poor. Our memory of how the war was fou that there were real alternatives, at least fo

Our present situation is more like that of their continental allies. How well wo situation? We were actually confronted War. The U.S. Navy's Maritime Stra way to fight a continental war. It is stil

The Accidental Arm

When the British entered World Wa Asquith expected the French and bulk of the forces on land; the B France was to be largely symbol French to hold the German army steamroller" smashed from the ally approved War Minister Lo to create massive "New Armie explain their rationale). The F est army in their history. On be denied to the French whe trouble in 1915. Once there Most of the 800,000 British died on the Western Front.

Were these horrific losses inevitable? Given the sheer depth of modern economies and the power of the defense, the war on land would surely have been a protracted bloodbath. Did it have to be a British bloodbath? Asquith was Prime Minister of the United Kingdom, not of some Franco-British combination. It was clearly in the interest of the French that the British army fought alongside theirs and helped preserve France. Was that in British interests, too? How deep should coalition partnership cut? Could the British have fought a more maritime war? In Vietnam, in Iraq, and in Afghanistan the United States has faced the question of how far to go in support of a coalition partner.

Perhaps the saddest feature of British prewar and wartime planning was Admiral Sir John Fisher's futile attempt to point out that although (as everyone agreed) no success on the Western Front could be decisive, the Germans were extraordinarily sensitive to threats to their Baltic coast—a place accessible by sea, albeit with considerable danger. Unfortunately, Fisher made his point, both before and during the war, in an obscure, even mystical way.[2] The often-denigrated Dardanelles operation was a remnant of the abortive British maritime strategy; it was intended to help sustain Russia. Fisher's great objection was that it would swallow forces he thought could have been used more effectively in the Baltic—again, to support the Russians on what he and others thought was the decisive front.

The deeper reason for British planning failure is that almost [up] to the declaration of war virtually no one in London believed [th]at there could ever be a war. It was widely accepted that, [be]cause the major economies were so closely intertwined, any [wa]r would be disastrous. The Britain of 1914 was a much more [mo]dern nation than its European partners. International finance [play]ed a larger part in the British economy than in any other. [The] financial sector still considers war futile: If one asks some[one] on Wall Street right now whether a war with China is possi[ble,] the answer is emphatically no, that would be ruinous. If the [poin]t of government is to maintain national prosperity, big wars [are ab]surd. The British government of the years before 1914 [did no]t, it seems, understand that those governing Germany had [quite] different ideas. How well do we understand how foreign [gover]nments think? Are big wars really obsolete?

[Eco]nomy as Weapon = Double-[Edg]ed Sword

[In fa]ct, those in London thought that what was much later [called m]utual assured destruction prevailed. War fighting and [prewar] war planning were of little account. The British army [commit]ment to France was much more symbolic than real, an [effort t]o show the French that the British would back them [in the ev]ent of a crisis. This plan was accepted (though not, [wholly] wholeheartedly) largely because it was far more

important that prewar War Minister Richard Haldane led an influential faction in the governing Liberal Party than that the army's favored plan for deployment in France made much military sense.

The British government naturally became interested in economic attack as a means of quickly concluding any war that broke out. The Admiralty became an advocate of such warfare as a natural extension of the traditional naval economic weapon of blockade. In 1908 a prominent British economist pointed out that in a crisis the British banks, which were central to the world economic system, could attack German credit with devastating results.[3] Somewhat later the British banks pointed out that since Germany was Britain's most important trading partner, any damage would go both ways. Banking had to be omitted from the arsenal of economic weapons. It turned out that sanctions imposed on Britain's main trading partner were less than popular in the United Kingdom—and that they badly damaged the British economy which depended on trade. For example, a prohibition against trading with the enemy made it necessary to prove that every transaction was not with the enemy. It was not at all clear that the damage done to the British economy did not exceed that done to the German.

In pre-1914 Europe the single life-and-death problem for most governments was internal stability. Most thought in domestic terms. For example, the British Liberal Party resisted naval and military spending because it considered social spending vital for British stability. The tsarist government in Russia sought to create a strong peasant class as a bulwark against socialist workers (assuring grain exports, which would create the prosperous peasant class, required free access to the world grain market via the Dardanelles). However, the Austro-Hungarian government feared nationalist upheaval triggered from outside, most notably from Serbia (and was unable to promote internal reform).

German leaders thought they faced an imminent internal crisis.[4] The perceived crisis was the rise of a hostile majority in the Reichstag, the lower house of the German parliament. Although hardly comparable to the British Parliament, the Reichstag was responsible for the budget. In elections from 1890 on, the Social Democrats, whom the Kaiser and his associates considered dangerous revolutionaries, consistently won majorities of the vote, but because seats were gerrymandered they did not win a majority in the Reichstag until 1912. The German army's general staff considered itself and the army the bulwark of the regime. Although in theory the Kaiser ruled Germany, in fact he had been sidelined for several years. Army expansion, which might be associated with the sense of internal crisis, began in 1912.

The following year the nightmare became visible, as the Reichstag passed a vote of no confidence after the army exonerated an officer who had attacked a civilian in Alsace.[5] The vote

did not bring down the government, because Prime Minister Theobald von Bethmann-Hollweg was responsible to the Kaiser rather than to the Reichstag. The center-left coalition shrank from rejecting the year's budget. However, there was a sense of escalating internal crisis. A member of the German General Staff told a senior Foreign Ministry official that his task for the coming year was to foment a world war, and to make it defensive for Germany so that the Reichstag would support the war.[6]

In this light, the event that precipitated the war—the assassination of the Austrian crown prince Franz Ferdinand—seems to have been much more a useful pretext than the reason the world blew up. The Kaiser was largely on the periphery of rapidly unfolding events during the crisis. He kept asking why the army was attacking France when the crisis was about Russia and Serbia. Do we understand who actually rules countries that may be hostile to us?

Internal Motivations, External Aggression

In 1912–14 the German army general staff could look back to 1870. By drawing France into a war at that time, Prussia had created the German Empire. The spoils of that war were a way of showing that it had been worthwhile, but the war was really about the internal political needs of the German state. In 1914, the general staff doubtless expected that victory would shrivel the Social Democrats (a 1907 military victory over the Hottentots in Africa had reversed their rise, though only briefly). No other military seems to have had a record of deliberately instigating war as a specific way of gaining an internal political end. After World War I, there was a general sense that the German general staff had been responsible for the war, but not to the extent that now seems apparent.

At one time a standard explanation for enmity between Britain and Germany, leading to war, was commercial rivalry. It was taken so seriously that interwar U.S. Navy war planners used British-U.S. rivalry to explain why a war might break out between the two countries. Similarly, one might see Chinese-U.S. trade rivalry as a possible cause of war. However, those concerned with commerce are too aware of how ruinous war can be. Wall Street really does prefer commercial competition to blowing apart its rivals. It has too clear an idea of what war might mean. Naval wars connected with commercial rivalry were fought before commercial and financial interests came to dominate governments. The perceived need to keep the state alive is a very different matter, and it seems to have been what propelled Germany in 1914. Do we see similar motives at work now, or in the near future? The lesson of 1914 is that others' decision to fight is far more often about internal politics than about what we may do.

The Vital Importance of Coalitions

British strategy in 1914–15 may not seem odd in itself, but it is decidedly odd in the context of other wars the British fought on the continent. Everyone in the 1914 Cabinet knew something of the Napoleonic Wars, though probably not from a strategic point of view. That was unfortunate, because they might have benefited from seeing the new war in terms of the earlier one. The British fought Napoleon as a member of a coalition. They watched their coalition partners collapse, to the point where they alone resisted Napoleon. They were forced to agree to a peace in 1801, which they rightly considered nothing more than a pause in the war—and they used that peace to consolidate what advantages they could.

Once the war against Napoleon resumed, the British wisely made it their first step to insure against invasion by blocking and then neutralizing the French and their allied fleets. Once they had been freed from the threat of invasion by the victory at Trafalgar, they could mount high-risk, high-gain operations around the periphery of Napoloeon's empire. Ultimately that meant Wellington's war on the Iberian Peninsula. Napoleon realized that he could not tolerate British resistance. Since he could not invade, he was forced into riskier and riskier operations intended to crush Britain economically. His disastrous 1812 invasion of Russia was in this category (it was intended to cut off Russian trade with Britain). The British limited their own liability on the Continent. Knowing that they could not be invaded (hence defeated), they could afford to be patient—and they won. Victory was a coalition achievement, which is why it did not matter that so many of the troops at Waterloo were not British.

World War I was shaped by the fact that Britain entered it. Until that moment, the German army staff could envisage a quick war which would end in the West with the hoped-for defeat of the French army. Once Britain was in the war, no German victory on land could be complete. Ironically, the Germans guaranteed that Britain would enter the war by building a large fleet specifically directed against it. Some current British historians have asked whether it was really worthwhile for the British of 1914 to have resisted the creation of a unified Europe under German control. They have missed the maritime point. In 1914 the British saw the Germans as a direct threat to their lives, because the Germans had been building their massive fleet. By 1914 most Britons well understood that their country lived or died by its access to the sea and to the resources of the world. The Royal Navy had worked hard for nearly 30 years to bring that message home. It resonated because it was true. In 1914 the British government would have had to fight public opinion to keep the country out of a war the Germans started.

The German decision to build a fleet seems, in retrospect, to have been remarkably casual. The fleet was completely disconnected from the war plan created by the army's general staff; it had no initial role whatsoever. The German navy came into its own only when it became clear that the army could not achieve a decision on land. Then it was not so much the big fleet (that had caught British attention prewar) but the U-boats that Admiral Alfred Tirpitz, the fleet's creator, grudgingly built. The British government might well have decided to oppose Germany in 1914 to preserve the balance of power in Europe—a historic British policy—but without the obvious threat of the German fleet its decision would not have enjoyed anything like the same level of support.

In 1939 the British again faced a continental war. Everyone in the British government had experienced World War I as a horrific bloodbath. This time the British consciously limited their liability. It helped that by 1939 they believed that the Germans could not destroy the United Kingdom by air attack (thanks to radar and modern fighters), so that as in World War I, Britain was a defensible island. Winston Churchill, who had a far more strategic viewpoint than most, certainly did not intend to surrender when the British were ejected from the continent in 1940. He understood that the overseas Empire and the overseas world could and would support Britain against Germany (which is why the Battle of the Atlantic was his greatest concern). He also understood that it would take a coalition to destroy Hitler.

During the Cold War, NATO faced a continental threat not entirely unlike that the British had faced in 1914 and in 1939. Attention was focussed on the Central Front, unfortunately so named because it was in the center between the alliance's northern and southern flanks. The U.S. Navy offered a maritime alternative, both in the 1950s and in the 1980s. Captain Peter Swartz, U.S. Navy (Retired), who chronicled the U.S. Navy's Maritime Strategy, summarized the way that a maritime power deals with a land power: It combines a coalition with its own land partner and it exploits maritime mobility to cripple the enemy army.

"Hard Thinking about the Object of War."

Not being able to end a war may seem to be a tame sort of disadvantage to the land power sweeping all before it in Europe. However, both in Napoleon's time and during World War I, the land power (France and Germany, respectively) found that it could not stop fighting. Its effort to knock the British out of the war eventually brought in enemies the land power could not handle. In Napoleon's time that was the Russians, whose territory absorbed the French army, and whose limitless mass of troops eventually helped invade France. Obviously there were many other contributions to French defeat, including Wellington's campaign in Spain, but the point is that none of that would have mattered had Napoleon been able to end the war as he liked.

In World War I the Germans found that their only leverage against the British was to attack their overseas source of strength, either at source in the United States or at sea en route to Britain. Either move was risky. Unrestricted submarine warfare against shipping led to angry reactions from the United States; in 1915–16 the German Foreign Ministry convinced the government (i.e. the general staff) to pull back. As an alternative, in 1916 the Germans organized the sabotage of munitions plants supplying the Allies, most notably Black Tom in New York Harbor. Although the U.S. government almost immediately discovered that the Germans had caused the Black Tom explosion, President Woodrow Wilson badly wanted to stay out of the war. That was not enough for the German general staff. Against Foreign Ministry opposition, it turned again in February 1917 to unrestricted submarine warfare as a way of strangling the Allies.

It was understood that resumption of such warfare would probably bring the United States into the war. With this possibility in mind, the Germans authorized their diplomats in Mexico to offer an alliance under which Mexico would regain the territory it had lost to the United States 60 years earlier: California, New Mexico, Arizona, Nevada, and Texas. Revelation of this Zimmermann Telegram helped bring the United States into the war on the Allied side. U.S. naval and industrial resources helped neutralize the German U-boat campaign in the Atlantic. The U.S. Army and Marines Corps tipped the balance of power in Europe, though it was at least as important that the British and the French became adept at all-arms warfare.

It is also possible that, in the end, the Western Front, where so much blood was spilled, was not decisive in itself. In 1918 the defense still enjoyed considerable advantages. The Germans told themselves that they could shore up their defense in the West, but in September and October 1918 their position in the south, the area in which maritime power had made Allied action possible, collapsed. Whatever they could do on the Western Front, the Germans could not spare troops to cover their southern and eastern borders. In this sense the collapse in the south (of Austria-Hungary, Turkey, and Bulgaria) may have been far more important than is generally imagined.

Maritime never meant purely naval. Success came from using land and sea forces in the right combinations. Maritime did demand hard thinking about the object of the war. In 1914, was it to preserve France or above all to defeat Germany? Because the prewar British government believed in deterrence, it never thought through this kind of question, and by the time it might have been asked, there was a huge British army in France. Withdrawal would have been difficult at best. After the disaster on the Somme in 1916, many in the British government began to ask what the British should do if they were forced to accept an unsatisfactory peace, as in 1801. Part of their answer was that phase two of the war should concentrate more on the east. That is why the British had such large forces in places like the Caucasus and the Middle East when the war ended in November 1918.[7]

A century later, we are in something like the position the British occupied in 1914. We are the world's largest trading nation, and we live largely by international trade—much of which has to go by sea. We do not have a formal empire like the British, but they and we are at the core of a commercial commonwealth which is our real source of economic strength. In a crisis our trade—our lifeblood—would be guaranteed by the U.S. and allied navies, the U.S. Navy dwarfing the others. That we depend on imports means that we have vital interests in far corners of the world. It happens that relatively few Americans understand as much, or see what happens in the Far East as central to their own prosperity. Access to our trading partners there is crucial to us, just as access to overseas trading partners (and the Empire) was a life-or-death matter for the British in 1914. Like the British in 1914, we regard war as too ruinous to be worthwhile, and we often assume that other governments take a similar view. Like the British, we are not very sensitive to the possibility that other governments' views may not match ours. A long look back at 1914 may be well worth our while.

Notes

1. Michael and Eleanor Brock, eds., H. H. Asquith: *Letters to Venetia Stanley* (Oxford, UK: Oxford University Press, 1982).

2. Holger M. Herwig, *"Luxury Fleet:" The German Imperial Navy 1888–1918* (London: Allen & Unwin, 1980).

3. Nicholas A. Lambert, *Planning Armageddon: British Economic Warfare and the First World War* (Cambridge, MA: Harvard University Press, 2012).

4. V. R. Berghahn, *Germany and the Approach of War in 1914,* second ed. (New York: St. Martin's Press, 1993).

5. Jack Beatty, *The Lost History of 1914: How the Great War Was Not Inevitable* (London: Bloomsbury, 2012).

6. David Fromkin, *Europe's Last Summer: Who Started the Great War in 1914?* (New York: Knopf, 2004).

7. Brock Millman, *Pessimism and British War Policy, 1916–1918* (London: Frank Cass, 2001).

Critical Thinking

1. Do you think that a great war is possible between the United States and China? Why or why not?

2. Why was World War I called the Great War?

3. Why did the United Kingdom declare war against Germany?

Create Central

www.mhhe.com/createcentral

Internet References

Centenary News
http://www.centenarynews.com

National Army Museum Website
http://www.nam.ac.ukwwI

The Great War Centenary
http://www.greatwar.co.uk/events/2014-2018-www1-centenary-events-htm

Trenches on the Web
http://www.worldwar1.com

DR. FRIEDMAN, whose "World Naval Developments" column appears monthly in *Proceedings,* is the author of *The Naval Institute Guide to World Naval Weapons Systems,* Fifth Edition, *The Fifty-Year War: Conflict and Strategy in the Cold War,* and other works. This article is based on his new book, *Fighting the Great War at Sea: Strategy, Tactics, and Technology,* forthcoming in September from the Naval Institute Press.

Article   Prepared by: Robert Weiner, *University of Massachusetts, Boston*

Russia's Latest Land Grab: How Putin Won Crimea and Lost Ukraine

JEFFREY MANKOFF

Learning Outcomes

After reading this article, you will be able to:

- Understand Russia's strategy in dealing with post-Soviet states.

- Understand the effects of Russian strategy on Ukraine.

- Understand why Russian strategy has upset the Cold War settlement.

- Understand why the European Union is opposed to Russia strategy toward the Ukraine.

Russia's occupation and annexation of the Crimean Peninsula in February and March have plunged Europe into one of its gravest crises since the end of the Cold War. Despite analogies to Munich in 1938, however, Russia's invasion of this Ukrainian region is at once a replay and an escalation of tactics that the Kremlin has used for the past two decades to maintain its influence across the domains of the former Soviet Union. Since the early 1990s, Russia has either directly supported or contributed to the emergence of four breakaway ethnic regions in Eurasia: Transnistria, a self-declared state in Moldova on a strip of land between the Dniester River and Ukraine; Abkhazia, on Georgia's Black Sea coast; South Ossetia, in northern Georgia; and, to a lesser degree, Nagorno-Karabakh, a landlocked mountainous region in southwestern Azerbaijan that declared its independence under Armenian protection following a brutal civil war. Moscow's meddling has created so-called frozen conflicts in these states, in which the splinter territories remain beyond the control of the central governments and the local de facto authorities enjoy Russian protection and influence.

Until Russia annexed Crimea, the situation on the peninsula had played out according to a familiar script: Moscow opportunistically fans ethnic tensions and applies limited force at a moment of political uncertainty, before endorsing territorial revisions that allow it to retain a foothold in the contested region. With annexation, however, Russia departed from these old tactics and significantly raised the stakes. Russia's willingness to go further in Crimea than in the earlier cases appears driven both by Ukraine's strategic importance to Russia and by Russian President Vladimir Putin's newfound willingness to ratchet up his confrontation with a West that Russian elites increasingly see as hypocritical and antagonistic to their interests.

Given Russia's repeated interventions in breakaway regions of former Soviet states, it would be natural to assume that the strategy has worked well in the past. In fact, each time Russia has undermined the territorial integrity of a neighboring state in an attempt to maintain its influence there, the result has been the opposite. Moscow's support for separatist movements within their borders has driven Azerbaijan, Georgia, and Moldova to all wean themselves off their dependence on Russia and pursue new partnerships with the West. Ukraine will likely follow a similar trajectory. By annexing Crimea and threatening deeper military intervention in eastern Ukraine, Russia will only bolster Ukrainian nationalism and push Kiev closer to Europe, while causing other post-Soviet states to question the wisdom of a close alignment with Moscow.

Frozen Conflicts Playbook

These frozen conflicts are a legacy of the Soviet Union's peculiar variety of federalism. Although Marxism is explicitly internationalist and holds that nationalism will fade as class solidarity develops, the Soviet Union assigned many of its territorial units to particular ethnic groups. This system was largely the work of Joseph Stalin. In the first years after the Bolshevik Revolution, Stalin headed the People's Commissariat for Nationality Affairs, the Soviet bureaucracy set up in 1917 to deal with citizens of non-Russian descent. Stalin's commissariat presided over the creation of a series of ethnically defined territorial units. From 1922 to 1940, Moscow formed the largest of these units into the 15 Soviet socialist republics; these republics became independent states when the Soviet Union dissolved in 1991.

Although designed as homelands for their titular nationalities, the 15 Soviet socialist republics each contained their own minority groups, including Azeris in Armenia, Armenians in Azerbaijan, Abkhazians and Ossetians in Georgia, Uzbeks in Kyrgyzstan, and Karakalpaks in Uzbekistan, along with Russians scattered throughout the non-Russian republics. Such diversity was part of Stalin's plan. Stalin drew borders through ethnic groups' historical territories (despite the creation of Uzbekistan, for example, the four other Central Asian Soviet republics were left with sizable Uzbek minorities) and included smaller autonomous enclaves within several Soviet republics (such as Abkhazia in Georgia and Nagorno-Karabakh in Azerbaijan). From Azerbaijan to Uzbekistan, the presence of concentrated minorities within ethnically defined Soviet republics stoked enough tension to limit nationalist mobilization against Moscow. The Ukrainian Soviet Socialist Republic already had sizable Russian and Jewish populations, but Soviet Premier Nikita Khrushchev's decision to give the republic the Crimean Peninsula in 1954 added a large, territorially concentrated Russian minority. (Crimean Tatars, who are the peninsula's native population, composed close to a fifth of the population until 1944, when most of them were deported to Central Asia for allegedly collaborating with the Nazis. According to the last census, from 2001, ethnic Russians compose about 58 percent of Crimea's population, Ukrainians make up 24 percent, and Crimean Tatars, around 12 percent. The remaining 6 percent includes Belarusians and a smattering of other ethnicities.)

For a long time, the strategy of ethnic division worked. During the 1980s, most of these minority groups opposed the nationalist movements that were pressing for independence in many of the Soviet republics, viewing the continued existence of the Soviet Union as the best guarantee of their protection against the larger ethnic groups that surrounded them. As a result, local officials in Abkhazia, South Ossetia, and Transnistria largely supported the August 1991 coup against Mikhail Gorbachev, who they believed was speeding the dissolution of the Soviet Union. In Crimea, only 54 percent of voters supported Ukrainian independence in a December 1991 referendum—by far the lowest figure anywhere in Ukraine.

As the Soviet Union dissolved, many of these divisions sparked intercommunal violence, which Moscow exploited to maintain a foothold in the new post-Soviet states. In 1989, as part of a national project to promote a shared linguistic identity with Romania, its neighbor to the west, the Moldavian Soviet Socialist Republic voted to reinstitute the Latin alphabet and adopt Moldovan as its only official language, downgrading Russian. Feeling threatened, the ethnic Russian and Ukrainian populations of Transnistria declared the area's independence in 1990, and, in an eerie preview of recent events in Crimea, pro-Russian paramilitary units took over Moldovan government buildings in the territory. Later, in 1992, when fighting broke out between Transnistrian separatists and a newly independent Moldova, Russia's 14th Army, which was still stationed in Transnistria as a holdover from Soviet times, backed the separatists. A cease-fire signed in July of that year created a buffer zone between the breakaway region and Moldova, enforced by the Russian military, which has remained in Transnistria ever since.

Similar scenes unfolded in Georgia. In 1989, the Georgian Soviet Socialist Republic, on its way to declaring independence, established Georgian as the official state language, angering Abkhazia and South Ossetia, which had enjoyed autonomy in Soviet Georgia. In 1990, clashes broke out after Georgian authorities voted to revoke South Ossetia's autonomy in response to the region's efforts to create a separate South Ossetian parliament. After Abkhazia declared its independence from the new Georgian state in 1992, Georgia's army invaded, sparking a civil war that killed 8,000 people and displaced some 240,000 (mostly ethnic Georgians). In both conflicts, the Soviet or Russian military intervened directly on the side of the secessionists. The 1992 cease-fire in South Ossetia and the 1994 cease-fire in Abkhazia both left Russian troops in place as peacekeepers, cementing the breakaway regions' de facto independence.

Tensions were renewed in 2004, when Mikheil Saakashvili, a brash, pro-Western 36-year-old, was elected president of Georgia. Saakashvili sought to bring Georgia into NATO and recover both breakaway republics. In response, Moscow encouraged South Ossetian forces to carry out a series of provocations, eventually triggering, in 2008, a Georgian military response and giving Russia a pretext to invade Georgia and formally recognize Abkhazian and South Ossetian independence.

In Nagorno-Karabakh, which was an autonomous region in Soviet Azerbaijan populated primarily by ethnic Armenians, intercommunal violence in the late 1980s grew, in the early 1990s, into a civil war between, on the one side, separatists backed by the newly independent state of Armenia and, on the other, the newly independent state of Azerbaijan. Although Soviet and then Russian forces were involved on both sides throughout the conflict, the rise of a hard-line nationalist leadership in Baku in 1992 encouraged Moscow to tilt toward Armenia, leading to the separatists' eventual victory. In 1994, after as many as 30,000 people had been killed, a truce left Nagorno-Karabakh in the hands of the ethnic Armenian separatists, who have since built a small, functional statelet that is technically inside Azerbaijan but aligned with Armenia—an entity that no UN member recognizes, including, paradoxically, Russia. As energy-rich Azerbaijan has subsequently grown wealthier and more powerful, Armenia—and, by extension, Nagorno-Karabakh—has cemented its alliance with Russia.

Back in the USSR?

In each of those cases, Russia intervened when it felt its influence was threatened. Russia has consistently claimed in such instances that it has acted out of a responsibility to protect threatened minority groups, but that has always been at best a secondary concern. The moves have been opportunistic, driven more by a concern for strategic advantage than by humanitarian or ethnonational considerations. Pledges to defend threatened Russian or other minority populations outside Russia may play well domestically, but it was the Azerbaijani, Georgian, and Moldovan governments' desire to escape Russia's geopolitical orbit—more than their real or alleged persecution of minorities—that led Moscow to move in. Russia has never intervened militarily to defend ethnic minorities, including Russians, in the former Soviet republics of Central Asia, who have often suffered much more than their co-ethnics in other former Soviet republics, probably because Moscow doesn't assign the same strategic significance to those Central Asian countries, where Western influence has been limited.

Leading up to the annexation of Crimea, Putin and his administration were careful to talk about protecting "Russian citizens" (anyone to whom Moscow has given a passport) and "Russian speakers" (which would include the vast majority of Ukrainian citizens), instead of referring more directly to "ethnic Russians." Moscow has also used the word "compatriots" (sootechestvenniki), a flexible term enshrined in Russian legislation that implies a common fatherland and gives Putin great latitude in determining just whom it includes. In announcing Crimea's annexation to Russia's parliament, however, Putin noted that "millions of Russians and Russian-speaking citizens live and will continue to live in Ukraine, and Russia will always defend their interests through political, diplomatic, and legal means." The Kremlin is walking a narrow line, trying to garner nationalist support at home and give itself maximum leeway in how it acts with its neighbors while avoiding the troubling implications of claiming to be the protector of ethnic Russians everywhere. But in Ukraine, once again, Moscow has intervened to stop a former Soviet republic's possible drift out of Russia's orbit and has justified its actions as a response to ethnic persecution, the claims of which are exaggerated.

It is important to note that although Russia has felt free to intervene politically and militarily in all these cases, until Crimea, it had never formally annexed the territory its forces occupied, nor had it deposed the local government (although, by many

accounts, Moscow did contemplate marching on Tbilisi in 2008 to oust Saakashvili). Instead, Russia had been content to demand changes to the foreign policies of Azerbaijan, Georgia, and Moldova, most notably by seeking to block Georgia's NATO aspirations. The annexation of Crimea is thus an unprecedented step in Russia's post-Soviet foreign policy. Although in practice the consequences may not be that different from in the other frozen conflicts (assuming Russia does not precipitate a wider war with Ukraine), Moscow's willingness to flout international norms in the face of clear warnings and the Obama administration's search for a diplomatic way out of the crisis hints at other motivations. More than in the conflicts of the early 1990s or even in Georgia in 2008, the Kremlin conceived of the invasion and annexation of Crimea as a deliberate strike against the West, as well as Ukraine. Putin apparently believes that he and Russia have more to gain from open confrontation with the United States and Europe—consolidating his political position at home and boosting Moscow's international stature—than from cooperation.

Mother Russia

Despite the differences in the case of Crimea, what has not changed in the Kremlin's tactics since the fall of the Soviet Union is Russia's paternalistic view of its post-Soviet neighbors. Russia continues to regard them as making up a Russian sphere of influence, where Moscow has what Russian Prime Minister Dmitry Medvedev, in 2008, termed "privileged interests." In the early 1990s, Russian officials described the former Soviet domains as Russia's "near abroad." That term has since fallen out of favor. But the idea behind it—that post-Soviet states in eastern Europe and Eurasia are not fully sovereign and that Moscow continues to have special rights in them—still resonates among the Russian elite. This belief explains why Putin and other Russian officials feel comfortable condemning the United States for violating the sovereignty of faraway states such as Iraq and Libya while Russia effectively does the same thing in its own backyard.

Such thinking plays another role as well. These days, Russia has little to justify its claims to major-power status, apart from its seat on the UN Security Council and its massive nuclear stockpile. Maintaining Russia's influence across the former Soviet Union helps Russian leaders preserve their image of Russia's greatness. Under Putin, the Kremlin has sought to reinforce this influence by pushing economic and political integration with post-Soviet states, through measures such as establishing a customs union with Belarus and Kazakhstan and forming the Eurasian Union, a new supranational bloc that Putin claims is directly modeled on the EU and that he hopes to unveil in 2015 (Belarus and Kazakhstan have already signed on; Armenia, Kyrgyzstan, and Tajikistan have expressed their interest).

Putin hopes to turn this Eurasian bloc into a cultural and geopolitical alternative to the West, and he has made clear that it will amount to little unless Ukraine joins. This Eurasian dream is what made the prospect of Kiev signing an association agreement with the EU back in November—one that would have permanently excluded Ukraine from the Eurasian Union—so alarming to Putin and led him, at the last minute, to bribe President Viktor Yanukovych with Russian loan guarantees to Ukraine, so that he would reject the deal with Brussels. Thus far, Putin's tactic has failed: not only did Yanukovych's refusal to sign the association agreement spawn the protests that eventually toppled him, but on March 21, the new, interim government in Kiev signed the agreement anyway.

Although Moscow has a variety of tools it can use to exert regional influence—bribes, energy exports, trade ties—supporting separatist movements remains its strongest, if bluntest, weapon. Dependent on Russian protection, Abkhazia, South Ossetia, Transnistria, and now Crimea serve as outposts for projecting Russian political and economic influence. (In this sense; Moscow doesn't back Nagorno-Karabakh directly, but backs Armenia.) Abkhazia, South Ossetia, and Transnistria all permit Russia to base troops on their territory, as does Armenia. Abkhazia and South Ossetia each host roughly 3,500 Russian troops, along with 1,500 Federal Security Service personnel; Transnistria has some 1,500 Russian soldiers on its territory; and Armenia has around 5,000 Russian troops. One of the principal reasons Moscow has regarded Crimea as so strategically valuable is that the peninsula already hosted Russia's Black Sea Fleet.

But Russia's tactics are not cost-free. By splitting apart internationally recognized states and deploying its military to disputed territories, Moscow has repeatedly damaged its economy and earned itself international condemnation. The bigger problem, however, is that Moscow's coercive diplomacy and support of separatist movements diminish Russian influence over time—that is, these actions achieve the exact opposite of what Russia hopes. It is no coincidence that aside from the Baltic countries, which have joined NATO and the EU, the post-Soviet states that have worked hardest to decrease their dependence on Russia over the past two decades are Azerbaijan, Georgia, and Moldova.

These states have moved westward directly in reaction to Russian meddling. During the 1990s, Azerbaijan responded to Russia's intervention over Nagorno-Karabakh by seeking new markets for its oil and gas reserves in the West. It found a willing partner in Georgia, leading to the construction of an oil pipeline from Baku through Tbilisi to the Turkish port of Ceyhan, which started operations in 2005. A parallel gas pipeline in the southern Caucasus opened the next year. Both freed Azerbaijan's and Georgia's economies from a reliance on Russia. Since 2010, Azerbaijan has also secured regional security guarantees from

Turkey, which would complicate any future Russian intervention. Georgia, meanwhile, continues to pursue membership in NATO, and even if it never makes it, Tbilisi will be able to count on some support from the United States and other Western powers if threatened. And Moldova, despite its fractious domestic politics, has also made great strides in aligning itself with Europe, committing to its own EU association agreement last November, just as Yanukovych backed out.

Russia's invasion and annexation of Crimea, especially if it is followed by incursions into eastern Ukraine, will have the same effect. Far from dissuading Ukrainians from seeking a future in Europe, Moscow's moves will only foster a greater sense of nationalism in all parts of the country and turn Ukrainian elites against Russia, probably for a generation. The episode will also make Ukraine and other post-Soviet states, including those targeted for membership in the Eurasian Union, even more reluctant to go along with any Russian plans for regional integration. Russia may have won Crimea, but in the long run, it risks losing much more: its once-close relationship with Ukraine, its international reputation, and its plan to draw the ex-Soviet states back together.

Critical Thinking

1. What is the strategy posed by Russia toward the post-Soviet states?

2. Has Russia achieved its objectives in Ukraine? Why or why not?

3. Is ethnic nationalism an important factor in Russian policy toward the successor states?

Create Central

www.mhhe.com/createcentral

Internet References

Association for Borderland Studies
   http://absborderlands.org
Eurasian Economic Commission
   http://www.eurasiancommission.org/en/Pages/default.aspx
Russian Foreign Ministry
   http://www.mid.ru/BRP_4.nsf/main_eng
Ministry of Foreign Affairs of Ukraine
   http://www.mfa.gov.ua/en

JEFFREY MANKOFF is Deputy Director of and a Fellow in the Russia and Eurasia Program at the Center for Strategic and International Studies.

Article   Prepared by: Robert Weiner, *University of Massachusetts, Boston*

The Utility of Cyberpower

Kevin L. Parker

Learning Outcomes

After reading this article, you will be able to:

- Understand what is meant by cyberspace.

- Understand what is the relationship between realism and the defense of U.S. national interest in cyberspace.

After more than 50 years, the Korean War has not officially ended, but artillery barrages seldom fly across the demilitarized zone.[1] U.S. forces continue to fight in Afghanistan after more than 10 years, with no formal declaration of war.[2] Another conflict rages today with neither bullets nor declarations. In this conflict, U.S. adversaries conduct probes, attacks, and assaults on a daily basis.[3] The offensives are not visible or audible, but they are no less real than artillery shells or improvised explosive devices. This conflict occurs daily through cyberspace.

To fulfill the U.S. military's purpose of defending the nation and advancing national interests, today's complex security environment requires increased engagement in cyberspace.[4] Accordingly, the Department of Defense (DOD) now considers cyberspace an operational domain.[5] Similar to other domains, cyberspace has its own set of distinctive characteristics. These attributes present unique advantages and corresponding limitations. As the character of war changes, comprehending the utility of cyberpower requires assessing its advantages and limitations in potential strategic contexts.

Defining Cyberspace and Cyberpower

A range of definitions for cyberspace and cyberpower exist, but even the importance of establishing definitions is debated.

Daniel Kuehl compiled 14 distinct definitions of cyberspace from various sources, only to conclude he should offer his own.[6] Do exact definitions matter? In bureaucratic organizations, definitions do matter because they facilitate clear division of roles and missions across departments and military services. Within DOD, some duplication of effort may be desirable but comes at a high cost; therefore, definitions are necessary to facilitate the rigorous analyses essential for establishing organizational boundaries and budgets.[7] In executing assigned roles, definitions matter greatly for cross-organizational communication and coordination.

No matter how important, precise definitions to satisfy all viewpoints and contexts are elusive. Consider defining the sea as all the world's oceans. This definition lacks sufficient clarity to demarcate bays or riverine waterways. Seemingly inconsequential, the ambiguity is of great consequence for organizations jurisdictionally bound at a river's edge. Unlike the sea's constant presence for millennia, the Internet is a relatively new phenomenon that continues to expand and evolve rapidly. Pursuing single definitions of cyberspace and cyberpower to put all questions to rest may be futile. David Lonsdale argued that from a strategic perspective, definitions matter little. In his view, "what really matters is to perceive the infosphere as a place that exists, understand the nature of it and regard it as something that can be manipulated and used for strategic advantage."[8] The definitions below are consistent with Lonsdale's viewpoint and suffice for the purposes of this discussion, but they are unlikely to satisfy practitioners who wish to apply them beyond a strategic perspective.

> Cyberspace: the domain that exists for inputting, storing, transmitting, and extracting information utilizing the electromagnetic spectrum. It includes all hardware, software, and transmission media used, from an initiator's input (e.g., fingers making keystrokes, speaking into microphones, or feeding documents into scanners)

to presentation of the information for user cognition (e.g., images on displays, sound emitted from speakers, or document reproduction) or other action (e.g., guiding an unmanned vehicle or closing valves).
Cyberpower: The potential to use cyberspace to achieve desired outcomes.[9]

Advantages of Wielding Cyberpower

With these definitions being sufficient for this discussion, consider the advantages of operations through cyberspace.

Cyberspace provides worldwide reach. The number of people, places, and systems interconnecting through cyberspace is growing rapidly.[10] Those connections enhance the military's ability to reach people, places, and systems around the world. Operating in cyberspace provides access to areas denied in other domains. Early airpower advocates claimed airplanes offered an alternative to boots on the ground that could fly past enemy defenses to attack power centers directly.[11] Sophisticated air defenses developed quickly, increasing the risk to aerial attacks and decreasing their advantage. Despite current cyberdefenses that exist, cyberspace now offers the advantage of access to contested areas without putting operators in harm's way. One example of directly reaching enemy decision makers through cyberspace comes from an event in 2003, before the U.S. invasion of Iraq. U.S. Central Command reportedly emailed Iraqi military officers a message on their secret network advising them to abandon their posts.[12] No other domain had so much reach with so little risk.

Cyberspace enables quick action and concentration. Not only does cyberspace allow worldwide reach, but its speed is unmatched. With aerial refueling, air forces can reach virtually any point on the earth; however, getting there can take hours. Forward basing may reduce response times to minutes, but information through fiber optic cables moves literally at the speed of light. Initiators of cyberattacks can achieve concentration by enlisting the help of other computers. By discretely distributing a virus trained to respond on command, thousands of co-opted botnet computers can instantly initiate a distributed denial-of-service attack. Actors can entice additional users to join their cause voluntarily, as did Russian "patriotic hackers" who joined attacks on Estonia in 2007.[13] With these techniques, large interconnected populations could mobilize on an unprecedented scale in mass, time, and concentration.[14]

Cyberspace allows anonymity. The Internet's designers placed a high priority on decentralization and built the structure based on the mutual trust of its few users.[15] In the decades since, the number of Internet users and uses has grown exponentially beyond its original conception.[16] The resulting system makes it very difficult to follow an evidentiary trail back to any user.[17] Anonymity allows freedom of action with limited attribution.

Cyberspace favors offense. In Clausewitz' day, defense was stronger, but cyberspace, due to the advantages listed above, currently favors the attack.[18]

Historically, advantages from technological leaps erode over time.[19] However, the current circumstance pits defenders against quick, concentrated attacks, aided by structural security vulnerabilities inherent in the architecture of cyberspace.

Cyberspace expands the spectrum of nonlethal weapons. Joseph Nye described a trend, especially among democracies, of antimilitarism, which makes using force "a politically risky choice."[20] The desire to limit collateral damage often has taken center stage in NATO operations in Afghanistan, but this desire is not limited to counterinsurgencies.[21] Precision-guided munitions and small-diameter bombs are products of efforts to enhance attack capabilities with less risk of collateral damage. Cyberattacks offer nonlethal means of direct action against an adversary.[22] The advantages of cyberpower may be seductive to policymakers, but understanding its limitations should temper such enthusiasm. The most obvious limitation is that your adversary may use all the same advantages against you. Another obvious limitation is its minimal influence on nonnetworked adversaries. Conversely, the more any organization relies on cyberspace, the more vulnerable it is to cyberattack. Three additional limitations require further attention.

Cyberspace attacks rely heavily on second order effects. In Thomas Schelling's terms, there are no brute force options through cyberspace, so cyberoperations rely on coercion.[23] Continental armies can occupy land and take objectives by brute force, but success in operations through cyberspace often hinges on how adversaries react to provided, altered, or withheld information. Cyberattacks creating kinetic effects, such as destructive commands to industrial control systems, are possible. However, the unusual incidents of malicious code causing a Russian pipeline to explode and the Stuxnet worm shutting down Iranian nuclear facility processes were not ends.[24] In the latter case, only Iranian leaders' decisions could realize abandonment of nuclear technology pursuits. Similar to strategic bombing's inability to collapse morale in World War II, cyberattacks often rely on unpredictable second order effects.[25] If Rear Adm. Wylie is correct in that war is a matter of control, and "its ultimate tool . . . is the man on the scene with a gun," then operations through cyberspace can only deliver a lesser form of control.[26] Evgeny Morozov quipped, "Tweets, of course, don't topple governments; people do."[27]

Cyberattacks risk unintended consequences. Just as striking a military installation's power system may have cascading ramifications on a wider population, limiting effects through interconnected cyberspace is difficult. Marksmanship instructors teach shooters to consider their maximum range and what lies beyond their targets. Without maps for all systems, identifying maximum ranges and what lies beyond a target through cyberspace is impossible.

Defending against cyberattacks is possible. The current offensive advantage does not make all defense pointless. Even if intrusions from sophisticated, persistent attacks are inevitable, certain defensive measures (e.g., physical security controls, limiting user access, filtering and antivirus software, and firewalls) do offer some protection. Redundancy and replication are resilience strategies that can deter some would-be attackers by making attacks futile.[28] Retaliatory responses via cyberspace or other means can also enhance deterrence.[29] Defense is currently disadvantaged, but offense gets no free pass in cyberspace.

Expectations and Recommendations

The advantages and limitations of using cyberpower inform expectations for the future and several recommendations for the military.

Do not expect clear, comprehensive policy soon.[30] Articulating a comprehensive U.S. strategy for employing nuclear weapons lagged 15 years behind their first use, and the timeline for clear, comprehensive cyberspace policy may take longer.[31] Multiple interests collide in cyberspace, forcing policy makers to address concepts that traditionally have been difficult for Americans to resolve. Cyberspace, like foreign policy, exposes the tension between defaulting to realism in an ungoverned, anarchic system, and aspiring to the liberal ideal of security through mutual recognition of natural rights. Cyberspace policy requires adjudicating between numerous priorities based on esteemed values such as intellectual property rights, the role of government in business, bringing criminals to justice, freedom of speech, national security interests, and personal privacy. None of these issues is new. Cyberspace just weaves them together and presents them from unfamiliar angles. For example, free speech rights may not extend to falsely shouting fire in crowded theaters, but through cyberspace all words are broadcast to a global crowded theater.[32]

Beyond the domestic front, the Internet access creates at least one significant foreign policy dilemma. While it can help mobilize and empower dissidents under oppressive governments, it also can provide additional population control tools to authoritarian leaders.[33] The untangling of these sets of overlapping

issues in new contexts is not likely to happen quickly. It may take several iterations, and it may only occur in crises. Meanwhile, the military must continue developing capabilities for operating through cyberspace within current policies.

Defend in Depth—Inner Layers

Achieving resilience requires evaluating dependencies and vulnerabilities at all levels. Starting inside the firewall and working outward, defense begins at the lowest unit level. Organizations and functions should be resilient enough to sustain attacks and continue operating. In a period of declining budgets, decision makers will pursue efficiencies through leveraging technology.[34] Therefore, prudence requires reinvesting some of the savings to evaluate and offset vulnerabilities created by new technological dependencies.[35] Future war games should not just evaluate what new technologies can provide, but also they should consider how all capabilities would be affected if denied access to cyberspace.

Beyond basic user responsibilities, forces providing defense against cyberattacks require organizations and command structures particular to their function. Martin van Creveld outlined historical evolutions of command and technological developments. Consistent with his analysis, military cyberdefense leaders should resist the technology-enabled urge to centralize and master all available information at the highest level. Instead, their organizations should act semi-independently, set low decision thresholds, establish meaningful regular information reporting, and use formal and informal communications.[36] These methods can enhance "continuous trial-and-error learning essential to collectively make sense of disabling surprises" and shorten response times.[37] Network structures may be more appropriate for this type of task than traditional hierarchical military structures.[38] Whatever the structure, military leaders must be willing to subordinate tradition and task-organize their defenses for effectiveness against cyberattacks.[39] After all, weapons "do not triumph in battle; rather, success is the product of man-machine weapon systems, their supporting services of all kinds, and the organization, doctrine, and training that launch them into battle."[40]

Defend in Depth—Outer Layers

Defending against cyberattacks takes more than firewalls. Expanding defense in depth requires creatively leveraging influence. DOD has no ownership or jurisdiction over the civilian sectors operating the Internet infrastructure and developing computer hardware and software. However, DOD systems are vulnerable to cyberattack through each of these avenues beyond their control.[41] Richard Clarke recommended federal regulation starting with the Internet backbone as the best way to overcome systemic vulnerabilities.[42] Backlash over potential legislation regulating Internet activity illustrates the problematic nature of regulation.[43] So, how can DOD effect change seemingly beyond its control? Label it "soft power" or "friendly conquest of cyberspace," but the answer lies in leveraging assets.[44]

One of DOD's biggest assets to leverage is its buying power. In 2011, DOD spent over \$375 billion on contracts.[45] The military should, of course, use its buying power to insist on strict security standards when purchasing hardware and software. However, it also can use its acquisition process to reduce vulnerabilities through its use of defense contractors. Similar to detailed classification requirements, contracts should specify network security protocols for all contract firms as well as their suppliers, regardless of the services provided. Maintaining stricter security protocols than industry standards would become a condition of lucrative contracts. Through its contracts, allies, and position as the nation's largest employer, DOD can affect preferences to improve outer layer defenses.[46]

Develop an Offensive Defense

Even in defensive war, Clausewitz recognized the necessity of offense to return enemy blows and achieve victory.[47] Robust offensive capabilities can enhance deterrence by affecting an adversary's decision calculus.[48] DOD must prepare for contingencies calling for offensive support to other domains or independent action through cyberspace.

The military should develop offensive capabilities for potential scenarios but should purposefully define its preparations as defense. Communicating a defensive posture is important to avoid hastening a security-dilemma-inspired cyberarms race that may have already started.[49] Over 20 nations reportedly have some cyberwar capability.[50] Even if it is too late to slow others' offensive development, controlling the narrative remains important.[51] Just as the name Department of Defense sends a different message than its former name—War Department—developing defensive capabilities to shut down rogue cyberattackers sounds significantly better than developing offensive capabilities that "knock [the enemy] out in the first round."[52]

Do not expect rapid changes in international order or the nature of war. Without question, the world is changing, but world order does not change overnight. Nye detailed changes due to globalization and the spread of information technologies, including diffusion of U.S. power to rising nations and nonstate actors. However, he claimed it was not a "narrative of decline" and wrote, "The United States is unlikely to decay like ancient Rome or even to be surpassed by another state."[53] Adapting to current trends is necessary, but changes in the strategic climate are not as dramatic as some proclaim.

Similarly, some aspects of war change with the times while its nature remains constant. Clausewitz advised planning should account for the contemporary character of war.[54] Advances in cyberspace are changing war's character but not totally eclipsing traditional means. Sir John Slessor noted, "If there is one attitude more dangerous than to assume that a future war will be just like the last one, it is to imagine that it will be so utterly different that we can afford to ignore all the lessons of the last one."[55] Further, Lonsdale advised exploiting advances in cyberspace but not to "expect these changes to alter the nature of war."[56] Wars will continue to be governed by politics, affected by chance, and waged by people even if through cyberspace.[57]

Do not Overpromise

Advocates of wielding cyberpower must bridle their enthusiasm enough to see that its utility only exists within a strategic context. Colin Gray claimed airpower enthusiasts "all but invited government and the public to ask the wrong questions and hold air force performance to irrelevant standards of superheroic effectiveness."[58] By touting decisive, independent, strategic capabilities, airpower advocates often failed to meet such hyped expectations in actual conflicts. Strategic contexts may have occurred where airpower alone could achieve strategic effects, but more often, airpower was one of many tools employed.

Cyberpower is no different. Gray claimed, "When a new form of war is analyzed and debated, it can be difficult to persuade prophets that prospective efficacy need not be conclusive."[59] Cyberpower advocates must recognize not only its advantages, but also its limitations applied in a strategic context.

Conclusion

If cyberpower is the potential to use cyberspace to achieve desired outcomes, then the strategic context is key to understanding its utility. As the character of war changes and cyberpower joins the fight alongside other domains, military leaders must make sober judgments about what it can contribute to achieving desired outcomes. Decision makers must weigh the opportunities and advantages cyberspace presents against the vulnerabilities and limitations of operations in that domain. Sir Arthur Tedder discounted debate over one military arm or another winning wars single-handedly. He insisted, "All three arms of defense are inevitably involved, though the correct balance between them may and will vary."[60] Today's wars may involve more arms, but Tedder's concept of applying a mix of tools based on their advantages and limitations in the strategic context still stands as good advice.

Notes

1. See Chico Harlan, "Korean DMZ troops exchange gunfire," *Washington Post,* 30 October 2010, <http://www.washingtonpost.com/wp-dyn/content/article/2010/10/29/AR2010102906427.html>. Bullets occasionally fly across the demilitarized zone, but occurrences are rare.

2. See Authorization for Use of Military Force, Public Law 107–40, 107th Cong., 18 September 2001, <http://www.gpo.gov/fdsys/pkg/PLAW-107publ40/html/PLAW-107publ40.htm>. The use of military force in Afghanistan was authorized by the U.S. Congress in 2001 through Public Law 107–40, which does not include a declaration of war.

3. "DOD systems are probed by unauthorized users approximately 250,000 times an hour, over 6 million times a day." Gen. Keith Alexander, director, National Security Agency and Commander, U.S. Cyber Command (remarks, Center for Strategic and International Studies Cybersecurity Policy Debate Series: US Cybersecurity Policy and the Role of US Cybercom, Washington, DC, 3 June 2010, 5), <http://www.nsa.gov/public_info/_files/speeches_testimonies/100603_alexander_transcript.pdf>.

4. "The purpose of this document is to provide the ways and means by which our military will advance our enduring national interests . . . and to accomplish the defense objectives in the 2010 Quadrennial Defense Review." Joint Chiefs of Staff, *The National Military Strategy of the United States of America, 2011: Redefining America's Military Leadership* (Washington, DC: United States Government Printing Office [GPO], 8 February 2011), i.

5. DOD, *DOD Strategy for Operating in Cyberspace* (Washington, DC: GPO, July 2011), 5.

6. Daniel T. Kuehl, "From Cyberspace to Cyberpower: Defining the Problem," in *Cyberpower and National Security,* eds. Franklin D. Kramer, Stuart H. Starr, and Larry K. Wentz (Dulles, VA: Potomac Books, 2009): 26–28.

7. *Staff Report to the Senate Committee on Armed Services, Defense Organization: The Need for Change,* 99th Cong., 1st sess., 1985, Committee Print, 442–44.

8. David J. Lonsdale, *The Nature of War in the Information Age: Clausewitzian Future* (London: Frank Cass, 2004), 182.

9. See Joseph S. Nye, Jr., *The Future of Power* (New York: PublicAffairs, 2011), 123. This definition is influenced by the work of Nye.

10. "From 2000 to 2010, global Internet usage increased from 360 million to over 2 billion people," DOD Strategy for Operating in Cyberspace, 1.

11. Giulio Douhet, *The Command of the Air* (Tuscaloosa, AL: University of Alabama Press, 2009), 9.

12. Richard A. Clarke and Robert K. Knake, *Cyber War: The Next Threat to National Security and What to Do about It* (New York: HarperCollins Publisher, 2010), 9–10.

13. Nye, 126.

14. Audrey Kurth Cronin, "Cyber-Mobilization: The New Levée en Masse," *Parameters* (Summer 2006): 77–87.

15. Clarke and Knake, 81–84.

16. See Clarke and Knake, 84–85. Trends in the number of Internet-connected devices threaten to use up all 4.29 billion available addresses based on the original 32-bit numbering system.

17. Clay Wilson, "Cyber Crime," in *Cyberpower and National Security,* eds. Franklin D. Kramer, Stuart H. Starr, Larry Wentz (Washington, DC: NDU Press, 2009), 428.

18. Carl von Clausewitz, *On War,* ed. and trans. Michael Howard and Peter Paret (Princeton, NJ: Princeton University Press, 1976), 357; John B. Sheldon, "Deciphering Cyberpower: Strategic Purpose in Peace and War," *Strategic Studies Quarterly* (Summer 2011): 98.

19. Martin van Creveld, *Command in War* (Cambridge, MA: Harvard University Press, 1985), 231.

20. Nye, 30.

21. Dexter Filkins, "US Tightens Airstrike Policy in Afghanistan," *New York Times,* 21 June 2009, <http://www.nytimes.com/2009/06/22/world/asia/22airstrikes.html>.

22. "We will improve our cyberspace capabilities so they can often achieve significant and proportionate effects with less cost and lower collateral impact." Chairman of the Joint Chiefs of Staff *The National Military Strategy of the United States of America 2011: Redefining America's Military Leadership* (Washington, DC: GPO, 2011), 19.

23. Thomas C. Schelling, *Arms and Influence* (New Haven, CT: Yale University, 2008), 2–4.

24. For Russian pipeline, see Clarke and Knake, 93; for Stuxnet, see Nye, 127.

25. Lonsdale, 143–45.

26. Rear Adm. J.C. Wylie, *Military Strategy: A General Theory of Power Control* (Annapolis, MD: Naval Institute Press, 1989), 74.

27. Evgeny Morozov, *The Net Delusion: The Dark Side of Internet Freedom* (New York: PublicAffairs, 2011), 19.

28. Nye, 147.

29. Richard L. Kugler, "Deterrence of Cyber Attacks," *Cyberpower and National Security,* eds. Franklin D. Kramer, Stuart H. Starr, and Larry K. Wentz (Washington, DC: NDU Press, 2009), 320.

30. See United States Office of the President, *International Strategy for Cyberspace: Prosperity, Security, and Openness in a Networked World,* May 2011.

31. See Clarke and Knake, 155. International strategy for cyberspace addresses diplomacy, defense, and development in cyberspace but fails to outline relative priorities for conflicting policy interests. 31.

32. First Amendment free speech rights and their limits have been a contentious issue for decades. "Shouting fire in a crowded theater" comes from a 1919 U.S. Supreme Court case, *Schenck v. United States.* Justice Oliver Wendell Holmes' established context as relevant for limiting free speech. An "imminent lawless action" test superseded his "clear and present danger"

test in 1969, <http://www.pbs.org/wnet/supremecourt/capitalism/landmark_schenck.html>.

33. Morozov, 28.

34. "Today's information technology capabilities have made this vision [of precision logistics] possible, and tomorrow's demand for efficiency has made the need urgent." Gen. Norton Schwartz, chief of staff, U.S. Air Force, "Toward More Efficient Military Logistics," address on 29 March 2011, to the 27th Annual Logistics Conference and Exhibition, Miami, FL, <http://www.af.mil/shared/media/document/AFD-110330-053.pdf>.

35. Chris C. Demchak, *Wars of Disruption and Resilience: Cybered Conflict, Power, and National Security* (Athens, GA: University of Georgia Press, 2011), 44.

36. Van Creveld, 269–70.

37. Demchak, 73.

38. Antoine Bousquet, *The Scientific Way of Warfare: Order and Chaos on the Battlefields of Modernity* (New York: Columbia University Press, 2009), 228–29.

39. See R.A. Ratcliff, *Delusions of Intelligence: Enigma, Ultra, and the End of Secure Ciphers* (Cambridge, UK: Cambridge University Press, 2006), 229–30. Allied World War II Enigma code-breaking offers a successful example of creatively task-organizing without rigid hierarchy.

40. Colin S. Gray, *Explorations in Strategy* (Westport, CT: Praeger, 1996), 133.

41. *DOD Strategy for Operating in Cyberspace*, 8.

42. Clarke and Knake, 160.

43. Geoffrey A. Fowler, "Wikipedia, Google Go Black to Protest SOPA," *Wall Street Journal*, 18 January 2012, <http://online.wsj.com/article/SB10001424052970204555904577167873208040252.html?mod=WSJ_Tech_LEADTop>; Associated Press, "White House objects to legislation that would undermine 'dynamic' Internet," Washington Post, 14 January 2012, <http://www.washingtonpost.com/politics/courts-law/white-house-objects-to-legislation-that-would-undermine-dynamic-internet/2012/01/14/gIQAJsFcyP_story.html>.

44. "Soft power," see Nye, 81–82; "friendly conquest," see Martin C. Libicki, *Conquest in Cyberspace: National Security and Information Warfare* (Cambridge, UK: Cambridge University Press, 2007), 166.

45. U.S. Government, USASpending.gov official Web site, "Prime Award Spending Data," <http://www.usaspending.gov/explore?carryfilters=on> (18 January 2012). "2011" refers to the fiscal year.

46. DOD Web site, "About the Department of Defense," <http://www.defense.gov/about> (18 January 2012). DOD employs 1.4 million active, 1.1 million National Guard/Reserve, 718,000 civilian personnel.

47. Clausewitz, 357.

48. Kugler, "Deterrence of Cyber Attacks," 335.

49. "Many observers postulate that multiple actors are developing expert [cyber] attack capabilities." Ibid., 337.

50. Clarke and Knake, 144.

51. "Narratives are particularly important in framing issues in persuasive ways." Nye, 93–94.

52. Quote from Gen. Robert Elder as commander of Air Force Cyber Command. See Clarke and Knake, 158; Defense Tech, "Chinese Cyberwar Alert!" 15 June 2007, <http://defensetech.org/2007/06/15/chinese-cyberwar-alert>.

53. Nye, 234.

54. Clausewitz, 220.

55. John Cotesworth Slessor, *Air Power and Armies* (Tuscaloosa, AL: University of Alabama Press, 2009), iv.

56. Lonsdale, 232.

57. Clausewitz, 89.

58. Gray, 58.

59. Colin S. Gray, *Modern Strategy* (Oxford, UK: Oxford University Press, 1999), 270.

60. Arthur W. Tedder, *Air Power in War* (Tuscaloosa: University of Alabama Press, 2010), 88.

Critical Thinking

1. What is the greatest threat to U.S. cybersecurity?

2. Why is it so difficult to defend U.S. cyberspace?

3. What recommendations would you make to defend U.S. cyberspace?

Create Central

www.mhhe.com/createcentral

Internet References

Department of Defense Strategy for Operating in Cyberspace
http://www.defense.gov/news/d20110714.cyber.pdf

Economist debates cyberwar
http://www.economist.com/debate/overview/256

National Security Agency
https://www.nsa.gov

KEVIN L. PARKER, U.S. Air Force, is the commander of the 100th Civil Engineer Squadron at RAF Mildenhall, United Kingdom. He holds a BS in civil engineering from Texas A&M University, an MA in human resource development from Webster University, and an MS in military operational art and science and an MPhil in military strategy from Air University. He has deployed to Saudi Arabia, Kyrgyzstan, and twice to Iraq.

Lt. Col. Kevin Parker, "The Utility of Cyberpower," *Military Review*, May/June 2014, pp 26–33. HQ. Department of the Army, US Army Combined Arms Center.

Article   Prepared by: Robert Weiner, *University of Massachusetts, Boston*

U.N. Treaty Is First Aimed at Regulating Global Arms Sales

Neil MacFarquhar

Learning Outcomes

After reading this article, you will be able to:

- Describe the scope of the new treaty.

- Discuss the opposition to this treaty and their reasons.

- Consider the issues surrounding the prospects for ratification in the U.S. Senate.

United Nations—The United Nations General Assembly voted overwhelmingly on Tuesday to approve a pioneering treaty aimed at regulating the enormous global trade in conventional weapons, for the first time linking sales to the human rights records of the buyers.

Although implementation is years away and there is no specific enforcement mechanism, proponents say the treaty would for the first time force sellers to consider how their customers will use the weapons and to make that information public. The goal is to curb the sale of weapons that kill tens of thousands of people every year—by, for example, making it harder for *Russia* to argue that its arms deals with *Syria* are legal under international law.

The treaty, which took seven years to negotiate, reflects growing international sentiment that the multibillion-dollar weapons trade needs to be held to a moral standard. The hope is that even nations reluctant to ratify the treaty will feel public pressure to abide by its provisions. The treaty calls for sales to be evaluated on whether the weapons will be used to break humanitarian law, foment genocide or war crimes, abet terrorism or organized crime or slaughter women and children.

"Finally we have seen the governments of the world come together and say 'Enough!' " said *Anna MacDonald,* the head of arms control for Oxfam International, one of the many rights groups that pushed for the treaty. "It is time to stop the poorly regulated arms trade. It is time to bring the arms trade under control."

She pointed to the Syrian civil war, where 70,000 people have been killed, as a hypothetical example, noting that Russia argues that sales are permitted because there is no arms embargo.

"This treaty won't solve the problems of Syria overnight, no treaty could do that, but it will help to prevent future Syrias," Ms. MacDonald said. "It will help to reduce armed violence. It will help to reduce conflict."

Members of the General Assembly voted 154 to 3 to approve the Arms Trade Treaty, with 23 abstentions—many from nations with dubious recent human rights records like Bahrain, Myanmar and Sri Lanka.

The vote came after more than two decades of organizing. Humanitarian groups started lobbying after the 1991 Persian Gulf war to curb the trade in conventional weapons, having realized that Iraq had more weapons than France, diplomats said.

The treaty establishes an international forum of states that will review published reports of arms sales and publicly name violators. Even if the treaty will take time to become international law, its standards will be used immediately as political and moral guidelines, proponents said.

"It will help reduce the risk that international transfers of conventional arms will be used to carry out the world's worst crimes, including terrorism, genocide, crimes against humanity and war crimes," Secretary of State John Kerry said in a statement after the United States, the biggest arms exporter, voted with the majority for approval.

But the abstaining countries included China and Russia, which also are leading sellers, raising concerns about how many countries will ultimately ratify the treaty. It is scheduled to go into effect after 50 nations have ratified it. Given the overwhelming vote, diplomats anticipated that it could go into effect in two to three years, relative quickly for an international treaty.

Proponents said that if enough countries ratify the treaty, it will effectively become the international norm. If major sellers like the United States and Russia choose to sit on the sidelines while the rest of the world negotiates what weapons can be traded globally, they will still be affected by the outcome, activists said.

The treaty's ratification prospects in the Senate appear bleak, at least in the short term, in part because of opposition by the gun lobby. More than 50 senators signaled months ago that they would oppose the treaty—more than enough to defeat it, since 67 senators must ratify it.

Among the opponents is Senator John Cornyn of Texas, the second-ranking Republican. In a statement last month, he said that the treaty contained "unnecessarily harsh treatment of civilian-owned small arms" and violated the right to self-defense and United States sovereignty.

In a bow to American concerns, the preamble states that it is focused on international sales, not traditional domestic use, but the National Rifle Association has vowed to fight ratification anyway. The General Assembly vote came after efforts to achieve a consensus on the treaty among all 193 member states of the United Nations failed last week, with Iran, North Korea and Syria blocking it. The three, often ostracized, voted against the treaty again on Tuesday.

Vitaly I. Churkin, the Russian envoy to the United Nations, said Russian misgivings about what he called ambiguities in the treaty, including how terms like genocide would be defined, had pushed his government to abstain. But neither Russia nor China rejected it outright.

"Having the abstentions from two major arms exporters lessens the moral weight of the treaty," said Nic Marsh, a proponent with the Peace Research Institute in Oslo. "By abstaining they have left their options open."

Numerous states, including Bolivia, Cuba and Nicaragua, said they had abstained because the human rights criteria were ill defined and could be abused to create political pressure. Many who abstained said the treaty should have banned sales to all armed groups, but supporters said the guidelines did that effectively while leaving open sales to liberation movements facing abusive governments.

Supporters also said that over the long run the guidelines should work to make the criteria more standardized, rather than arbitrary, as countries agree on norms of sale in a trade estimated at $70 billion annually.

The treaty covers tanks, armored combat vehicles, large-caliber weapons, combat aircraft, attack helicopters, warships, missiles and launchers, small arms and light weapons. Ammunition exports are subject to the same criteria as the other war matériel. Imports are not covered.

India, a major importer, abstained because of its concerns that its existing contracts might be blocked, despite compromise language to address that.

Support was particularly strong among African countries—even if the compromise text was weaker than some had anticipated—with most governments asserting that in the long run, the treaty would curb the arms sales that have fueled many conflicts.

Even some supporters conceded that the highly complicated negotiations forced compromises that left significant loopholes. The treaty focuses on sales, for example, and not on all the ways in which conventional arms are transferred, including as gifts, loans, leases and aid.

"This is a very good framework to build on," said Peter Woolcott, the Australian diplomat who presided over the negotiations. "But it is only a framework."

Rick Gladstone contributed reporting from New York, and Jonathan Weisman from Washington.

This article has been revised to reflect the following correction:

Correction: April 4, 2013

An article on Wednesday about the United Nations General Assembly's overwhelming approval of an arms control treaty misspelled, in some copies, the surname of the head of arms control for Oxfam International, one of many rights groups that pushed for the treaty. She is Anna MacDonald, not McDonald.

Critical Thinking

1. What is the status of disarmanent talks between Russia and United States regarding nuclear arms?

2. How does the issue of disarmament relate to the issue of human rights?

3. How much influence in domestic politics do arms manufacturers have in the United States and elsewhere?

Create Central

www.mhhe.com/createcentral

Internet References

United Nations Office for Disarmament Affairs
www.un.org/disarmament/ATT

Oxfam International
www.oxfam.org

Human Rights Watch
www.hrw.org

Article   Prepared by: Robert Weiner, *University of Massachusetts, Boston*

Turkey at a Tipping Point

JENNY WHITE

Learning Outcomes

After reading this article, you will be able to:

- Understand the relationship between Kemalism and Turkish National Identity.
- Discuss the impact of Erdoğan on the Turkish political system.

Something substantially different is shaping up in today's Turkey. Given the many variables in play, no one can be sure what the country will look like in 10 years. The recent autocratic turn of the pious former prime minister and now president, Recep Tayyip Erdoğan, cannot be explained simply as a form of Islamic radicalization. After more than a decade of economic growth and social reform under the Justice and Development Party (AKP), Muslim and Turkish identities have been transformed to such an extent that it is nearly impossible to assign people to one end or the other of a secular-Islamist divide, particularly that half of the population that is under 30. Many young people have heterogeneous identities, composed of seemingly contradictory positions and affiliations. Turkey is now split along more complex lines, pitting Sunni against Sunni, Sunni against Alevi (a heterodox Shia sect that makes up more than 10 percent of the population), and both pious and secular nationalists against Kurds. It could be argued that a lust for power and profit on the part of one man and his inner circle, rather than a wider cohort, has driven recent events as much as religion. This is no novelty in the world of dictators, which may well be the direction Turkey is taking.

Part of the answer to what is happening in the present lies in the past, in Kemalist practices (the legacy of Mustafa Kemal Atatürk, who founded the modern Turkish state in 1923) that still powerfully shape social and political life today. Erdoğan, threatened by recent street protests and the actions of a rival Islamic movement, has returned to the fearmongering and aggressive political paternalism that were ingrained in the Turkish psyche for much of the twentieth century, making them powerful tools for social manipulation. Kemalism has been largely dethroned, but the levers of power it developed remain in place. In the absence of Kemalist symbolism, AKP rule has taken on an Ottoman and Sunni Muslim veneer.

What is fundamentally different, though, is that Erdoğan has begun, for the first time, to dismantle the democratic structures that, creaky and biased though they were, provided a balance of power among institutions. Under Erdoğan, these institutions, from universities and the media to police, prosecutors, and judges, have been forced to answer not to a party, but essentially to one man who has taken control of most mechanisms of rule. This is a new and worrisome development, out of step with the AKP's (and Erdoğan's) accomplishments over the previous decade. Those who claim to have seen this coming could have done so only by closing their eyes to what the party accomplished—and what these newest developments put at risk.

David or Goliath?

From 2002 until 2011, the AKP attracted a wide variety of voters, drawn to its economic program, global outlook, revival of Turkey's European Union accession process, and introduction of much-needed reforms, which included placing the military under civilian control. The party profited from a reservoir of public sympathy and support after the military in 2007 and the Constitutional Court in 2008 threatened to bring the government down for alleged anti-secular activities. The AKP represented David against the military Goliath that had ousted several governments since 1960.

Once in power, the AKP reached out to minorities and former national enemies like Greece and Armenia. It broke nationalist taboos by acknowledging, to some degree, the 1915 Armenian massacres and the slaughter of Alevis at Dersim in 1937 and 1938, while pursuing a solution to the division of Cyprus, Kurdish cultural rights, and peace with the separatist Kurdistan Workers' Party (PKK).

Per capita income doubled on the AKP's watch, although unemployment remained near 10 percent, with youth unemployment much higher and women's labor force participation just 29 percent. An improved economy, social welfare, and new roads and subways brought votes, while opposition parties were ineffectual. This combination continues to be successful: About half of the population consistently votes for the AKP (43 percent in March local elections and 52 percent in the August presidential election). In other words, the AKP appears to have done well by the country, and there is no other party voters trust to keep the train on the rails.

A noticeable change in direction occurred in 2011. In a general election that June, the AKP won just under 50 percent of the vote, giving it a majority of 326 seats in the 550-seat parliament, and empowering Erdoğan to centralize power. He replaced independent thinkers in the party with loyalists who often lacked the requisite experience or expertise. The military was brought to heel through a series of trials (known as the Ergenekon and Sledgehammer cases) and the subsequent imprisonment of hundreds of high-ranking officers accused of plotting coups. In July 2011, the chief of the general staff and the commanders of the land, sea, and air forces resigned en masse; they were replaced by more tractable men. Once the threat of a military coup and dissenting voices within the party were removed, the AKP's message became narrower, focused on a romanticized notion of Ottoman Sunni brotherhood, and more intolerant. Erdoğan began to see enemies and threats everywhere, mistaking dissent and protest against government policies for coup attempts.

For most of its rule, the AKP had worked in tandem with the Hizmet movement led by the Muslim cleric Fethullah Gülen, who has lived in self-imposed exile in Pennsylvania since 1997. Hizmet excelled at setting up well-regarded schools and businesses in Turkey and abroad, with the aim of developing what Gülen called a "golden generation" of youth equipped with business and science skills and Muslim ethics, who could staff state agencies. For every embassy the AKP government opened abroad—dozens in sub-Saharan Africa alone—Hizmet would set up local schools and businesses. But relations between the AKP and Hizmet began to fray several years ago.

Hizmet is widely thought to have a heavy presence in the Turkish police and security services. In December 2013, Erdoğan accused it of being behind prosecutors and police who tried to arrest close members of his circle on corruption charges. He claimed that the investigation was a coup attempt, and that Hizmet had created a "parallel state." He transferred or fired thousands of police officers and prosecutors in a successful attempt to derail the charges. The AKP also closed down Hizmet's lucrative prep schools in Turkey and brought Bank Asya, which is associated with Gülen, to its knees by orchestrating a massive withdrawal of deposits. Each side, proclaiming its Sunni piety, has vowed to destroy the other.

The constitution is designed to protect the rights of the state, not the individual.

Out of Touch

In response to this perceived coup attempt, the AKP curtailed civil liberties, banning YouTube and Twitter after they were used to circulate taped evidence from the corruption investigation. Recently passed laws allow intrusive government surveillance and arrests of citizens for thought crimes. Given the jailing and harassment of journalists and protesters, and the impunity of the police in using violence, little is now possible in the way of freedom of speech. Erdoğan has revived the Kemalist threat paradigm, using the same language, railing against outside and inside enemies, and presenting himself in his campaign ads and speeches as the heroic savior of the nation, the patriarchal father protecting the honor of his national family and keeping the dangerous chaos of liberalism at bay.

By pulling the levers of suspicion and social polarization, Erdoğan appeals to the conservative nationalist core of his supporters, but he is out of touch with a large part of the population. There is a growing disconnect between the twenty-first-century aspirations of both pious and secular youth, who grew up in the AKP environment of great promise, and the twentieth-century values and practices of Turkey's leadership, which cannot bend to meet that promise and is preoccupied with serving its own interests. The AKP raked in enormous profits through rampant development all over the country, despoiling environments, neighborhoods, and archaeological sites. The 2013 demonstrations began as a peaceful sit-in to save Gezi Park in Istanbul's Taksim Square, then grew into a nationwide protest against the disproportionate police violence used to break it up.

The Gezi events occurred around the same time that enormous crowds filled Cairo streets to show their approval of the Egyptian army's coup against President Mohamed Morsi. Erdoğan, who felt a kinship with the Muslim Brotherhood and Morsi, clearly viewed the Gezi protests in light of the events in Egypt, convinced that the protesters were plotting to overthrow him. He responded with an all-out crackdown, including arrests of protesters under draconian terrorism laws. It is not only secular youth, however, who have taken up the call of environmentalism and other social justice issues. There has been a convergence in lifestyle and aspirations between secular and pious youth, who have developed a taste for making their own choices and demanding accountability.

Erdoğan's increasing volatility and consolidation of power have opened fissures in the AKP edifice. Party members uncomfortable with his policies dare not speak up. Many hoped that Abdullah Gül, when he stepped down from the presidency in August, would capitalize on his popularity and legitimacy by leading a moderate branch of the party, but he has disappeared from the headlines.

Even the conservative provincial folk who make up a large part of the AKP's core constituency have recoiled from the gloves-off exercise of raw power by Erdoğan and his circle, which even religious pretexts can no longer disguise. Earlier this year, many citizens were shocked by the callousness with which Erdoğan and his advisers treated family members waiting for news of their missing relatives after a mine disaster in the western town of Soma, in which 301 miners were killed. Despite media censorship, a photo of an Erdoğan aide kicking a miner went viral, as did a video of a large crowd booing the prime minister. Erdoğan was forced to take refuge in a market, where he was caught on camera punching another miner.

Another wild card is the recruitment of Turks by Islamic State (ISIS) jihadists to join the group's fighters in Iraq and Syria. Many of its recruits hail from nearby countries like Iraq and Saudi Arabia; they have moved freely across Turkey's borders and taken up residence in its cities and border towns. Turkey's largely Sunni and Alevi population has no affinity with ISIS's puritan Salafist creed and in the past has been suspicious of foreigners, including Arabs. But the weakening of physical borders as a result of the AKP's dream of a Muslim union of states in former Ottoman lands, and the breakdown of firm national and Muslim identities and proliferation of alternative practices beyond "Turkish Islam," have opened cracks in Turkish society in which ISIS can establish roots.

Sèvres Syndrome

Over the past decade, Washington slowly and somewhat reluctantly came to the realization that Turkey was no longer the pliant, army-led Kemalist ally of Cold War years, but had become a self-possessed nation with a booming economy, proactive foreign policy, global political and economic reach, and a headstrong and openly pious Muslim prime minister. Pundits initially warned that the Islam-rooted AKP was moving the country away from the West and toward the Islamic East, but that view dissipated when it became clear that Turkey was pursuing interests in Europe, sub-Saharan Africa, South America, and Asia, not just the Middle East. The new Turkish leaders imagined themselves walking in the footsteps not of Atatürk, the war hero and first president of the nation, but of the Ottomans, lords of a world empire. When the Middle East imploded in the 2011 Arab uprisings and their turbulent aftermath, Turkey seemed to be the one stable Muslim-majority country left standing in the region.

This new brand of Turkey emerged in sharp contrast to the crisis-ridden country of earlier decades. Although the Kemalist state oversaw free and fair elections that became the expected standard, the country was micromanaged socially and politically by elites positioned in state institutions and by the military, which carried out several coups when it felt that national unity was threatened by nonconforming identities and ideologies. This aggressive defensiveness, which some scholars call Turkey's Sèvres Syndrome, is a century-long hangover from the dismemberment of the Ottoman Empire by the Europeans, formalized by the 1923 Treaty of Sèvres.

Since then, in schoolbooks and a variety of rituals from grade school to adulthood, Turks have learned to be militant, to know who their enemies are, and to be suspicious of outsiders. Polls show that a majority of Turks not only lead the world in disliking the United States, but they dislike pretty much everyone else too, Muslim countries included. That hostility extends to next-door neighbors with different religious beliefs or lifestyles. A continual drumbeat of acts of intolerance against Armenians, Greek Christians, Protestants, Kurds, Alevis, Roma, Jews, and others has left deep tears in the social fabric.

Citizenship, in the sense of a contract between the nation-state and its people, was poorly developed. Schoolchildren were taught that the ideal quality was unquestioning obedience to the state, the highest expression of which would be to sacrifice their lives for it. There was little mention of what the state would provide for its citizens, aside from protection against the ever-present threat posed by what were called inside and outside enemies, the bogeymen of the nation-state. The current constitution, written under military oversight following a 1980 coup, is designed to protect the rights of the state, not the individual. Kemalism's message was one of unceasing embattlement, buttressed by conspiracy theories, and nurturing a deep-seated belief that a strong patriarchal state (*Devlet Baba*, or Father State, in popular parlance) and army were necessary in order to protect the national family and its citizen children from outsiders still hell-bent on destroying them.

Muslim Nationalism

Non-Muslim citizens and other ethnic minorities like the Kurds suffered greatly under Kemalist nationalist policies that defined them as pawns manipulated by outside powers to undermine Turkish national unity. Although Kemalists promoted a secular lifestyle, their policies were based on a religio-racial understanding of Turkishness that was contingent on being Muslim. Yet Kemalist Islam did not require piety and, indeed, eyed it with suspicion; for many years, the headscarf was barred from government offices and universities (the ban was lifted in 2013). Until the 1990s, the headscarf and other overt demonstrations of piety were associated with the rural poor and urban migrants from the countryside, both romanticized and disdained.

The Kemalist state ran a tight Islamic ship. The Presidency of Religious Affairs controlled mosques, religious teaching, and public expressions of faith. State laicism was not secularism so much as state-controlled Sunni Islam. Other faiths and forms of Islamic worship, such as the heterodox Alevi sect and officially banned but proliferating Sufi orders, coexisted in the shadows and gained adherents, including some politicians. In the 1980s, under the leadership of Necmettin Erbakan, political parties with a clear Islamist bent began to make headway in elections, but were continually closed down by the courts, only to reopen under other names. Erdoğan, Gül, and other dissidents broke away from Erbakan's Welfare Party after his government was forced out in 1997, and in 2001 they founded the AKP, which they claimed was not Islamic, but rather a secular (not laicist) party run by pious Muslims. That is, Muslimhood was a personal attribute of individual politicians, not a party ideology. The party would make policy based on pragmatic considerations, not Islam. It aimed to represent all sectors of Turkish society. And for a time, it did.

Erdoğan began to see enemies and threats everywhere, mistaking dissent and protest for coup attempts.

In the mid-1980s, Prime Minister Turgut Özal had opened Turkey's economy to the world market, unleashing provincial entrepreneurs who had been left out of state-supported industrial development. These businessmen tended to be pious, and their newly acquired wealth and dominance in social and political networks led to the rise of an Islamic bourgeoisie. Under the AKP, they have developed alternative definitions of the nation and the citizen based on a post-Ottoman rather than a republican model.

Such changes have allowed the new pious elites to experiment with expressions of Muslimhood and national identity that would not have been possible before. Muslim nationalism is based on a cultural ideal of Turkishness, rather than blood-based Turkish ethnicity. It imagines the nation with more flexible Ottoman postimperial boundaries, instead of the historically embattled republican borders. The founding moment for this ideology is not the 1923 establishment of the nation-state, but the 1453 conquest of Constantinople by the Turks, which is reenacted, visually depicted in public places, and commemorated in festivities, sometimes displacing Kemalist national rituals.

This shift has created quite a different understanding of Turkish national interests, freeing the AKP to engage with Turkey's non-Muslim minorities, open borders to Arab states by waiving visa requirements, and make global alliances and pursue economic and political interests without concern for the ethnic identity of its interlocutors or the role they played in republican

history—for instance, in relations with former enemies Greece and Armenia. When it was first elected, the AKP systematically began to break down military tutelage and reach out to non-Muslims and Kurds, returning confiscated properties and allowing use of previously banned non-Turkish languages. Erdoğan began to negotiate a peace deal with the Kurdish PKK, which the government classifies as a terrorist organization, after three decades of fighting and more than 40,000 dead.

The ban on three letters of the alphabet used in Kurdish—q, w, and x—was eliminated. Education in the Kurdish language was allowed in private institutions, though not in public ones. Place names of villages and regions were restored to their Kurdish or Alevi originals. Tunceli, for instance, would once again become Dersim, reminding everyone of the state massacre of Alevis that occurred there in the 1930s (Erdoğan blamed it on the secular, Kemalist Republican People's Party, which was in power at the time).

There has been a convergence in lifestyle and aspirations between secular and pious youth.

New Identities

Kemalism as a nationalist ideology has been pushed to the margins, although nationalism itself is alive and thriving in new forms. The concept of what it means to be Turkish, which was shaped by ideological indoctrination in schools, has become more malleable in recent years, up for reinterpretation in a marketplace of identities browsed by a burgeoning middle class that is young, globalized, and desires to be modern. For the first time in republican history, an Islamic identity is associated with upward mobility. Islam is a faith, but also a lifestyle choice with its own fashions, leisure options, musical styles, and media that mirror secular society. If they choose to work, pious young women can now find jobs and arenas of activism and professional development open to them, especially since the lifting of the headscarf ban.

The 2013 protests began in response to the government's attempt to turn Gezi Park, one of central Istanbul's last parks in a city with less than 2 percent public green space, into a mall. The police violently put down the protests, but instead of making them fade away, this response provoked a spontaneous, nationwide series of mass demonstrations. Mostly young and secular, and including many women, the Gezi protesters are another product of the changes in Turkish society since the 1980s. They are global, playful, and consumerist. Turkishness is a personal attribute for them, just as the AKP suggested that Muslimhood was a personal attribute. They represent themselves, not an ideological position, a party, or a scheming foreign power. It was the first time in Turkish history that such masses of people—many with contradictory or competing interests—came together without any ideological or party organization.

The emergence of these new publics, even if only briefly, heralded an important step in Turkey's transformation away from twentieth-century values and incomplete political structures, toward a more tolerant democratic order and a civic nationalism based on citizenship rather than blood or group membership. But young people and women have little place in a political system dominated by older males. They find outlets in a civil society and in the street, but are unlikely for at least the next decade to have an impact on the system that Erdoğan is consolidating under himself—unless that system changes dramatically to permit independent voices, which at this juncture seems doubtful.

The rigidity of the political system is heightened by a widely shared majoritarian understanding of democracy in which the electoral winners get to determine what is allowed and what is banned in social life according to the norms of their community, with no room for nonconforming practices or ideas. This is true whether the issue is banning alcohol consumption or banning the veil. As Erdoğan told the Gezi protesters: If you don't agree with my decisions, win an election.

Kurdish Crisis

In October, nationwide protests by Kurdish citizens broke out against the government's refusal to help protect the Kurdish town of Kobani, just across the border in Syria, against an ISIS onslaught. The protests turned violent, leaving 40 people dead. The reluctance to act reflected Turkish perceptions that Syrian President Bashar al-Assad's survival and the strengthening of Kurdish nationalist aspirations in Syria are greater dangers to Turkey's national integrity than ISIS. The Turkish government (as well as many of its nationalist constituents who will be casting votes in the June 2015 general election) perceives the PKK as an existential threat, though Ankara is in peace talks with jailed PKK leader Abdullah Öcalan and on good terms with the Kurdistan Democratic Party (PDK) in Iraq. Indeed, Iraqi Kurdistan has become a lucrative trade partner.

If Kobani falls, the peace negotiations may be a dead letter; but one could argue that they are already on life support. The PKK appears to be experiencing a struggle for supremacy between the still-popular Öcalan and top military commander Cemil Bayık. On Ankara's side, nationalist factions in government and the military may be pushing against any accommodation with the Kurds, while others advocate continuing the talks. In October, the negotiations were proceeding in Ankara at the same moment as Turkish planes were bombing PKK militants in eastern Turkey in retaliation for the killing of three soldiers.

Turkey sought to enlist a Syrian Kurdish group, the Democratic Union Party (PYD), to help topple Assad, but was rebuffed. If there is to be an autonomous Kurdish region in Syria (which could benefit Turkey by buffering it from the Syrian war), Ankara would prefer that it not be run by the unpredictable PYD, an ally of the PKK. Ankara's recent decision to allow *peshmerga* fighters from Iraqi Kurdistan to cross into Kobani via Turkey, while rejecting international pressure to arm the PYD, is an awkward compromise. Turkey trusts the peshmerga, but Iraqi Kurds and the PYD/PKK are rivals for power, not friends.

Nevertheless, Turkey had to do something to avert another wave of refugees. In the first week after ISIS assaulted Kobani, 140,000 Syrians fled into Turkey in two days alone—a 10 percent increase in the refugee population of 1.4 million. Turkey feels it does not get enough international aid or respect for carrying this burden. Officials fear that any further influx, combined with rising unrest among the Kurds and increasing anti-refugee sentiment, could lead to major social instability. ISIS is a threat, but Ankara sees no good outcome from confronting it. The international coalition fighting ISIS seems to have no strategic goals to resolve the situation in Syria. Turkish public opinion outside of the Kurdish areas is strongly against involvement in Syria, and suspicion of the PKK is widespread. ISIS is fighting both Assad and the PYD, which seem to be the more immediate evils.

Turkey's broader foreign policy is in tatters, as illustrated in October by its humiliatingly decisive loss in a bid for a non-permanent seat on the United Nations Security Council. The AKP's support for the Muslim Brotherhood and Hamas, both considered threats by Saudi Arabia, the Gulf states (with the exception of Qatar), Egypt, and other regimes in the region, has led not to a Sunni *Pax Ottomana*, but rather to an attenuation of diplomatic ties with these countries.

Open Wounds

Turkey is at a tipping point, held in the balance between those seeking to loosen the reins of heavy-handed paternalistic governance and those unsettled by the chaos of liberalism and desiring order and prosperity (the AKP demonized the Gezi protesters as hoodlums destroying property). Pulling the sectarian lever, however, nourishes extremism.

Within the new context of Muslim nationalism, these tensions have dangerous implications. ISIS penetration of Turkish borders is made possible partly because geographic boundaries in practice have become nearly irrelevant. Although Turkish opinion polls show widespread revulsion against ISIS, it could be argued that part of the population might be vulnerable to recruitment because boundaries of identity are also in flux. In the new post-Ottoman, globalized, commercialized environment of today's Turkey, a choosing Muslim does not have to see himself as a Turkish Muslim, and being a Turk no longer means being bounded by the borders of the nation-state. ISIS recruits are primed to embrace jihadist life by the deep structure of Turkish society, which requires obedience to a patriarchal hierarchy and submergence of selfhood, casting the citizen as self-sacrificing hero.

All of this is destabilizing Turkey internally, ripping open wounds that had partly healed after a decade of reforms. Those wounds are now vulnerable to infection by outside ideologies and actors. Erdoğan, in the meantime, is dismantling Turkey's checks and balances. Surrounded by yes-men, he has moved into his newly constructed thousand-room presidential palace in Ankara. Recently he railed against "those Lawrences" (of Arabia) in the Middle East who, he claimed, are trying to do again today what they did with the Treaty of Sèvres after World War I. Preoccupied with imagined enemies, Turkey's leader is blind to the real threat inside the gates.

Critical Thinking

1. What is the relationship between the Turkish government and the Kurds?

2. How has Turkey's foreign policy changed recently?

3. How has the national identity of Turkey changed, if at all?

Internet References

Justice and Development Party
www.akparti.org.tr/english

People's Democratic Party
https://hdpenglish.wordpress.com

Republic of Turkey, Ministry of Foreign Affairs
www.mfa.gov.tr/default.en.mfa

JENNY WHITE is a professor of anthropology at Boston University. Her latest book is *Muslim Nationalism and the New Turks* (Princeton University Press, 2012).

Article

Prepared by: Robert Weiner, *University of Massachusetts, Boston*

Kurdish Nationalism's Moment of Truth?

Michael Eppel

Learning Outcomes

After reading this article, you will be able to:

- Discuss the effects of the conflicts in Iraq and Syria on Kurdish nationalism.

- Understand the position of Turkey and the Arab states toward Kurdish nationalism.

The moment long awaited by Kurdish nationalists may have arrived. The conditions that arose in Iraq following the fall of Saddam Hussein's regime in 2003, the deterioration of the state in Syria as a result of the civil war raging there since 2011, and changes in traditional Turkish positions regarding the Kurds in Turkey and the autonomous Kurdish region in Iraq have created a historic opportunity. The Kurds are now closer than ever to establishing an independent state or new autonomous regions within the framework of the states that control the territories they inhabit.

The creation of a large independent state is the ultimate vision of the Kurdish national movement. Nonetheless, the leaders of the main Kurdish political forces in Iraq, Syria, Turkey, and Iran have repeatedly declared that their objective is the establishment of autonomous Kurdish regions, disavowing any intention of seceding from the existing states. In Iraq, where the Kurdistan Regional Government (KRG) rules the autonomous region that has existed since 1991, Kurdish leaders profess their loyalty to Baghdad while at the same time making that loyalty contingent on the preservation of broad autonomy. Since 2003, their strategy has aimed to fortify a de facto "state within a state."

In Syria, Salih Muslim Muhammad, the leader of the Democratic Union Party (PYD)—the strongest and most active Kurdish political force in that country—stated several times in 2013

and 2014 that the Kurds' objective is the establishment of an autonomous area within the Syrian state. The principal Kurdish nationalist political forces in Turkey, including the radical leftist organization known as the Kurdistan Workers' Party (PKK), also speak of autonomy and equality for the Kurds within the Turkish state. Mustafa Karasu, one of the PKK's senior leaders, declared in May 2014 that the party had abandoned the aim of creating an independent state and is instead seeking the democratization of Turkey, meaning recognition of the Kurds' right to self-rule. In the same month, leaders of the Free Life Party of Kurdistan (PJAK), the main Kurdish nationalist group in Iran, declared that their new policy aimed to achieve "democratic autonomy for Iranian Kurds."

Since the summer of 2014, an offensive by the jihadist Islamic State in Iraq and Syria (ISIS) has threatened the Kurdish areas in both countries, and complicated their relations with Turkey. But the Kurds' central role in the international military response to ISIS has also strengthened the nationalists' position.

A Stateless People

The Kurds, numbering 25 to 35 million, are the world's largest population group with a developed modern national movement but without a state. Although Kurdish distinctiveness and the signifiers *kurd* and *akrad* have existed in the discourse of the Kurds and among their neighbors since ancient times and certainly since the beginnings of Islam and the Arab conquest, there has never been an independent Kurdish state in Kurdistan at any time in recorded history.

The development of social strata with a modern education and the emergence of a national movement among the Kurds were slow and limited during the nineteenth and early twentieth centuries. This sluggish pace resulted from historical as well as geopolitical conditions. Kurdistan was a landlocked area divided among strong states, the centers of which were in Anatolia, Iran, and Mesopotamia. Others inhibiting factors included the tribal-feudal fragmentation of Kurdish society, which prevented the formation of a state; the absence of a hegemonic Kurdish dialect and the slow development of the written language due to the historical dominance of Arabic, Turkish, and Persian; and a relative detachment from direct modern Western influences, which primarily reached Kurdistan in the context of the Ottoman state's modernization in the nineteenth century.

At the end of World War I, when the map of the Middle East was redrawn, the Kurdish national movement, then in its infancy, did not succeed in obtaining an independent state. British interests (despite the support of then—Colonial Secretary Winston Churchill for a Kurdish state), as well as the weakness and fragmentation of the Kurdish political forces, led to the division of Kurdistan within the boundaries of Turkey, Iran, and the new Arab states of Iraq and Syria created by Britain and France.

Throughout the twentieth century the chances of establishing a Kurdish state remained slim, given strong objections by the states that controlled Kurdish lands, and the preference shown by world powers for cultivating relations with those states rather than supporting the national movement. Ever since World War I, the possibility of an independent Kurdish state, or even of Kurdish autonomy, had been the nightmare of the states among which Kurdistan was divided. Admittedly, in cases where conflicts developed between those states, their regimes did not hesitate to support Kurdish tribal revolts and political forces that acted against their rivals.

Turkey, Iraq, Iran, and Syria shared a common interest in the prevention of a Kurdish state. Their fears of Kurdish nationalism on the one hand, and of Russia and communism on the other, were the undeclared motives for two regional agreements: the Saadabad Pact of 1937 and the Baghdad Pact of 1955. They objected even to Kurdish autonomy within the confines of any of their territories, fearing that this would be the first step toward an independent state, strengthening separatist and irredentist trends throughout the region. The creation of a Kurdish state would have torn extensive territories away from those countries, weakening their status and regimes, and it likely would have incited other minorities to demand autonomy or independence. Turkey and Turkish nationalism since Mustafa Kemal Atatürk went so far as to deny the very existence of a Kurdish people and identity, and made violent efforts to suppress or even eradicate the Kurdish culture and language.

The weakening of nation-states throughout the Arab world since 2011 is creating auspicious conditions for Kurdish nationalism.

Various countries around the world pursued their own interests in maintaining relations with the states that were hostile toward the Kurds. Many of those countries, with minorities that had separatist or autonomist movements of their own, feared the precedent that might be set by the establishment of a Kurdish state.

Encouraging Signs

Since the end of the global Cold War in 1990–91, significant changes have taken place in the conditions that, throughout the twentieth century, prevented the creation of a Kurdish state or of autonomous Kurdish areas. Some of the world's multinational states have crumbled, and minorities have established independent states or autonomous regions within existing states. Developments such as the disintegration of the Soviet Union, Czechoslovakia, and Yugoslavia encouraged the Kurds. They followed with great interest the establishment of South Sudan, which broke away from Sudan and achieved independence in 2011, facing no real opposition from the Arab states or any other country.

The strengthening of movements for autonomy or independence in Catalonia and the Basque region in Spain, in Scotland and Belgium, in Quebec in Canada, and in the Philippines, and the legitimacy that these trends have achieved, reflect a change in the ideological climate, in the characteristics of the international arena, and in the relationship between society and the state in the age of globalization. These trends strengthen the Kurdish claims to autonomy. The Kurds have closely watched the separatist movements in Scotland and Catalonia.

The opposition of Turkey, Iran, Iraq, and the Arab world to the establishment of a Kurdish state continues to constitute a severe obstacle. However, the chances for a state or for more autonomous regions have increased following the consolidation of strong, stable Kurdish autonomy in northern Iraq, as well as growing nationalist sentiment among the Kurds in Turkey, the beginnings of a dialogue between the Turkish government and the Kurds, and the rise of Kurdish self-rule amid civil war in Syria. The weakening of nation-states throughout the Arab world since 2011, together with the strengthening of worldwide autonomist and federalist trends since the end of the Cold War, is creating auspicious conditions for Kurdish nationalism.

Autonomy in Iraq

The Kurdistan Regional Government (KRG) in Iraq has nurtured connections with autonomous regions and federalist movements around the world. At the same time, the Kurds are encouraging those Sunni and Shia forces that seek their own autonomous regions and a federal system in Iraq. An autonomous Shia region in southern Iraq and a Sunni area to the north and west of Baghdad, as well as the transformation of Iraq into a federal or confederal state, would reinforce the autonomy of the Kurdish region relative to the central government in Baghdad.

When Hussein's regime was overthrown in 2003, it appeared that the Kurds' historic opportunity to declare an independent state, at least in Iraqi Kurdistan, had come. However, the Kurdish national leadership in Iraq—Massoud Barzani, head of the Kurdistan Democratic Party (KDP), and Jalal Talabani, leader of the Patriotic Union of Kurdistan (PUK)—felt that such a move would provoke a harsh and uncompromising reaction by Turkey and Iran. Kurdistan, lacking an outlet to the sea, would not be able to export oil or receive foreign aid without the consent and cooperation of at least one of the surrounding countries that opposed an independent Kurdish state: Turkey, Syria, and Iran.

Thus, despite broad-based popular support for independence, the Kurdish leadership in Iraq took the realistic course of building autonomy within the framework of the Iraqi state. But it has also kept open the option of seceding from Iraq and establishing an independent state. So as not to arouse a Turkish and Iranian response, KRG spokespersons have made sure to declare their support for the continued unity of Iraq. In April 2014, Barzani affirmed that the Kurdish aim is broad autonomy within Iraq. However, he warned that if the Iraqi government tried to limit this autonomy, the Kurds would consider an independent state or confederation.

ISIS has enabled the Kurds to achieve international recognition and cemented their status as an independent political and military force.

Ties with Turkey

The KRG is forging closer economic and political ties with Turkey in order to reinforce its status vis-à-vis the central government in Baghdad, but also to assuage Turkish fears of the Kurdish national movement and to strengthen Ankara's interest in the preservation of the autonomous region in Iraq. Turkey's opposition to Kurdish autonomy in Iraq and its threats to invade the region had posed a major obstacle to its transformation into an

independent state, but since 2005 they have worked on improving their relationship. In light of the growing Iranian influence on Baghdad and the tightening of economic bonds between Iraq and Iran, Turkey's interest in improving its position in Iraq and its economic and political ties with the KRG has increased.

More than 1,000 Turkish companies are operating in Iraqi Kurdistan and benefiting from rapid development and construction. The region is likely to become Turkey's principal source of gas and oil, and Turkey will be the main route for KRG energy exports to markets worldwide. At the end of 2013, the KRG and Ankara reached an agreement on Kurdish exports of oil and gas. In 2014, they inaugurated a new oil pipeline to the Turkish port of Ceyhan. Both the KRG and Turkey are exerting pressure on the Iraqi government to accept the agreement and recognize it as commensurate with Iraq's constitution and sovereignty.

Turkey continues to oppose Kurdish secession from Iraq and to regard the establishment of a Kurdish state as contradictory to its strategic interests. In practice, though, the Kurds see progress toward independent oil and gas exports as a significant move toward building a thriving economy and reducing their dependence on the Iraqi government, a trend that reinforces their autonomy and preserves the option of independence. The strengthening of the economic relationship and interdependence between the Iraqi Kurdish region and Turkey constrains Baghdad's ability to exert pressure on the KRG.

The Kurdish leadership in Iraq has an interest in successful negotiations between the Turkish government and the Kurds in Turkey. A renewed struggle would confront the KRG with the dilemma of maneuvering between popular desire to support the Kurdish national struggle in Turkey and the KRG's interests in preserving the relationship with Ankara.

Following the establishment of the Atatürk regime after World War I, Turkey negated the very existence of Kurdish nationality and suppressed the Kurds' culture and language. The PKK launched an armed separatist campaign that has left more than 40,000 dead since the 1980s. But over the past decade, the Turkish government has shifted under Prime Minister (now President) Recep Tayyip Erdoğan toward dialogue with the Kurds. The government initiated informal talks with Abdullah Öcalan, the imprisoned PKK leader. Erdoğan's attempt to increase the flexibility of Turkey's position has inspired the Kurds to hope they are within reach of their objectives of equal rights in Turkey and a regional autonomy that will reflect their national culture and identity.

An arrangement acceptable to the Kurds will require Turkey to make far-reaching compromises, including profound changes in the discourse of Turkish nationalism and the Kemalist ideology on which the Turkish state has been based since World War I. To reach its goal of achieving the status of a regional power, Turkey must find a way to resolve its conflict with the Kurds. This process depends on Turkey's ability not only to accept Kurdish autonomy within its borders, but also to cultivate its relations with the Kurdish regions in Iraq and Syria, or with a Kurdish state, if and when such a state is established in northern Iraq.

Syrian Rivalries

Amid the civil war in Syria that began in 2011, the Kurds have achieved control of northern areas along the border with Turkey, which they call Rojava. The armed wing of the PYD, the People's Protection Units (YPG), is the strongest Kurdish force in Syria. The PYD's main rival is the Kurdistan Democratic Party in Syria (KDP-S), affiliated with Barzani's KDP in Iraq. The KDP-S leads the Kurdish National Council, a coalition of Kurdish parties in Syria that oppose the PYD.

In July 2012, Barzani made efforts to foster reconciliation and coordination among the rival Kurdish forces in Syria. He convened a conference in Erbil, the capital of Iraqi Kurdistan, at which the Syrian parties agreed to establish a Supreme Council. This move would have given Barzani and the KDP the prestige of pan-Kurdish leadership. However, the rivalry among the Syrian groupings has flared up again, becoming violent at times.

The PYD has strong connections with the PKK in Turkey. PYD activists in Syria fly the PKK's flag and openly express their support for the group. They have a close ideological relationship, sharing radical leftist tendencies. Whereas the vital interest of the KRG and especially the KDP lies in cultivating close relations with Turkey and allaying its anxiety over the Kurdish national movement, the relationship between Turkey and the PYD is tense and fraught with suspicion, due to the latter's close ties with the PKK.

In November 2013, the PYD declared self-rule in Syria's Kurdish regions, and in the beginning of 2014 it announced the establishment of three autonomous cantons. The Kurds in Syria control three separate regions, in which they constitute the majority, along the Turkish and Iraqi borders. The fact that the Syrian Kurds as of October 2014 had not achieved territorial contiguity creates a strategic and geopolitical problem for them. The PYD declaration of autonomy met with resistance and criticism from other parties in Syria that belong to the Kurdish National Council. The KRG also rejected the declaration. But the other main forces in Iraqi Kurdistan—the Change Movement (Gorran) and the PUK—voiced their support for the autonomous government declared by the PYD.

Different Paths

Since World War I, the Kurdish nationalist political movements have developed in different ways in each of the states that divide Kurdistan. The Kurdish region in Iraq is the most advanced in terms of political autonomy, and its leadership has aspirations for power or at least seniority in the wider Kurdish nation. It has the advantage of oil profits, whereas the Turkish, Syrian, and Iranian parts of Kurdistan suffer from poverty, underdevelopment, and discrimination. However, the Iraqi Kurds have a population of only 5 million to 5.5 million, compared with the 12 to 18 million Kurds in Turkey, 6 to 8 million in Iran, and 1.8 to 2.5 million in Syria. The Iraqi Kurdish leadership, which acquired a central status within the Kurdish national movement over the course of the twentieth century, will have to become accustomed to a situation in which other power centers—each with its own viewpoint and perception of interests—arise and flourish in Turkey, Syria, and Iran.

These differences contribute to the complexity and unpredictability of political relations among the Kurdish regions. In the development of the national movement in Iraq, tribal or clannish solidarities and affiliations have been interwoven with modern political patterns. The political leadership of the Kurds in Turkey is less influenced by these tribal affinities, given the Turkish suppression of any expression of Kurdishness, of the Kurdish language and culture, and of Kurdish nationalism.

The Turkish state's co-optation of the tribal landowners, or *aghas,* in the post—World War I economic system caused social alienation between them and the peasants, deepening the landowners' investment in the state and turning them against modern Kurdish nationalism. This socioeconomic situation was the background to the revival of Kurdish nationalism in Turkey in the 1980s by a new, nontribal element—educated, modernized, radical leftist activists, many of them from poor rural families. They grew up amid social ferment aroused by discrimination and repression of Kurdish identity and national aspirations.

Such divergent political perspectives, interests, sociopolitical backgrounds, and social visions shape the relations among the KDP in Iraq, the PKK in Turkey, and the PYD in Syria. Rojava, the Syrian Kurdistan, became a prize for which the KDP and its Kurdish Syrian allies contended against the PKK and its Syrian ally, the PYD. It is up to the leadership of the various factions to reserve their domestic rivalries for the political field and avoid sliding into intra-Kurdish violence. Whatever the outcome of the civil war that is now raging in Syria, the Kurds will have to struggle for the establishment of their autonomy and its recognition by the state. A shift toward a federal regime in Syria could prove extremely beneficial for the Kurds.

If Ankara pursues further reform on the Kurdish issue, that could create conditions favoring the establishment of autonomy for the Kurds in Turkey. But any transformation in deeply rooted anti-Kurdish attitudes will require significant changes in the conception of Turkish nationalism that has prevailed since Atatürk's time.

Identity Politics

If they can make such progress, the Kurds will build a system of "states within states" and establish relations among themselves on the basis of the Kurdish national identity. The creation of Kurdish autonomous regions with confederative status in the framework of existing states could serve as a catalyst for the decentralization and democratization of the regimes in Syria, Iraq, Turkey, and Iran, and for the possible development of federalist systems in these states.

A Kurdish state would be the fulfillment of the nationalist movement's vision. The overwhelming majority of participants in a referendum conducted in Iraqi Kurdistan in 2005 supported the creation of an independent Kurdish state. Nonetheless, in light of the political, social, and linguistic splits and rivalries among the Kurds, and given the separate development of the Kurdish political forces in the various parts of Kurdistan, the establishment of a large state encompassing the Kurdish areas in Turkey, Syria, Iraq, and Iran will require a complex process, fraught with tension, which could easily escalate into violence among the various Kurdish forces.

In 2012 and 2013, relations deteriorated between the KRG and the government of Iraq, headed by Prime Minister Nuri Kamal al-Maliki, primarily with regard to the powers granted to the Kurds for oil production and export. Maliki ordered an end to monthly fiscal transfers to the KRG after it announced that it would continue independent oil exports. During this crisis, Kurdish statements in favor of establishing an independent state grew more forceful. Barzani declared in early July 2014 that the Kurds wanted independence and asked the Kurdistan parliament to prepare a referendum. The Kurds see the increased flexibility of Turkish policy with regard to oil exports, starting in 2013 and especially since May 2014, as a means of reducing their economic dependence on the Iraqi government. However, the KRG's efforts to sell its oil independently in world markets have met obstacles. The Iraqi government used legal claims to block transactions. Most nations and oil companies, including American ones, have refrained from buying Kurdish oil.

Meanwhile, Turkey, Iran, and the United States have made it clear to the Kurdish leaders that they still oppose secession from Iraq. Notwithstanding Turkey's increased receptiveness to Kurdish autonomy in Iraq, and despite some Turkish analysts who advocate reconsidering the objection in principle to the establishment of an independent Kurdish state, Ankara remained steadfast in its opposition.

The Jihadist Onslaught

In June 2014, an offensive by ISIS captured Mosul, the biggest Sunni city in Iraq, demonstrating the weakness of the Iraqi Army and bringing the country to the verge of collapse. It also created an opportunity for the Kurds to annex the northern city of Kirkuk, along with its oil fields, and other disputed areas. In its early stages, the ISIS offensive was interpreted by the KRG leadership, and particularly by the KDP, as just one more chapter in Iraq's long-running Shia-Sunni sectarian struggle.

At the beginning of August, however, ISIS attacked the Kurds. On August 2, ISIS opened a murderous offensive against the Yazidis, a Kurdish minority sect (whereas most Kurds are Sunni, the Yazidis practice a syncretic religion based on a blend of Zoroastrianism and mystical elements of Islam, Christianity, and Assyrian traditions), who fled to Mount Sinjar. The Kurdish military force, the *peshmerga,* suffered heavy losses and retreated from the isolated mountain. On August 6, ISIS captured the towns of Makhmour and Gwer, the entry points to Erbil, and advanced toward the city's airport. ISIS fighters used American arms that they had seized from the Iraqi Army in Mosul. As it became clear that ISIS posed a severe danger to the Kurds, the United States began airstrikes against the group's forces outside Erbil.

There has never been an independent Kurdish state at any time in recorded history.

With the collapse of the Iraqi Army, the peshmerga were the only force holding back the advancing Islamists. Until August, the United States had declined to supply the Kurds with arms, insisting that such military assistance would have to come through the Iraqi government in Baghdad—which vetoed any supply of arms and matériel to the peshmerga forces by third parties, while refusing to provide them itself. But the weakness of Iraq's army and the threat posed by ISIS led the United States to change its position and begin to supply arms to the Kurds, over the protests of the Baghdad government. In August, France became the first European state to send military aid and weapons to the peshmerga.

The regional and global threat represented by ISIS enabled the Kurds to achieve international recognition and cemented their status as an independent political and military force. French President François Hollande, Italian Prime Minister Matteo Renzi, and foreign ministers and emissaries of many nations visited Erbil and held meetings with KRG leaders. Britain, Germany, Denmark, the Czech Republic, Canada, Australia, and other countries have also expressed their support for the Kurds and sent them arms and military equipment.

Nonetheless, despite the improvement in the status of the Kurds, who have proved themselves the strongest and most efficient force capable of taking on ISIS despite setbacks in the August battles, there has been no change in the positions of

Turkey, the United States, and Iran with regard to preserving the integrity of Iraq—that is, preventing the establishment of an independent Kurdish state. In view of the need for cooperation with the Shia and Sunni forces in Iraq, the Kurds have tempered overly enthusiastic statements about independence.

Given this state of affairs, the realistic alternative open to the Kurds is to continue to promote their de facto independence, without seceding from Iraq, and to pursue efforts toward the establishment of a confederal regime. KRG Prime Minister Nachirwan Barzani declared in September 2014 that the Kurds prefer coexistence in a federal Iraq. The declaration of an independent Kurdish state will have to wait until international conditions are more favorable.

On the Brink

In the beginning of July, the ISIS forces in Syria attacked the small city of Kobani and the surrounding Kurdish villages, near the Turkish border. Although People's Protection Units (YPG) fighters repelled the first offensive against the city, ISIS occupied the villages. The Kurdish population of the area, some 150,000 to 200,000, fled to Turkey and to the Kurdish region in Iraq. ISIS besieged the city in September and October, but Turkey was reluctant to come to Kobani's aid because of its fear that supporting the PYD would strengthen the PKK. Ankara's inaction aroused anger among the Kurds in Turkey, leading to protests demanding that volunteers be allowed to cross the border to reinforce the Kurdish fighters in Kobani. Turkish police opened fire on the Kurdish demonstrators, and the Turkish Air Force attacked PKK bases in the mountains of the southeast.

In October, Ankara allowed peshmerga fighters from the Kurdish region in Iraq to cross through Turkey in order to join the battle in Kobani. The Turkish decision resulted from mounting international pressure on Ankara to take a more active role against ISIS. Continuing to refuse to assist Kobani in any way would also damage the peace process between Erdoğan and the Kurds in Turkey.

Massoud Barzani sponsored negotiations between the rival Kurdish forces in Syria, achieving an agreement calling for the PYD and the KDP-S to establish a common council to run the Kurdish cantons in Syria, to cooperate in the fighting against ISIS, and to allow the peshmerga to join the defense of Kobani. The role of the Kurds in fighting ISIS, along with the ongoing Syrian civil war, is setting the stage for Kurdish autonomy in Syria.

These developments are creating an opportunity for historic change in the political and diplomatic situation of the Kurds. Prolonged instability and violence or state collapse in Syria and Iraq, as well as major changes in Turkish attitudes, may produce the right conditions for the establishment of a Kurdish state or new autonomous regions. The new threat of ISIS and the central role of the Kurds in the international struggle against the group are facilitating the growth of confederal relations between the Kurdish region in Iraq and the government in Baghdad. The Iraqi Kurds are using the opportunity to strengthen their state within a state, to gain wide international recognition of their quasi-independent status in Iraq, and to retain the option of declaring a fully independent state when favorable conditions prevail in the international arena.

The spread of federalizing, decentralizing political change across the Middle East would allow for the existence of Kurdish regions enjoying broad autonomy or de facto independence within the framework of federal or confederal states. In any case, a political solution for Kurdish national aspirations, whether an independent state or autonomous federal regions, is indispensable for the stabilization and peaceful development of the Middle East.

Critical Thinking

1. What are the differences and similarities between the various Kurdish political parties?

2. What recent changes have occurred in the policies of the Turkish government toward the Kurds?

3. Why doesn't the United States support an independent Kurdistan?

Internet References

Kurdish Democratic Party
www.kpp.se

Kurdistan Regional Government-Iraq
new.krg.us

Patriotic Union of Kurdistan
www.aha.ru/~said/puk.htm

MICHAEL EPPEL is a professor of Middle Eastern history at the University of Haifa.

Article   Prepared by: Robert Weiner, *University of Massachusetts, Boston*

The New Russian Chill in the Baltic

MARK KRAMER

Learning Outcomes

After reading this article, you will be able to:

- Understand the effect of the conflict in Ukraine on the Baltic States.

- Understand the Russian military provocations in the Baltics.

I n late February and March 2014, shortly after the violent overthrow of Ukrainian President Viktor Yanukovych, Russian President Vladimir Putin sent troops to occupy the Crimean Peninsula, which had long been part of Ukraine. Putin's subsequent annexation of Crimea sparked a bitter confrontation with Western governments and stoked deep anxiety in Central and Eastern Europe about the potential for Russian military encroachments elsewhere. Nowhere has this anxiety been more acute than in Poland and the three Baltic countries—Lithuania, Latvia, and Estonia—where fears have steadily mounted as Russia has helped to fuel a civil war in eastern Ukraine while undertaking a series of military provocations in the Baltic region.

Estonia, Latvia, and Lithuania were forcibly annexed by the Soviet Union in 1940 and remained an involuntary part of it until 1991, when they were finally able to regain their independence. Their relations with post-Soviet Russia have often been tense, even though Russian troops were withdrawn on schedule from Baltic territory by 1994. As a deterrent against possible threats from Russia, all three Baltic countries pressed hard to gain membership in NATO, a status that would entitle them to protection from the United States and other alliance members. Initially, the NATO governments were skeptical about bringing in the Baltic states, but in November 2002 the allied leaders invited Estonia, Latvia, and Lithuania to join. The three were formally admitted into the alliance in 2004.

Poland, for its part, had joined NATO several years earlier. After Communist rule came to an end in Poland in 1989, a broad consensus emerged among Polish elites and the public that membership in NATO would be crucial for the country's long-term security vis-à-vis Russia and other potential threats. In the early 1990s, leaders in Washington and other NATO capitals were wary of adding new members to an alliance that already included 16 countries. Over time, however, NATO shifted in favor of enlargement, and in 1997 the member states agreed to invite Poland and two other former Warsaw Pact countries (Hungary and the Czech Republic) to join. The three formally gained membership in 1999.

The subsequent entry of Estonia, Latvia, and Lithuania into NATO brought most of the Baltic region under the alliance's auspices. The only exceptions have been Finland and Sweden, both of which have chosen thus far to remain outside military alliances, as they have since 1945. However, one of the byproducts of the Russia-Ukraine confrontation and the recent spate of Russian military provocations in northern Europe has been a surge of public discussion in both Finland and Sweden about the need for closer links with NATO and even possible membership in the alliance—a step that once would have been unthinkable.

Prior Misgivings

Well before the conflict between Russia and Ukraine erupted in 2014, concerns had arisen in the Baltic region about Russia's intentions. The August 2008 Russia-Georgia war, which saw the Russian Army quickly overwhelm and defeat the much smaller Georgian military and then carve off two sizable parts of Georgia's territory (the self-declared independent republics of South Ossetia and Abkhazia), stirred doubts in the Baltic countries and Poland about the willingness of the United States and other key NATO members to defend them against Russian military pressure or intervention. Even though Georgia was not a NATO member and thus had not received any guarantee of allied protection, the televised images of Russian forces sweeping across Georgian territory and overrunning Georgian military positions came as a jolt to both elites and the wider public in the Baltic countries. Their misgivings were reinforced by the conspicuous maneuvers undertaken by Russian ground forces along the Russian-Estonian border in late 2008 and by the provocative nature of Russia's "Zapad 2009" military exercises with Belarus in September 2009, which involved simulations of rapid offensive operations against NATO.

To allay misgivings in the Baltic countries, NATO leaders in December 2009 authorized the preparation of contingency plans for the reinforcement and defense of the whole Baltic region against an unspecified enemy. Contingency planning known as Eagle Guardian already existed for the defense of Poland, but until 2009 neither the United States nor Germany had wanted to produce additional blueprints to defend the Baltic states, for fear that such an effort would damage relations with Russia if it became publicly known. Polish officials initially expressed concern that the decision to include the Baltic countries would dilute Eagle Guardian, but they were eventually willing to embrace the expanded contingency plan, provided that Poland was treated separately and that US bilateral military support would increase.

Russian actions came perilously close to provoking a military confrontation or a collision with a passenger aircraft.

The new plan designated a minimum of nine NATO divisions—from the United States, Britain, Germany, and Poland—for combat operations to repulse an attack against Poland or the Baltic countries. US and German policy makers tried to keep the revised contingency planning secret, but some details began leaking to the press in early 2010. Soon thereafter, the unauthorized release of thousands of classified US State Department documents on the WikiLeaks website, including many items pertaining to Eagle Guardian and the concerns that led to its expansion, revealed the alliance's planning for all to see.

The public disclosure of NATO's behind-the-scenes deliberations and revised Eagle Guardian plans in late 2010 and early 2011 spawned hyperbolic commentary in the Russian press and drew a harsh reaction from the Russian Foreign Ministry. High-ranking officials claimed to be "bewildered" and "dismayed" that NATO, after issuing countless "proclamations of friendship," would be treating Russia as "the same old enemy in the Cold War." The Russian ambassador to NATO, Dmitri Rogozin—a notorious hard-liner—denounced the "sinister manipulations and intrigues" of the allied governments and accused them of engaging in "warmongering," "odious discrimination," "hateful anti-Russian propaganda," and "flagrant hypocrisy."

The ensuing tensions, coming at an early stage of Barack Obama's presidency, tarnished his administration's

much-ballyhooed "reset" of relations with Moscow and eroded NATO's credibility in its dealings with Russia, including its repeated statements that "NATO does not view Russia as a threat." Perhaps if Dmitri Medvedev had stayed on as Russian president (as the Obama administration expected), the damage from the disclosures would have abated relatively quickly and would not have hindered closer ties via the NATO-Russia Council. But with the return of Putin as president in 2012 and the Russian government's growing predilection for flamboyant anti-Western rhetoric and policies, the adverse impact of the revelations persisted.

Among other things, the Russian Army stepped up its military exercises, including simulations of attacks against the Baltic countries and Poland. Russia's Zapad 2011 and Zapad 2013 exercises with Belarus, which were given wide publicity in the Russian and Belarusian media, featured simulated preventive nuclear strikes against Poland and large-scale thrusts toward the Baltic countries. In March 2012, Russian combat aircraft also began to conduct simulated attacks against military sites in Sweden. The Russian authorities' shift to a more belligerent posture throughout the Baltic region began when Medvedev was still president, and any prospect for a rapprochement between NATO and Russia disappeared after Putin returned to the presidency, lending an even sharper edge to the two sides' military rivalry vis-à-vis the Baltic countries and Poland.

NATO sought to offset Russian military activities in the region by carrying out major maneuvers of its own in 2012 and 2013, especially Exercise Steadfast Jazz in the Baltic countries and Poland in early November 2013, which included more than 1,000 mechanized infantry, 2,000 other troops, 3,000 command-and-control personnel, 40 combat aircraft, 2 submarines, and 15 surface vessels. All the NATO countries as well as Finland, Sweden, and Ukraine (then still headed by Yanukovych) took part in Steadfast Jazz, which was tied to plans for a NATO Response Force capable of deploying thousands of allied troops to "defend all member states" in the Baltic region against external attack at very short notice. Before the exercise began, Russian Deputy Defense Minister Anatoly Antonov complained that it would mark a return to the "chill of the Cold War." Although the results of Steadfast Jazz and other joint exercises helped to calm nerves among Baltic leaders, apprehension in the region about Russia's intentions was mounting long before Put in authorized the annexation of Crimea or began fueling a civil war in eastern Ukraine.

Grave Threat

No sooner had the Russian government embarked on the takeover of Crimea in late February and early March 2014 than senior officials in Estonia, Latvia, and Lithuania began warning about the "grave threat" to their own countries. At an emergency meeting of European Union leaders in Brussels in early March, Lithuanian President Dalia Grybauskaite warned that "Russia today is dangerous.. . . They are trying to rewrite the borders established after the Second World War in Europe." The vice speaker of Lithuania's parliament, Petras Auštrevičius, concurred: "Russia is presenting a clear threat, and, knowing the Russian leadership, there is a great risk they might not stop with Ukraine. There is a clear risk of an extension of [Russia's military] activities." Estonian President Toomas Hendrik Ilves emphasized that "no one in [the Baltic] countries can safely assume that Russia's predatory designs will end with the seizure of Crimea."

No country would see much point in belonging to an alliance that refused to protect its members against external aggression.

Baltic leaders' concerns about the prospect of Russian aggression intensified after Putin delivered a bellicose speech before the Russian parliament on March 18, 2014, announcing the annexation of Crimea and proclaiming a duty to protect ethnic Russian populations in other countries. Senior Baltic officials and military commanders urged the United States and other leading NATO countries to reaffirm and strengthen Article 5 of the North Atlantic Treaty, which stipulates that "an armed attack against one or more of them in Europe or North America shall be considered an attack against them all" and obliges every NATO country to take "such action as it deems necessary, including the use of armed force," to repulse an attack against another NATO member state. The commander-in-chief of the Estonian Defense Forces, Major General Riho Terras, declared that although Estonia faced no immediate military threat, the events in Crimea demonstrated that the Russian Army has "a very credible capability" of "doing various things" elsewhere in Europe, especially in countries like Estonia with large ethnic Russian minorities. "It is very important," Terras warned, "that we [the members of NATO] now seriously think about defense plans based on Article 5."

Terras's comments were echoed a week later by Estonian Prime Minister Taavi Rõivas, who said that, in light of the annexation of Crimea and Russia's conspicuous military activities in the Baltic region, it would be "extremely important for the alliance," especially the United States, to deploy "boots on the ground" in the Baltic countries in order to "increase the NATO presence and defend all allies" against any possible encroachments. Only through a robust and lasting troop presence, he implied, could NATO counter "external threats" and ensure the "security and well-being" of alliance members.

Lithuanian Defense Minister Juozas Olekas expressed much the same view, arguing that the "very active" Russian troop movements in Kaliningrad Oblast (the Russian exclave on the Baltic Sea) along the border with Lithuania necessitated the deployment of "NATO ground forces in the [Baltic] region and visits from NATO navies" to deter "aggression from the East." Terras's predecessor as Estonia's commander-in-chief, General Ants Laaneots, addressed the issue even more bluntly in an interview with the Estonian press a few weeks later: "Putin has brought sense back to European minds regarding military dangers. I am happy that NATO and above all the EU members [of the alliance] have woken up after twenty years of self-delusion in the field of security."

These comments by Baltic officials were in line with broader public opinion in the Baltic countries and Poland. A survey in Poland in March 2014, as Russia's takeover of Crimea was unfolding and Russian military forces in and around the Baltic Sea were engaging in unscheduled large-scale maneuvers, indicated that 59 percent of Polish adults viewed Russia as a "threat to Poland's security." Surveys in Estonia and Lithuania in late March turned up even higher shares of respondents—74 percent of Estonians and 68 percent of Lithuanians—who saw Russia as the "greatest threat" to their countries and to the "whole of Europe."

After a civil war erupted in eastern Ukraine in the late spring of 2014 with Russia's active support of separatist rebels, anxiety in the Baltic countries and Poland steadily intensified. Polish Foreign Minister Radoslaw Sikorski warned in August 2014 that Putin "has moved beyond all civilized norms" and "thinks he's facing a bunch of degenerate weaklings [in NATO]. He thinks we wouldn't go to war to defend the Baltics. You know, maybe he's right." Sikorski was hardly alone in this view. Nearly every senior official in Poland and the Baltic countries expressed great unease. Political leaders and military commanders in the region increasingly warned that Putin was intent on undermining NATO's resolve to protect their countries. The Latvian and Estonian governments noted with consternation that Russian diplomats had stepped up efforts to give Russian passports and higher pensions to ethnic Russians in Latvia and Estonia. They urged the United States and other NATO countries to take full account of the "overriding danger" posed by Russia and the "evident desire by Moscow authorities to reestablish domination over their former empire in Europe."

Surging Provocations

Russia's annexation of Crimea and sponsorship of an armed insurgency in eastern Ukraine were accompanied by a surge of Russian military provocations in and around the Baltic region. Some of these actions were targeted at the Baltic countries and Poland, whereas others were aimed at Sweden, Finland, and

Norway. Although most of the incidents were little more than shows of strength and bravado, some proved highly dangerous. In a few cases, Russian actions came perilously close to provoking a military confrontation between Russia and one or more NATO countries or a collision with a passenger aircraft.

For more than a decade after the Soviet Union collapsed, Russian military forces engaged in relatively few exercises and kept a very low profile. Long training flights for Russian combat aircraft nearly ceased, and sea patrols by naval vessels and submarines were drastically curtailed. The situation began to change gradually in the early 2000s as Russia's economy started to recover from the steep output decline that followed the disintegration of the Soviet economy. By 2006, Russian military forces had returned to higher levels of readiness and resumed activities beyond Russia's borders, including lengthy training flights through international airspace. The extent of these activities was not quite as sizable as in the Soviet era, but on a gradually increasing number of occasions from 2006 through 2013 NATO fighter aircraft intercepted Russian military planes as they approached Polish and Baltic airspace. According to NATO's Combined Air Operations Center in Üdem, Germany, allied aircraft scrambled to intercept Russian combat planes roughly 45 times a year in the Baltic region from 2011 to 2013.

This pattern changed dramatically in 2014, as tensions mounted over Crimea and eastern Ukraine. Russian military activities of all sorts in northern Europe, especially harassing NATO countries and Sweden, precipitously increased. Russian warships intruded into Baltic countries' territorial waters, including one occasion when Russian vessels engaged in live-firing exercises that severely disrupted civilian shipping throughout the region. Russian fighter and bomber aircraft repeatedly buzzed warships and other naval vessels from NATO countries and Sweden in the Baltic and North Seas, carried out simulated attacks against NATO countries and Sweden (as well as a simulated volley of air-launched cruise missile targeting North America), intruded into NATO countries' airspace, and undertook armed missions against US and Swedish reconnaissance aircraft, forcing them to take evasive maneuvers.

NATO fighter aircraft in the region were scrambled to intercept Russian planes more than 130 times in 2014, roughly triple the number of interceptions in 2013. On many occasions, NATO fighters intercepted formations of Russian bombers and tanker aircraft as they approached or entered Baltic and Norwegian airspace. After the largest such incident, in late October 2014, NATO's Allied Operations Command reported that "the bomber and tanker aircraft from Russia did not file flight plans or maintain radio contact with civilian air traffic control authorities and they were not using onboard transponders. This poses a potential risk to civil aviation as civilian air traffic control cannot detect these aircraft or ensure there is no interference with civilian air traffic."

The quantity and provocative nature of Russian aerial incursions over northern Europe in 2014 marked a sharp departure from the pattern of earlier years. A report published in late 2014 by the European Leadership Network (a London-based think tank) highlighted the magnitude of the difference in the Baltic region from January to September 2014, when "the NATO Air Policing Mission conducted 68 'hot' identification and interdiction missions along the Lithuanian border alone, and Latvia recorded more than 150 incidents of Russian planes approaching its airspace." Also, "Estonia recorded 6 violations of its airspace in 2014, as compared to 7 violations overall for the entire period between 2006 and 2013." This pattern continued in late 2014 and 2015, far exceeding the number of incidents since the height of the Cold War.

Reckless Endangerment

Of particular concern to Polish and Baltic leaders were the seemingly deliberate efforts by Russian military forces to provoke armed clashes or to endanger civilian passenger aircraft. Provocations directed against NATO countries, Sweden, and Finland occurred throughout 2014 and early 2015 but were particularly frequent in the aftermath of the controversy surrounding the downing of Malaysia Airlines flight MH17 by Russian-backed insurgents in eastern Ukraine. In early September 2014, two days after Obama traveled to Estonia and pledged strong support to the three Baltic countries, Russian state security forces kidnapped at gunpoint an Estonian Internal Security Service officer, Eston Kohver, from a border post on Estonian territory and spirited him to Moscow. As of February 2015, Kohver was still being held without trial in Moscow's notorious Lefortovo Prison on charges of espionage.

Apprehension in the region about Russia's intentions was mounting long before Putin authorized the annexation of Crimea.

The same month Kohver was abducted, Russian strategic nuclear bombers carried out simulated cruise missile attacks against North America; Russian fighters buzzed a Canadian frigate in the Black Sea while Russian naval forces engaged in maneuvers nearby; Russian medium-range bombers intruded into Swedish airspace to test the reactions of air defense forces; and Russian warships seized a Lithuanian fishing vessel in international waters of the Barents Sea and brought it to Murmansk in defiance of the Lithuanian government's protests. The next month, Russian military forces not only kept up their aerial incursions in the Baltic region (including a mission against Swedish surveillance aircraft that was deemed "unusually provocative") but also dispatched submarines on a prolonged series of intrusions into Swedish territorial waters, causing sharp bilateral tensions and nearly provoking an armed confrontation at sea. The Swedish navy received authorization to use force if necessary to bring the submarines to the surface, but a 10-day search for the intruders proved unsuccessful.

Equally disturbing was the apparent willingness—indeed eagerness—of the Russian authorities to deploy military forces in ways that endangered civilian air traffic in northern Europe. The most egregious incident of this sort occurred in March 2014, when a Russian reconnaissance plane that was not transmitting its position nearly collided with an SAS passenger airliner carrying 132 people. A fatal collision was avoided only because of the alertness and skillful reaction of the SAS pilots. Other such incidents occurred in the spring and early summer. Even after the MH17 incident in July 2014 drew opprobrium from around the world, Russian military aircraft continued to pose dangers to civilian airliners. The frequency and audacity of the incidents left no doubt that they were deliberate.

By the end of 2014, it had become clear that, as the European Leadership Network report stated, the "Russian armed forces and security agencies seem to have been authorized and encouraged to act in a much more aggressive way toward NATO countries, Sweden, and Finland." Against a backdrop of large-scale Russian military exercises and force redeployments in the Baltic region and elsewhere in 2014 and early 2015, the long series of Russian military provocations has raised troubling questions about Moscow's intentions.

Boots on the Ground?

The surge of tensions over Ukraine and Russian military provocations in the Baltic region spurred officials from Poland and the three Baltic countries to push for a strong show of resolve by NATO and a concrete reaffirmation of Article 5. The US government moved relatively quickly to allay some of these concerns, announcing in March 2014 that it would send six F-15C fighters and two KC-135 tanker aircraft to the headquarters of NATO's Baltic Air Policing Mission at the Šiauliai air base in Lithuania, joining the four F-15Cs that had been on patrol since the mission was established in 2004 when the Baltic countries (which lack their own combat aircraft) entered the alliance. The United States also deployed twelve F-15s and F-16s to Poland to assist air defense operations there and augmented the US naval presence in the Baltic Sea. Subsequently, Denmark, France, and Britain sent additional fighter planes to Šiauliai to expand the air policing mission further and relieve some of the aircraft already on patrol. NATO also expanded its surveillance of the Baltic region with extra flights of allied Airborne Warning and Control Systems planes, which provided broad coverage around the clock.

These incremental increases of NATO's military presence in the Baltic region, and a decision by alliance foreign ministers in April 2014 to "suspend all practical civilian and military cooperation between NATO and Russia," were welcomed by the Baltic and Polish governments, but they urged the United States and other large NATO countries to go further with defense preparations. Officials in Warsaw, Tallinn, Riga, and Vilnius sought the permanent stationing of allied ground and air forces on Polish and Baltic territory. Estonian Prime Minister Rõivas's plea for "boots on the ground" was echoed by other leaders in the region, who hoped that their requests would be endorsed by the NATO governments at a summit meeting in Wales on September 4–5, 2014. For years, such a step had been precluded by the NATO-Russia Founding Act, signed by Russian and NATO leaders at a Paris summit in May 1997. To mitigate Moscow's aversion to the enlargement of NATO, the Founding Act established conditions for the deployment of allied troops on the territory of newly admitted member states:

> NATO reiterates that in the current and foreseeable security environment, the Alliance will carry out its collective defense and other missions by ensuring the necessary interoperability, integration, and capability for reinforcement rather than by additional permanent stationing of substantial combat forces. Accordingly, it will have to rely on adequate infrastructure commensurate with the above tasks. In this context, reinforcement may take place, when necessary, in the event of defense against a threat of aggression.

In accordance with this provision, the United States and other allied countries had always eschewed any prolonged deployment of "substantial" military forces on the territory of new NATO members in the Baltic region.

In the lead-up to the Wales summit, the Polish and Baltic governments argued that international circumstances had fundamentally changed since 1997 and that NATO should no longer be bound by anything in the Founding Act. Referring to the clause stating that NATO and Russia no longer regarded each other as enemies, General Terras of Estonia contended in May 2014 that the whole document had become obsolete: "Russia sees NATO as a threat, and therefore NATO should not view Russia as a friendly, cooperative country. That is very clear. The threat assessment of NATO needs to fit the current realistic circumstances." Terras returned to this theme a few months later, just before the Wales summit:

> No one now believes that friendly relations between NATO and Moscow can be reestablished. Russia today regards NATO as an enemy, and this must facilitate changes [in NATO's force posture]. Some changes have already taken place, and the NATO summit must give out a clear message to the allies and to Russia that NATO

is the world's most powerful military organization and is willing to do everything to protect its member states, including increasing its presence in areas bordering Russia.

Terras and other senior military and political officials in the Baltic region also argued that in light of Russia's actions in Ukraine and elsewhere, NATO should move ahead as expeditiously as possible with concrete military preparations for "defense against a threat of aggression," as stipulated in the Founding Act.

Limited Measures

Many officials in the United States and other NATO countries were sympathetic to the arguments of Polish and Baltic leaders, but the US government ultimately decided not to proceed with long-term deployments of "substantial" military forces in the Baltic region. US officials at the Wales summit did make an effort to address Baltic and Polish concerns, not least by joining with all the other NATO allies in vowing to uphold Article 5: "The Alliance poses no threat to any country. But should the security of any Ally be threatened we will act together and decisively, as set out in Article 5 of the Washington Treaty." The summit declaration made clear that this warning was meant for Russia.

In the military sphere, however, the summit mostly just endorsed and extended the relatively limited measures that had been adopted earlier in the year to expand the Baltic Air Policing Mission and to bolster Western naval forces in the Baltic Sea. Although the Wales summit participants welcomed the fact that more than 200 military exercises had been held in Europe in 2014, the reality was that few of these exercises were of any appreciable size. Moreover, although they endorsed the deployment of "ground troops in the eastern parts of the Alliance for training and exercises," they made clear that these troops were stationed there solely "on a rotational basis," not permanently. The allied leaders did adopt a Readiness Action Plan to enlarge the long-planned NATO Response Force (from 13,000 to 20,000 troops) and to put it on a higher state of readiness, with a "Spearhead Force" of up to several thousand troops and reinforcements that could be deployed to the Baltic region within a few days. Whether those projections will actually materialize in 2015 and 2016 remains to be seen, however. Even if the proposals are fully implemented, they fall well short of what the Baltic countries and Poland had been seeking.

In the months following the Wales summit, the NATO governments tried to fulfill several of the pledges they had adopted, most notably with the establishment of multinational command-and-control centers in the Baltic countries, Poland, Bulgaria, and Romania, consisting of "personnel from Allies on a rotational basis" who are to "focus on planning and exercising collective defense." At a February 2015 meeting in Brussels, NATO defense ministers pledged to increase the size of the Response Force to 30,000, including a Spearhead Force of 5,000. NATO military planners and individual governments took other concrete steps, including the upgrading of infrastructure and the prepositioning of weaponry and support equipment, to enhance the alliance's capacity to uphold Article 5 in the Baltic region.

Nevertheless, in the absence of large, permanent deployments of US and other NATO ground and air forces in the Baltic countries and Poland, doubts about the collective defense of the region are bound to persist. Terras highlighted this problem when he noted that although he himself did not doubt NATO's willingness to carry out its defense commitments, "the real question is whether Putin believes that Article 5 works." He warned, "We should not give any option of miscalculation for President Putin."

Gloomy Scenarios

Russia's actions in Crimea and eastern Ukraine, and the risky nature of Russian military operations and exercises in and around the Baltic region in 2014 and 2015, have sparked acute unease in Poland and the Baltic countries. Although the available evidence does not indicate that the Russian authorities will attack a NATO member state, Putin does seem intent on undermining NATO by raising doubts about the credibility of Article 5. Despite the strong pledges of support offered at the Wales summit, some uncertainty remains about what would happen if Russia undertook a limited military probe against one or more of the Baltic states. Certain European members of NATO might hinder a timely response, but if that were to happen the United States and some other NATO member states would likely act outside the alliance's command structure to defend the Baltic states, as envisaged in NATO's contingency defense planning. They would undoubtedly try to avoid escalation to all-out war against Russia, not least because that would require the NATO countries to fight in a region in which they would be at a serious geographic disadvantage.

However, if the United States and its allies failed to uphold Article 5 in the Baltic region and refrained from intervening against Russian military forces, this would gravely damage the credibility of all of NATO's defense commitments. No country would see much point in belonging to an alliance that refused to protect its members against external aggression. If Putin were fool-hardy enough to risk all-out war by embarking on military action in the Baltic region, NATO would have no fully reliable or attractive military and diplomatic options. But the worst option of all would be to do nothing and allow Russian military expansion to proceed unchecked.

These gloomy scenarios seem improbable for now, but the very fact that they are being discussed seriously in NATO circles as well as in Warsaw and the Baltic capitals is a sign of how gravely Russia's actions in Ukraine and elsewhere have affected the post—Cold War European security order. Peace and security in the Baltic region, which only a decade ago appeared more robust than ever, now seem all too precarious.

Critical Thinking

1. Why would Sweden and Finland consider joining NATO?
2. Should Poland and the Baltics be defended separately or together from external attacks?
3. What can the U.S. do to protect the Baltics from Russia?

Internet References

Latvian Foreign Ministry
mfa.gov.lv

Lithuanian Ministry of Foreign Affairs
urm.lt

Republic of Estonia, Ministry of Foreign Affairs
vm.el

Russian Ministry of Foreign Affairs
government.ru

MARK KRAMER is director of the Cold War Studies program and a senior fellow of the Davis Center for Russian and Eurosian Studies at Harvard University.

Prepared by: Robert Weiner, *University of Massachusetts/Boston*

Article

The Cuban Missile Crisis at 50: Lessons for U.S. Foreign Policy Today

GRAHAM ALLISON

Learning Outcomes

After reading this article, you will be able to:

- Explain how the lessons learned from the Cuban missile crisis help in dealing with Iran, North Korea, and China.

- Explain why the Cuban missile crisis can serve as a guide on how to defuse crises.

Fifty years ago, the Cuban missile crisis brought the world to the brink of nuclear disaster. During the standoff, U.S. President John F. Kennedy thought the chance of escalation to war was "between 1 in 3 and even," and what we have learned in later decades has done nothing to lengthen those odds. We now know, for example, that in addition to nuclear-armed ballistic missiles, the Soviet Union had deployed 100 tactical nuclear weapons to Cuba, and the local Soviet commander there could have launched these weapons without additional codes or commands from Moscow. The U.S. air strike and invasion that were scheduled for the third week of the confrontation would likely have triggered a nuclear response against American ships and troops, and perhaps even Miami. The resulting war might have led to the deaths of 100 million Americans and over 100 million Russians.

The main story line of the crisis is familiar. In October 1962, a U.S. spy plane caught the Soviet Union attempting to sneak nuclear-tipped missiles into Cuba, 90 miles off the United States' coast. Kennedy determined at the outset that this could not stand. After a week of secret deliberations with his most trusted advisers, he announced the discovery to the world and imposed a naval blockade on further shipments of armaments to Cuba. The blockade prevented additional materiel from coming in but did nothing to stop the Soviets from operationalizing the missiles already there. And a tense second week followed during which Kennedy and Soviet Premier Nikita Khrushchev stood "eyeball to eyeball," neither side backing down.

Saturday, October 27, was the day of decision. Thanks to secret tapes Kennedy made of the deliberations, we can be flies on the wall, listening to the members of the president's ad hoc Executive Committee of the National Security Council, or ExComm, debate choices they knew could lead to nuclear Armageddon. At the last minute, the crisis was resolved without war, as Khrushchev accepted a final U.S. offer pledging not to invade Cuba in exchange for the withdrawal of the Soviet missiles.

Every president since Kennedy has tried to learn from what happened in that confrontation. Ironically, half a century later, with the Soviet Union itself only a distant memory, the lessons of the crisis for current policy have never been greater. Today, it can help U.S. policymakers understand what to do-and what not to do-about Iran, North Korea, China, and presidential decision-making in general.

What Would Kennedy Do?

The current confrontation between the United States and Iran is like a Cuban missile crisis in slow motion. Events are moving, seemingly inexorably, toward a showdown in which the U.S. president will be forced to choose between ordering a military attack and acquiescing to a nuclear-armed Iran.

Those were, in essence, the two options Kennedy's advisers gave him on the final Saturday: attack or accept Soviet nuclear missiles in Cuba. But Kennedy rejected both. Instead of choosing between them, he crafted an imaginative alternative with three components: a public deal in which the United States pledged not to invade Cuba if the Soviet Union withdrew its missiles, a private ultimatum threatening to attack Cuba within 24 hours unless Khrushchev accepted that offer, and a secret sweetener that promised the withdrawal of U.S. missiles from Turkey within six months after the crisis was resolved. The sweetener was kept so secret that even most members of the ExComm deliberating with Kennedy on the final evening were in the dark, unaware that during the dinner break, the president had sent his brother Bobby to deliver this message to the Soviet ambassador.

Looking at the choice between acquiescence and air strikes today, both are unattractive. An Iranian bomb could trigger a cascade of proliferation, making more likely a devastating conflict in one of the world's most economically and strategically critical regions. A preventive air strike could delay Iran's nuclear progress at identified sites but could not erase the knowledge and skills ingrained in many Iranian heads. The truth is that any outcome that stops short of Iran having a nuclear bomb will still leave it with the ability to acquire one down the road, since Iran has already crossed the most significant "redline" of proliferation: mastering the art of enriching uranium and building a bomb covertly. The best hope for a Kennedyesque third option today is a combination of agreed-on constraints on Iran's nuclear activities that would lengthen the fuse on the development of a bomb, transparency measures that would maximize the likelihood of discovering any cheating, unambiguous (perhaps secretly communicated) threats of a regime changing attack should the agreement be violated, and a pledge not to attack otherwise. Such a combination would keep Iran as far away from a bomb as possible for as long as possible.

The Israeli factor makes the Iranian nuclear situation an even more complex challenge for American policymakers than the Cuban missile crisis was. In 1962, only two players were allowed at the main table. Cuban Prime Minister Fidel Castro sought to become the third, and had he succeeded, the crisis would have become significantly more dangerous. (When Khrushchev announced the withdrawal of the missiles, for example, Castro sent him a blistering message urging him to fire those already in Cuba.) But precisely because the White House recognized that the Cubans could become a wild card, it cut them out of the game. Kennedy informed the Kremlin that it would be held accountable for any attack against the United States emanating from Cuba, however it started. His first public announcement said, "It shall be the policy of this Nation to regard my nuclear missile launched from Cuba against any nation in the Western Hemisphere as an attack by the Soviet Union on the United States, requiring a full retaliatory response upon the Soviet Union."

Today, the threat of an Israeli air strike strengthens U.S. President Barack Obama's hand in squeezing Iran to persuade it to make concessions. But the possibility that Israel might actually carry out a unilateral air strike without U.S. approval must make Washington nervous, since it makes the crisis much harder to manage. Should the domestic situation in Israel reduce the likelihood of an independent Israeli attack, U.S. policymakers will not be unhappy.

Carrots Go Better with Sticks

Presented with intelligence showing Soviet missiles in Cuba, Kennedy confronted the Soviet Union publicly and demanded their withdrawal, recognizing that a confrontation risked war. Responding to North Korea's provocations over the years, in contrast, U.S. presidents have spoken loudly but carried a small stick. This is one reason the Cuban crisis was not repeated whereas the North Korean ones have been, repeatedly.

In confronting Khrushchev, Kennedy ordered actions that he knew would increase the risk not only of conventional war but also of nuclear war. He raised the U.S. nuclear alert status to defcon 2, aware that this would loosen control over the country's nuclear weapons and increase the likelihood that actions by other individuals could trigger a cascade beyond his control. For example, nato aircraft with Turkish pilots loaded active nuclear bombs and advanced to an alert status in which individual pilots could have chosen to take off, fly to Moscow, and

drop a bomb. Kennedy thought it necessary to increase the risks of war in the short run in order to decrease them over the longer term. He was thinking not only about Cuba but also about the next confrontation, which would most likely come over West Berlin, a free enclave inside the East German puppet state. Success in Cuba would embolden Khrushchev to resolve the Berlin situation on his own terms, forcing Kennedy to choose between accepting Soviet domination of the city and using nuclear weapons to try to save it.

During almost two dozen face-offs with North Korea over the past three decades, meanwhile, U.S. and South Korean policymakers have shied away from such risks, demonstrating that they are deterred by North Korea's threat to destroy Seoul in a second Korean war. North Korean leaders have taken advantage of this fear to develop an effective strategy for blackmail. It begins with an extreme provocation, blatantly crossing a redline that the United States has set out, along with a threat that any response will lead to a "sea of fire." After tensions have risen, a third party, usually China, steps in to propose that "all sides" step back and cool down. Soon thereafter, side payments to North Korea are made by South Korea or Japan or the United States, leading to a resumption of talks. After months of negotiations, Pyongyang agrees to accept still more payments in return for promises to abandon its nuclear program. Some months after that, North Korea violates the agreement, Washington and Seoul express shock, and they vow never to be duped again. And then, after a decent interval, the cycle starts once more.

If the worst consequence of this charade were simply the frustration of being bested by one of the poorest, most isolated states on earth, then the repeated Korean crises would be a sideshow. But for decades, U.S. presidents have declared a nuclear-armed North Korea to be "intolerable" and "unacceptable." They have repeatedly warned Pyongyang that it cannot export nuclear weapons or technology without facing the "gravest consequences." In 2006, for example, President George W. Bush stated that "the transfer of nuclear weapon or material by North Korea to state or nonstate entities would be considered a grave threat to the United States, and North Korea would be held fully accountable for the consequences." North Korea then proceeded to sell Syria a plutonium producing reactor that, had Israel not destroyed it, would by now have produced enough plutonium for Syria's first nuclear bomb. Washington's response was to ignore the incident and resume talks three weeks later.

One lesson of the Cuban missile crisis is that if you are not prepared to risk war, even nuclear war, an adroit adversary can get you to back down in successive confrontations. If you do have redlines that would lead to war if crossed, then you have to communicate them credibly to your adversary and back them up or risk having your threats dismissed. North Korea's sale of a nuclear bomb to terrorists who then used it against an American target would trigger a devastating American retaliation. But after so many previous redlines have been crossed with impunity, can one be confident that such a message has been received clearly and convincingly? Could North Korea's new leader, Kim Jong Un, and his advisers imagine that they could get away with it?

The Rules

A similar dynamic may have emerged in the U.S. economic relationship with China. The Republican presidential candidate Mitt Romney has announced that "on day one of my presidency I will designate [China] a currency manipulator and take appropriate counteraction." The response from the political and economic establishment has been a nearly unanimous rejection of such statements as reckless rhetoric that risks a catastrophic trade war. But if there are no circumstances in which Washington is willing to risk a trade confrontation with China, why would China's leaders not simply take a page from North Korea's playbook? Why should they not continue, in Romney's words, "playing the United States like a fiddle and smiling all the way to the bank" by undervaluing their currency, subsidizing domestic producers, protecting their own markets, and stealing intellectual property through cybertheft?

Economics and security are separate realms, but lessons learned in one can be carried over into the other. The defining geopolitical challenge of the next half century will be managing the relationship between the United States as a ruling superpower and China as a rising one. Analyzing the causes of the Peloponnesian War more than two millennia ago, the Greek historian Thucydides argued that "the growth of the power of Athens, and the alarm which this inspired in Sparta, made war inevitable." During the Cuban missile crisis, Kennedy judged that Khrushchev's adventurism violated what Kennedy called the "rules of the precarious status quo" in relations between two nuclear superpowers. These rules had evolved during previous crises, and the resolution of the standoff in Cuba helped restore and reinforce them, allowing the Cold War to end with a whimper rather than a bang.

The United States and China will have to develop their own rules of the road in order to escape Thucydides' trap. These will need to accommodate both parties' core interests, threading a path between conflict and appeasement. Overreacting to perceived threats would be a mistake, but so would ignoring or papering over unacceptable misbehavior in the hope that it will not recur. In 1996, after some steps by Taipei that Beijing considered provocative, China launched a series of missiles over Taiwan, prompting the United States to send two aircraft carrier battle groups into harm's way. The eventual result was a clearer understanding of both sides' redlines on the Taiwan issue and a calmer region. The relationship may need additional such clarifying moments in order to manage a precarious transition as China's continued economic rise and new status are reflected in expanded military capabilities and a more robust foreign posture.

Do Process

A final lesson the crisis teaches has to do not with policy but with process. Unless the commander in chief has suffficient time and privacy to understand a situation, examine the evidence, explore various options, and reflect before choosing among them, poor decisions are likely. In 1962, one of the first questions Kennedy asked on being told of the missile discovery was, How long until this leaks? McGeorge Bundy, his national security adviser, thought it would be a week at most. Acting on that advice, the president took six days in secret to deliberate, changing his mind more than once along the way. As he noted afterward, if he had been forced to make a decision in the first 48 hours, he would have chosen the air strike rather than the naval blockade-something that could have led to nuclear war.

In today's Washington, Kennedy's week of secret deliberations would be regarded as a relic of a bygone era. The half-life of a hot secret is measured not even in days but in hours. Obama learned this painfully during his first year in office, when he found the administration's deliberations over its Afghanistan policy playing out in public, removing much of his flexibility to select or even consider unconventional options. This experience led him to demand a new national security decision-making process led by a new national security adviser. One of the fruits of the revised approach was a much more tightly controlled flow of information, made possible by an unprecedented narrowing of the inner decision-making circle. This allowed discussions over how to handle the discovery of Osama bin Laden's whereabouts to play out slowly and sensibly, with the sexiest story in Washington kept entirely secret for five months, until the administration itself revealed it after the raid on bin Laden's Abbottabad compound.

It has been said that history does not repeat itself, but it does sometimes rhyme. Five decades later, the Cuban missile crisis stands not just as a pivotal moment in the history of the Cold War but also as a guide for how to defuse conflicts, manage great-power relationships, and make sound decisions about foreign policy in general.

Critical Thinking

1. What is the relationship between Thucydides' history of the Peloponnesian war and the Cuban missile crisis?

2. Should the U.S. risk war to prevent Iran from developing the bomb?

3. What should U.S. policy be in dealing with North Korea?

Create Central

www.mhhe.com/createcentral

Internet References

World at the Brink
  http:microsites.jfklibrary.org/cmc/

National Archives and JFK Library Mark 50th Anniversary of Cuban Missile Crisis in October
  www.archives.gov/press/press-releases/2012/nr12-146.html.

GRAHAM ALLISON is Professor of Government and Director of the Belfer Center for Science and International Affairs at Harvard University's Kennedy School of Government.

Allison, Graham. From *Foreign Affairs*, July/August 2012, pp. 11–16. Copyright © 2012 by Council on Foreign Relations, Inc. Reprinted by permission of Foreign Affairs. www. ForeignAffairs.com

Unit 6

UNIT

Prepared by: Robert Weiner, *University of Massachusetts, Boston*

ETHICS AND VALUES

This unit contains several articles which deal with the question of the rule of law, both on the national and international level. The year 2015 marked 800 years since the signing of the first Magna Carta or Great Charter. The Magna Carta not only established the rule of law in England, but also established a model for the rule of law around the world. A definition of a rule of law state includes a liberal democracy which is characterized by free and fair elections, a competitive party system, a real rather than a nominal constitution, a free press, and an impartial judicial system which is appointed and allowed to function by the government without intimidation and political interference. An objective judicial system is critical in rooting out corruption in states like China. There have been a series of show trials in China, involving select high level officials, in an effort to maintain the political legitimacy of the ruling elite. Kleptocracy not only exists in China, but is a widespread phenomenon in a number of states, ranging from Ukraine to Nigeria. Kleptocracy invariably occurs in a repressive regime marked by human rights violations, and a loss of legitimacy of the government.

A rule of law state is also essential to ensure gender equality and justice for female victims of violent crimes in regions such as Latin America and the Caribbean. Furthermore, a rule of law state is necessary to eliminate racism in such liberal democracies as the United States, where unarmed black males fall victim to a discriminatory system, in a society which is becoming more diverse.

The year 2015 also marked 100 years since the Armenian genocide took place in Ottoman Turkey. Genocide is a term which was invented by the Polish lawyer and linguist Raphael Lemkin, who almost single-handedly persuaded the international community in 1948 to adopt the Convention on the Prevention and Punishment of the Crime of Genocide. The term genocide is based on the Greek word "gens" which means people, and the Latin word "cide" which means killing. Until 1948, genocide was known as the "crime with no name," as famously stated by Winston Churchill. Genocide was finally recognized as a crime under international law after the Holocaust of World War II. The Genocide Convention lists various acts of genocide, which are designed to destroy in whole or in part a group of people based on race, ethnicity, religion, or nationality. The Convention was criticized for only focusing on these four groups, and not including political, economic, or social groups. Moreover, the Convention did not contain a definition of "group" and left it up to subsequent international criminal courts to come up with a definition of group. International criminal courts have wrestled for years with figuring out whether membership in a group can be defined by objective or subjective factors, or both. Furthermore, the Convention did not explain what was meant by "in part." The courts have tried to determine whether "in part" refers to a substantial or significant part of a group, sometimes arriving at contradictory opinions. Finally, an international judicial body, as mentioned in the Convention, did not exist at the time the Convention was adopted, and took 50 years to appear as the International Criminal Court in 1998.

Although the international community hoped that genocide would never again take place, it has occurred again and again since the second World War. For example, in Rwanda in 1994 genocide resulted in the deaths of at least 800,000 to 1 million Tutsis and moderate Hutus. In Srebrenica, Bosnia in 1995, 7,000 to 8,000 Bosniak men and boys were massacred. Based on the genocides which occurred in Rwanda and Bosnia, a considerable amount of case law has been produced to provide a solid legal foundation for the prevention and punishment of this most heinous of crimes against humanity, as long as states have the political will to deal with such mass atrocities. The Genocide Convention has been virtually incorporated into the Statute of the International Criminal Court (ICC) which, after decades of negotiations, finally came into existence in 1998. A number of countries have also enacted national laws, using the Genocide Convention as a model. Although human rights advocates have been disappointed by the narrow legal basis of the Genocide Convention, the expansion of the definition of crimes against humanity has filled in the gap that has been left by the Convention.

Article   Prepared by: Robert Weiner, *University of Massachusetts, Boston*

Xi's Corruption Crackdown: How Bribery and Graft Threaten the Chinese Dream

James Leung

Learning Outcomes

After reading this article, you will be able to:

- Explain the scope of corruption in China.
- Understand the relationship between the Party and corruption.

In a series of speeches he delivered shortly after taking office in 2012, Chinese President Xi Jinping cast corruption as not merely a significant problem for his country but an existential threat. Endemic corruption, he warned, could lead to "the collapse of the [Chinese Communist] Party and the downfall of the state." For the past two years, Xi has carried out a sweeping, highly publicized anticorruption campaign. In terms of sheer volume, the results have been impressive: according to official statistics, the party has punished some 270,000 of its cadres for corrupt activities, reaching into almost every part of the government and every level of China's vast bureaucracy. The most serious offenders have been prosecuted and imprisoned; some have even been sentenced to death.

The majority of the people caught up in Xi's crackdown have been low- or midlevel party members and functionaries. But corruption investigations have also led to the removal of a number of senior party officials, including some members of the Politburo, the group of 25 officials who run the party, and, in an unprecedented move, to the expulsion from the party and arrest of a former member of the Politburo's elite Standing Committee.

Xi's campaign has proved enormously popular, adding a populist edge to Xi's image and contributing to a nascent cult of personality the Chinese leader has begun to build around himself. And it has the quiet support of the aristocratic stratum of "princelings," the children and grandchildren of revolutionary leaders from the Mao era. They identify their interests with those of the country and consider Xi to be one of their own. But there has been pushback from other elites within the system, some of whom believe the campaign is little more than a politically motivated purge designed to help Xi solidify his own grip on power. Media organizations in Hong Kong have reported that Xi's two immediate predecessors, Jiang Zemin and Hu Jintao, have asked him to dial back the campaign. And some observers have questioned the campaign's efficacy: in 2014, despite Xi's efforts, China scored worse on Transparency International's Corruption Perceptions Index than it had in 2013. Even Xi himself has expressed frustration, lamenting a "stalemate" in his fight to clean up the system while pledging, in grandiose terms, not to give up: "In my struggle against corruption, I don't care about life or death, or ruining my reputation," he reportedly declared at a closed-door Politburo meeting last year.

There is no doubt that Xi's campaign is in part politically motivated. Xi's inner circle has remained immune, the investigations are far from transparent, and Xi has tightly controlled the process, especially at senior levels. Chinese authorities have placed restrictions on foreign media outlets that have dared to launch their own investigations into corruption, and the government has detained critics who have called for more aggressive enforcement efforts.

But that doesn't mean the campaign will fail. The anticorruption fight is only one part of Xi's larger push to consolidate his authority by establishing himself as "the paramount leader within a tightly centralized political system," as the China expert Elizabeth Economy has written. So far, Xi seems capable of pulling off that feat. Although this power grab poses other risks, it puts Xi in a good position to reduce corruption significantly—if not necessarily in a wholly consistent, apolitical manner.

This might seem paradoxical: after all, too much central power has been a major factor in creating the corruption epidemic. That is why, in the long term, the fate of Xi's anticorruption fight will depend on how well Xi manages to integrate it into a broader economic, legal, and political reform program. His vision of reform, however, is not one that will free the courts, media, or civil society, or allow an opposition party that could check the ruling party's power. Indeed, Xi believes that Western-style democracy is at least as prone to corruption as one-party rule. Rather, Xi's vision of institutional reform involves maintaining a powerful investigative force that is loyal to an honest, centralized leadership. He seems to believe that, over the course of several years, consistent surveillance and regular investigations will change the psychology of bureaucrats, from viewing corruption as routine, as many now do, to viewing it as risky—and, finally, to not even daring to consider it.

Stamping out graft, bribery, and influence peddling could very well help China's leaders maintain the political stability they fear might slip away as economic growth slows and geopolitical tensions flare in Asia. But if Xi's fight against corruption becomes disconnected from systemic reforms, or devolves into a mere purge of political rivals, it could backfire, inflaming the grievances that stand in the way of the "harmonious society" the party seeks to create.

I'll Scratch Your Back . . .

One school of thought holds that corruption is a deeply rooted cultural phenomenon in China. Some political scientists and sociologists argue that when it comes to governance and business, the traditional Chinese reliance on guanxi—usually translated as "connections" or "relationships"—is the most important factor in explaining the persistence and scope of the problem. The comfort level that many Chinese citizens have with the guanxi system might help explain why it took so long for public outrage to build up to the point where the leadership was forced to respond. But all cultures and societies produce a form of guanxi, and China's version is not distinct enough to explain the depth and severity of the corruption that inflicts the Chinese system today. The main culprits are more obvious and banal: one-party rule and state control of the economy. The lack of firm checks and balances in a one-party state fuels the

spread of graft and bribery; today, no Chinese institution is free of them. And state control of resources, land, and businesses creates plenty of opportunities for corruption. In the past three decades, the Chinese economy has become increasingly mixed. According to Chinese government statistics, the private sector now accounts for around two-thirds of China's GDP and employs more than 70 percent of the labor force. And the Chinese economy is no longer isolated; it has been integrated into the global market. Nevertheless, the private sector is still highly dependent on the government, which not only possesses tremendous resources but also uses its regulatory and executive power to influence and even control private businesses.

When it comes to government purchasing and contracting and the sale of Chinese state assets (including land), bidding and auctioning processes are extremely opaque. Officials, bureaucrats, and party cadres exploit that lack of transparency to personally enrich themselves and to create opportunities for their more senior colleagues to profit in exchange for promotions. Midlevel officials who oversee economic resources offer their superiors access to cheap land, loans with favorable terms from state-owned banks, government subsides, tax breaks, and government contracts; in return, they ask to rise up the ranks. Such arrangements allow corruption to distort not just markets but also the workings of the party and the state.

Similar problems also exist in government organizations that do not directly control economic resources, such as China's military. To win promotion, junior military officers routinely bribe higher-ranking ones with gifts of cash or luxury goods. Last year, the authorities arrested Xu Caihou, a retired general who had served as a member of the Politburo and had been the vice chair of the Central Military Commission. In his house, they discovered enormous quantities of gold, cash, jewels, and valuable paintings—gifts, the party alleged, from junior officers who sought to advance up the chain of command. After the party expelled him, Xu confessed, according to Chinese state media; a few months later, he died, reportedly of cancer.

Direct state ownership, however, is hardly a prerequisite for self-dealing. The immense regulatory power that Chinese authorities hold over the private sector also helps them line their own pockets. In highly regulated industries, such as finance, telecommunications, and pharmaceuticals, relatives of senior government officials often act as "consultants" to private businesspeople seeking to obtain the licenses and approvals they need to operate. Zheng Xiaoyu, the former head of the State Food and Drug Administration, accepted around $850,000 in bribes from pharmaceutical companies seeking approval for new products. In 2007, after more than 100 people in Panama died after taking contaminated cough syrup that Zheng had approved, he was tried on corruption charges; he was found guilty and executed a few months later.

Corruption has also infected law enforcement and the legal system. Organized criminal groups pay police officers to protect their drug and prostitution rings. Criminal suspects and their relatives often bribe police officers to win release from jail or to avoid prosecution. If that fails, they can try their luck with prosecutors and judges. And of course, since China's judiciary is not independent, there are always party and government officials who might be able and willing to intervene in a case—for the right price. Authorities allege that Zhou Yongkang, a former member of the party's Standing Committee who oversaw legal and internal security affairs, personally intervened in many court cases after accepting bribes. Zhou was arrested, charged, and expelled from the party last year and is currently awaiting trial—the first time in decades that the state has pursued a criminal case against a former member of the Standing Committee.

As China's domestic markets have grown, multinational companies and banks have learned that getting access means knowing whose palms to grease. Many firms have taken to hiring the children of senior government officials, sometimes even paying their tuition at Western universities. Others have opted for a more direct route, paying hefty "consulting" fees to middlemen in order to participate in stock offerings or to win preferential treatment in bidding for government contracts. This environment has discouraged some multinational companies from investing and conducting business in China, especially those constrained by U.S. anticorruption laws.

Meanwhile, officials have taken advantage of loose financial controls and a lack of transparency to safeguard their illicit profits. Many officials hold a number of Chinese passports, often under different names but with valid visas, and use them to travel abroad and stash their money in foreign bank accounts.

But corruption is hardly limited to official circles and big business; every aspect of society feels its effects. Consider education. To give their child a shot at getting into one of the relatively small number of high-quality Chinese primary and secondary schools and universities, parents often have to bribe admissions officers or headmasters. Similarly, the scarcity of good hospitals and well-trained medical personnel has led to the practice of supplying doctors or medical administrators with a hongbao—a "red packet" of cash—to secure decent treatment.

Keep It Clean

Faced with this far-reaching problem, Xi has promised more than a mere Band-Aid, envisioning a long-term process of systemic reform. The first phase has been the heavily stage-managed crackdown of the past two years. So far, the campaign has contained an element of populism: it has targeted only officials, bureaucrats, and major business figures whom the party suspects of corrupt dealings; no ordinary Chinese people have felt the sting.

The campaign seeks not only to punish corruption but to prevent it as well. In late 2012, the party published a set of guidelines known as the "eight rules and six prohibitions," banning bureaucrats from taking gifts and bribes; attending expensive restaurants, hotels, or private clubs; playing golf; using government funds for personal travel; using government vehicles for private purposes; and so on.

The government has also required all officials and their immediate family members to disclose their assets and income, to make it harder to hide ill-gotten gains. At the same time, the party has sought to reduce incentives for graft by narrowing the income gaps within the system. In the last year, it raised the salaries and retirement benefits of military officers, law enforcement personnel, and other direct government employees, while sharply cutting the higher salaries enjoyed by top managers of state owned enterprises.

Still, to date, Xi's campaign has been chiefly an enforcement effort. Investigations are led by the party's Central Commission for Discipline Inspection (CCDI), which sends inspection teams to examine every ministry and agency and every large state-owned enterprise. The teams enjoy the unlimited power to investigate, detain, and interrogate almost anyone, but mainly government officials, the vast majority of whom are party members. Once the teams believe they have gathered sufficient evidence of wrongdoing, the CCDI expels suspects from the party and then hands them over to the legal system for prosecution.

Xi has declared that no corrupt official will be spared, no matter how high his position. In practice, however, the CCDI has chosen its targets very carefully, especially at senior levels. The decision to go after Zhou was heralded as setting a new precedent—since the late 1980s, the party has followed an unspoken rule against purging a member or former member of the Standing Committee. And yet Zhou's removal and prosecution remain unique; they appear to have been less a signal of things to come than a shot across the bow, intended to scare off any potential opposition to Xi within the leadership. Zhou was vulnerable because he was retired and no longer had direct control or power. Also, Zhou had backed a group of senior party officials who had challenged Xi's power and authority early in his tenure; among them was Bo Xilai, the influential party chief of Chongqing, who in 2013 was brought down by a scandal involving corruption and a murder plot in which his wife participated. Finally, Zhou and his immediate family members were particularly flagrant in their corrupt pursuits, which made him an easy target. Some media reports have indicated that authorities are investigating the family members of other retired Standing Committee members. But so far, no ranking member of the "red aristocracy" has yet been targeted, and all the highest-level targets, including Zhou and

Xu, have been part of a single loose political network. Apparently, there are still lines Xi is not willing to cross.

It is also worth noting that although Xi has allowed investigations of the country's key military institutions, he has yet to make any major personnel changes within the Commission for Discipline Inspection of the Central Military Commission, the armed forces' equivalent of the CCDI. Xi still needs more time to consolidate his control over the military and its institutions.

A number of other elements of Xi's campaign are also problematic, because they present opportunities for abuse and run contrary to the spirit of the legal reforms that Xi is pursuing. Xi claims that he wants to improve due process and reduce abusive police and judicial practices. But the CCDI itself does not always follow standard legal procedures. For example, Chinese law allows police to detain a suspect for only seven days without formally charging him, unless the police obtain express permission from legal authorities to extend the detention. The CCDI, on the other hand, has kept suspects in custody for far longer periods without seeking any approval and without issuing any formal charges, giving the appearance of a separate standard.

Meanwhile, with its newfound authority, the CCDI is gradually becoming the most powerful institution within the party system. Unless the party balances and limits the agency's power and influence, the CCDI could grow unaccountable and become a source of the very kinds of conduct it is supposed to combat.

Perhaps the biggest potential obstacle to the success of the campaign is strong resistance to it within the bureaucratic system. Xi has launched a direct attack on the interests of many entrenched bureaucrats and officials; even those who have escaped prosecution have watched their prosperity and privilege shrink. Many officials might also resent the idea that there is something fundamentally wrong with the way they are accustomed to conducting themselves. They may feel that they deserve the benefits they get through graft; without their work, after all, nothing would get done—the system wouldn't function.

Early in Xi's tenure, some officials seemed to believe that although the days of flagrant self-dealing were over, it would still be possible to exploit their positions for profit; they would just need to be a bit more subtle about it. In 2013, *The New York Times,* citing Chinese state media, reported that a new slogan had become popular among government officials: "Eat quietly, take gently, and play secretly." But that sense of confidence has evaporated as it has become clear that Xi is serious about cracking down. During the past two years, party members and state bureaucrats have become extremely cautious about running afoul of the new ethos, although many are quietly seething about the situation. This has interfered with the traditional wheel-greasing function of corruption and contributed to China's economic slowdown. If corruption no longer assists entrepreneurs in slipping past bureaucratic barriers, it will put additional pressure on Xi to institute economic reforms that genuinely reduce those obstacles.

The Politics of Anticorruption

Since the anticorruption campaign is just one of a number of major changes taking place in the Xi era, it's difficult to forecast what path it might take. In a pessimistic scenario, the campaign would end in failure after strong resistance within the top party leadership and the bureaucratic system forces Xi to back down. That outcome would be a catastrophe. Corruption would likely rise to pre-2012 levels (at the very least), destabilizing the economy, reducing investor confidence, and seriously eroding Xi's authority, making it difficult for him to lead.

In a more optimistic scenario, Xi would manage to overcome internal resistance and move on to broader economic, legal, and political reforms. Ideally, the campaign will strengthen Xi's power base enough and win him the support necessary to reduce the party's tight grip on policy and regulatory and administrative power, creating a favorable environment for the growth of a more independent private sector. Xi has no interest in creating a Western-style democratic system, but he does think that China could produce a cleaner and more effective form of authoritarianism. To better serve that goal, Xi should consider adding a number of more ambitious elements to the anti-corruption crusade, including a step that both Transparency International and the G-20 have called for: improving public registers to clarify who owns and controls which companies and land, which would make it harder for corrupt officials and businesspeople to hide their illicit profits.

At the moment, there is more reason for optimism than pessimism. Xi has already consolidated a great deal of control over the state's power structures and is determined and able to remove anyone who might resist or challenge his authority or policies. So far, within the senior leadership and the wider bureaucratic system, resistance to the anticorruption campaign has been passive rather than active: some bureaucrats have reportedly slowed down their work in a rather limited form of silent protest. Meanwhile, the anticorruption campaign continues to enjoy strong public support, especially from low- and middle-income Chinese who resent the way that corruption makes the Chinese system even more unfair than it already is. Anti-corruption thus represents a way for the party to ease the social tensions and polarization that might otherwise emerge as the economy slows, even as dramatic economic inequalities persist. To maintain this public support, the trick for Xi will be calibrating the scope and intensity of the campaign: not so narrow or moderate as to seem halfhearted, but not so broad or severe as to seem like a form of abuse itself.

Critical Thinking

1. What lines is president Xi willing to cross in the battle against corruption?

2. Why is corruption so endemic in China?

3. What path will the anti-corruption campaign take?

Internet References

Embassy of the People's Republic of China in the United States
china-embassy.org

Freedom House
https://freedom house.org

Kleptocracy Initiative
kleptocracyinitiative.org

JAMES LEUNG is a pseudonym for an economist with extensive experience in China, Europe, and the United States.

Article   Prepared by: Robert Weiner, *University of Massachusetts, Boston*

The G-Word: The Armenian Massacre and the Politics of Genocide

THOMAS DE WAAL

Learning Outcomes

After reading this article, you will be able to:

- Understand the historical background of the Armenian genocide.

- Understand the position of the Armenian community on the question today.

- Discuss the current Turkish position on the question.

O ne hundred years ago this April, the Ottoman Empire began a brutal campaign of deporting and destroying its ethnic Armenian community, whom it accused of supporting Russia, a World War I enemy. More than a million Armenians died. As it commemorates the tragedy, the U.S. government, for its part, still finds itself wriggling on the nail on which it has hung for three decades: Should it use the term "genocide" to describe the Ottoman Empire's actions toward the Armenians, or should it heed the warnings of its ally, Turkey, which vehemently opposes using the term and has threatened to recall its ambassador or even deny U.S. access to its military bases if the word is applied in this way? The first course of action would fulfill the wishes of the one-million-strong Armenian American community, as well as many historians, who argue that Washington has a moral imperative to use the term. The second would satisfy the strategists and officials who contend that the history is complicated and advise against antagonizing Turkey, a loyal strategic partner.

No other historical issue causes such anguish in Washington. One former State Department official told me that in 1992, a group of top U.S. policymakers sat in the office of Brent Scowcroft, then national security adviser to President George

H. W. Bush, and calculated that resolutions related to the topic were consuming more hours of their time with Congress than any other matter. Over the years, the debate has come to center on a single word, "genocide," a term that has acquired such power that some refuse to utter it aloud, calling it "the G-word" instead. For most Armenians, it seems that no other label could possibly describe the suffering of their people. For the Turkish government, almost any other word would be acceptable.

U.S. President Barack Obama has attempted to break this deadlock in statements he has made on April 24, the day when Armenians traditionally commemorate the tragedy, by evoking the Armenian language phrase Meds Yeghern, or "Great Catastrophe." In 2010, for example, he declared, "1.5 million Armenians were massacred or marched to their death in the final days of the Ottoman Empire. . . . The Meds Yeghern is a devastating chapter in the history of the Armenian people, and we must keep its memory alive in honor of those who were murdered and so that we do not repeat the grave mistakes of the past."

Armenian descendants seeking recognition of their grandparents' suffering could find everything they wanted to see there, except one thing: the word "genocide." That omission led a prominent lobbying group, the Armenian National Committee of America, to denounce the president's dignified statement as "yet another disgraceful capitulation to Turkey's threats," full of "euphemisms and evasive terminology."

In a sense, Obama had only himself to blame for this over-the-top rebuke. After all, during his presidential campaign, he had, like most candidates before him, promised Armenian American voters that he would use the word "genocide" if elected, but once in office, he had honored the relationship with Turkey and broken his vow. His 2010 address did go further than those of his predecessors and openly hinted that he had the G-word in mind when he stated, "My view of that history has not changed." But if he edged closer to the line, he stopped short of crossing it.

History as Battleground

Back in 1915, there was nothing controversial about the catastrophe suffered by ethnic Armenians in the Ottoman Empire. The Young Turkish government, headed by Mehmed Talat Pasha and two others, which ruled what was left of the empire, had entered World War I the year before on the side of Germany, fighting against its longtime foe Russia. The leadership accused Christian Armenians—a population of almost two million, most of whom lived in what is now eastern Turkey—of sympathizing with Russia and thus representing a potential fifth column. Talat ordered the deportation of almost the entire people to the arid deserts of Syria. In the process, at least half of the men were killed by Turkish security forces or marauding Kurdish tribesmen. Women and children survived in greater numbers but endured appalling depredation, abductions, and rape on the long marches.

Leading statesmen of the time regarded the deportation and massacre of the Armenians as the worst atrocity of World War I. One of them, former U.S. President Theodore Roosevelt, argued in a 1918 letter to the philanthropist Cleveland Dodge that the United States should go to war with the Ottoman Empire "because the Armenian massacre was the greatest crime of the war, and failure to act against Turkey is to condone it." Some of the best sources on the horrific events were American. Because the United States had remained neutral during the war's early years, dozens of its diplomatic officials and missionaries in the Ottoman Empire had stayed on the ground and witnessed what happened. In May 1915, Henry Morgenthau, the U.S. ambassador in Turkey, delivered a démarche from the Ottoman Empire's three main adversaries—France, Russia, and the United Kingdom—that denounced the deportation of the Armenians. The statement condemned the Ottoman government for "crimes against humanity," marking the first known official usage of that term. In July 1915, Morgenthau cabled to Washington, "Reports from widely scattered districts indicate systematic attempts to uproot peaceful Armenian populations." These actions, he wrote, involved arbitrary arrests, torture, and large-scale deportations of Armenians, "accompanied by frequent instances of rape, pillage, and murder, turning into massacre."

At the other corner of the Ottoman Empire, Jesse Jackson, the U.S. consul in Aleppo, watched as pitiful convoys of emaciated Armenians arrived in Syria. In September 1916, Jackson sent a cable to Washington that described the burial grounds of nearly 60,000 Armenians near Maskanah, a town in today's northern Syria: "As far as the eye can reach mounds are seen containing 200 to 300 corpses buried in the ground pele mele, women, children and old people belonging to different families."

By the end of World War I, according to most estimates of the time, around one million Armenians had died. Barely one-tenth of the original population remained in its native lands in the Ottoman Empire. The rest had mostly scattered to Armenia, France, Lebanon, and Syria. Many, in ever-greater numbers over the years, headed to the United States.

From the 1920s on, the events of the Great Catastrophe became more a matter of private grief than public record. Ordinary Armenians concentrated on building new lives for themselves. The main political party active in the Armenian diaspora, the Armenian Revolutionary Federation (which had briefly ruled an independent Armenia in 1918–20, before it became a Soviet republic), expended most of its efforts fighting the Soviet Union rather than Turkey. Only in the 1960s did Armenians seriously revive the memory of their grandparents' suffering as a public political issue. They drew inspiration from "Holocaust consciousness," the urge for collective remembrance and action that brought together the Jewish people after the 1961 trial of Adolf Eichmann for Nazi war crimes.

The Republic of Turkey, founded by Mustafa Kemal in 1923, was a state rooted in organized forgetting—not only of the crimes committed in the late Ottoman period against Armenians, Assyrians, and Greeks but also of the suffering of the Muslim population in a string of wars in Anatolia and the Balkans prior to 1923. As the new Turkish state developed, the vanishing of the Armenians became a political, historical, and economic fait accompli. In Turkey, only one substantial book addressing the issue was published between 1930 and the mid-1970s.

When Turkish historians finally returned to the topic in the late 1970s, they did so in response to a wave of terrorist attacks on Turkish diplomats in Western Europe, most of them carried out by Armenian militants based in Beirut. The campaign set off a war among nationalist historians. A simplistic Armenian narrative told of Turkish perpetrators, callous international bystanders, and innocent Armenian victims, downplaying the role that radical Armenian political parties had played in fueling the crackdown. Countering this story was an even cruder narrative spun by some pro-Turkish scholars, several of whom were receiving funding from the Turkish government. That story line portrayed the Armenians as traitors and Muslims as victims of scheming Christian great powers that sought to break up the Ottoman Empire.

The United States served as the main arena for these assertions and denials. In one book published in 1990, Heath Lowry, the head of the newly established Institute of Turkish Studies in Washington, D.C., pursued a common line of Turkish argument: casting doubt on the authenticity of Westerners' eyewitness testimonies. His account, *The Story Behind "Ambassador Morgenthau's Story,"* alleged that Morgenthau was an unreliable witness. Others argued that U.S. missionaries were untrustworthy sources because of their anti-Muslim bias. Over the years, efforts to discredit dozens of primary sources have grown increasingly tortuous. The U.S.-based Turkish website Tall Armenian Tale, for example, laboriously tries to cast doubt on every single one of the hundreds of eyewitness testimonies of the massacre.

A more legitimate line of historical inquiry has focused on the hitherto overlooked tribulations of Muslims in Anatolia and the Caucasus during World War I. These accounts have pointed out that the Armenians were not the only people to face persecution in eastern Turkey. The Kurdish and Turkish populations, too, suffered grievously at the hands of the Russian army, which contained several Armenian regiments, when these forces occupied swaths of eastern Turkey not long after the Armenian deportations. Later, in 1918–20, Muslim Azerbaijanis were deported from the briefly independent Republic of Armenia before it was conquered by the Bolsheviks.

The wartime context of the Armenian massacre and the multiple actors involved—in addition to Armenians and Turks: Assyrians, Azerbaijanis, Greeks, Kurds, British, Germans, and Russians—make it harder to tell the story in all its nuance. The history of the Armenian genocide lacks the devastating simplicity of the Holocaust's narrative. But a new generation of historians has finally taken up the challenge of explaining the full context of the tragedy. Some of them, such as Raymond Kevorkian, are Armenian, whereas others, including Donald Bloxham and Erik-Jan Zurcher, hail from Europe. Several come from Turkey, including Fikret Adanir, Taner Akcam, Halil Berktay, and Fuat Dundar.

At the heart of most of these histories lies a hard kernel of truth: although Muslims suffered enormously during World War I, in both Anatolia and the Caucasus, the Armenian experience was of a different order of pain. Along with the Assyrians, the Armenians were subjected to a campaign of destruction that was more terrible for being organized and systematic. And even though some Armenian nationalists helped precipitate the brutal Ottoman response, every single Armenian suffered as a result. As Bloxham has written, "Nowhere else during the First World War was the separatist nationalism of the few answered with the total destruction of the wider ethnic community from which the nationalists hailed. That is the crux of the issue."

Word as Weapon

If the issue of the experience of the Armenians in World War I were merely a matter of historical interpretation, a way forward would be clear. The huge volume of primary source material, combined with Armenian oral histories, authenticates the veracity of what Armenians recall—as does the plain fact that an entire people vanished from their historical homeland. All that historians have to do, it would seem, is fill out the context of the events and explain why the Young Turks treated the Armenians the way they did.

But what dominates the public discourse today is the word "genocide," which was devised almost three decades after the Armenian deportations to designate the destruction not just of people but also of an entire people. The term is closely associated with the man who invented it, the Polish-born Jewish lawyer Raphael Lemkin. Lemkin barely escaped the horror of the Holocaust, which wiped out most of his family in Poland after he immigrated to the United States. As he would later explain in a television interview, "I became interested in genocide because it happened so many times. It happened to the Armenians, and after the Armenians, Hitler took action."

Lemkin had a morally courageous vision: to get the concept of genocide enshrined in international law. His tireless lobbying soon paid off: in 1948, just four years after he invented the term, the United Nations adopted the Genocide Convention, a treaty that made the act an international crime. But Lemkin was a more problematic personality than the noble crusader depicted in modern accounts, such as Samantha Power's book *A Problem From Hell.* In his uncompromising pursuit of his goal, Lemkin allowed the term "genocide" to be bent by other political agendas. He opposed the Universal Declaration of Human Rights, adopted a week after the Genocide Convention, fearing that it would distract the international community from preventing future genocides—the goal that he thought should surpass all others in importance. And he won the Soviet Union's backing for the convention after "political groups" were excluded from the classes of people it protected.

The final definition of "genocide" adopted by the UN had several points of ambiguity, which gave countries and individuals accused of this crime legal ammunition to resist the charge. For example, Article 2 of the convention defines "genocide" as "acts committed with intent to destroy, in whole or in part, a national, ethnical, racial or religious group, as such." The meaning of the words "as such" is far from clear. And alleged perpetrators often deny that the destruction was "committed with intent"—an argument frequently made in Turkey.

Soon, however, only a careful few were bothering to refer to the UN convention in evoking the term. In the broader public's mind, the association with the Holocaust gave the word "genocide" totemic power, making it the equivalent of absolute evil. After 1948, the legal term that had initially been created to deter mass atrocities became an insult traded between nations and peoples accusing each other of past and present horrors. The United States and the Soviet Union each freely accused the other of genocide during the Cold War.

The Armenian diaspora saw the word as a perfect fit to describe what had happened to their parents and grandparents and began referring to the Meds Yeghern as "the Armenian genocide." The concept helped activate a new political movement. The year 1965 marked both the 50th anniversary of the massacre and the moment when the Armenian diaspora made seeking justice for the victims a political cause.

In the postwar United States, it was normal practice to put the words "Armenian" and "genocide" together in the same sentence. This usage came with the assumption that the UN convention—one of its first signatories was Turkey—had no retroactive force and therefore could not provide the basis for legal action related to abuses committed before 1948. For instance, in 1951, U.S. government lawyers submitted an advisory opinion on the Genocide Convention to the International Court of Justice, in The Hague, citing the Turkish massacre of the Armenians as an instance of genocide. In April 1981, in a proclamation on the Holocaust, U.S. President Ronald Reagan mentioned "the genocide of the Armenians before it, and the genocide of the Cambodians which followed it."

Political circumstances changed this thinking in the 1980s. Reagan himself performed an abrupt about-face following the 1982 assassination of Kemal Arikan, the Turkish consul general to the United States, by two young Armenian militants in Los Angeles. The death of a diplomat of a close NATO ally in Reagan's own home state enraged and embarrassed the president. He and his team concluded that on three of the foreign policy issues that concerned them the most—the Soviet Union, Israel, and terrorism—Turkey was staunchly on the U.S. side. Armenians, by contrast, were not.

Seven months after the killing of Arikan, the State Department's official bulletin published a special issue on terrorism, which included a piece titled "Armenian Terrorism: A Profile." A note at the end of the article said, "Because the historical record of the 1915 events in Asia Minor is ambiguous, the Department of State does not endorse allegations that the Turkish government committed a genocide against the Armenian people. Armenian terrorists use this allegation to justify in part their continuing attacks on Turkish diplomats and installations." In response to furious Armenian complaints, the bulletin ended up publishing not one but two clarifications of that statement. But from that point on, a new line had been drawn by the executive branch, and the term "Armenian genocide" was outlawed in the White House.

Deadlock on the Hill

Congress, meanwhile, was plowing its own furrow. By the 1970s, one million Armenians lived in the United States. Younger generations were no longer willing to limit the discussions of their ancestors' deaths to Sunday dinners, requiem services, and low-circulation newspapers. Many Armenian Americans who had political savvy and wealth, such as the Massachusetts businessman Stephen Mugar, began to lobby Congress. They found an ally in the Speaker of the House of Representatives, Tip O'Neill, whose congressional district included the de facto capital of the Armenian American community: Watertown, Massachusetts. In early 1975, urged on by Mugar and others, O'Neill managed to get the House to pass a resolution authorizing the president to designate April 24 of that year as the "National Day of Remembrance of Man's Inhumanity to Man" and observe it by honoring all victims of genocide, "especially those of Armenian ancestry who succumbed to the genocide perpetrated in 1915."

That occasion marked the only time Congress has passed any kind of resolution recognizing the Armenian genocide. In 1990, the Senate spent two days in fierce debate over whether April 24 should again be officially designated as a national day of remembrance, this time of the "Armenian Genocide of 1915–1923." Kansas Senator Bob Dole led the argument in favor of the motion, but opponents managed to block it. Ever since, with the White House opposed to officially recognizing the phrase "Armenian genocide," resolutions of this kind have failed. They have become an increasingly tired and predictable exercise: however much historical evidence the Armenian lobbyists produce to support their case, the Turks play the trump card of national security, lightly threatening that a yes vote would jeopardize the United States' continued use of the Incirlik Air Base, which is on Turkish territory, a key supply hub for U.S. military operations in the region. In 2007, when one genocide resolution appeared certain to pass the House, no fewer than eight former secretaries of state intervened with a joint letter advising Congress to drop the issue—which it ultimately did.

The fight for genocide recognition has now become the raison d'être for the two dominant Armenian American organizations, the Armenian Assembly of America and the Armenian National Committee of America. They do not conceal that the campaign helps them preserve a collective identity among the Armenian diaspora—an increasingly assimilated group that is losing other common bonds, such as the Armenian language and attendance at services of the Armenian Apostolic Church. But they do not like to admit that the campaign has also damaged their cause. For many Americans, the phrase "Armenian genocide" now evokes not a story of terrible human suffering but an exasperating, eye-roll-inducing tale of lobbying and congressional bargaining. Inevitably, the need to secure votes for any given resolution on the topic means that the memory of the Ottoman Armenians is cheapened by being tied to other items of congressional business. What results is routine horse-trading, as in, "You vote for the farm bill, and I'll back you on the genocide resolution."

A few thoughtful Armenians object to such genocide-recognition lobbying campaigns on the grounds that they turn the deaths of their grandparents into one big homicide case. They see that their fellow Armenians are less interested in grieving for the dead than in demonstrating outside the Turkish embassy with pictures of dead bodies—the more gruesome, the better—and struggling to prove something that they already know to be true. The obsession with genocide, argues the French Armenian philosopher Marc Nichanian, "forbids mourning." Armenian campaigners have a point when they contend that their pursuit of genocide recognition has had the benefit of focusing Turkey's mind on an issue that the country would rather have forgotten. But their campaign has also heightened Turkish passions, since their efforts have indirectly strengthened the Turkish nationalist story line of World War I. That partial, but not entirely inaccurate, account portrays the great powers of the time as conspirators plotting to undermine the Ottoman Empire. Consequently, any resolution passed by a modern great power condemning Turkey's historical crimes would only inflame a sore spot.

Fueling this paranoia, many Turkish policymakers have expressed their suspicion that a genocide resolution would pave the way for territorial concessions. These fears have little basis in reality. Although some radical groups, such as the Armenian Revolutionary Federation, continue to make territorial claims, the Republic of Armenia has all but officially recognized Turkey's current borders. Reestablishing full diplomatic relations between the two countries, which have been on hold since the Armenian-Azerbaijani war in the early 1990s, would make this recognition formal. No statements made by a political party that last ruled Armenia in 1920 can change that reality.

As for reparations, it is hard to see how Washington's adoption of the word "genocide" would make the case for them. Most international legal opinions are clear that the UN Genocide Convention carries no retroactive force and therefore could not be invoked to bring claims on dispossessed property. Such a scenario is all the more difficult to imagine because it would trigger a nightmarish relitigation of the whole of World War I, during which not only Armenians but also Azerbaijanis, Greeks, Kurds, and Turks were robbed of their possessions in Anatolia, the Balkans, and the Caucasus. Yet the invocation of the controversial word still fills Turkey with dread.

A Turkish Thaw

The only good news in this bleak historical tale comes from Turkey itself. Since the election in 2002 of the post-Kemalist government led by the Justice and Development Party (known as the AKP), in a process largely unconnected to outside pressure, Turkish society has begun to revisit some of the dark pages of its past, including the oppression of the non-Turkish populations of the late Ottoman Empire. This growing openness has allowed the descendants of forcibly Islamized Armenians to come out of the shadows, and a few Armenian churches and schools have reopened. Turkish historians have begun to write about the late Ottoman period without fear of retribution. And they have finally started to challenge the old dominant narrative, which the historian Berktay has called "the theory of the immaculate conception of the Turkish Republic."

From the Armenian standpoint, this opening has been too slow. But it could hardly have proceeded at a faster pace. As one of the key figures behind the thaw, the late Istanbul-based Armenian journalist Hrant Dink, pointed out, Turkey had been a closed society for three generations; it takes time and immense effort to change that. "The problem Turkey faces today is neither a problem of 'denial' or 'acknowledgement,'" Dink wrote in 2005. "Turkey's main problem is 'comprehension.' And for the process of comprehension, Turkey seriously needs an alternative study of history and for this, a democratic environment. . . . The society is defending the truth it knows."

In that spirit, Dink, a stalwart of the left and a confirmed antiimperialist, criticized genocide resolutions in foreign parliaments on the grounds that they merely replicated previous great-power bullying of Turkey. He saw his mission as helping Turks understand Armenians and the trauma they have passed down over generations, while helping Armenians recognize the sensitivities and legitimate interests of the Turks. Dink's stand broke both Turkish and Armenian taboos, and he paid the highest price for his courage: in 2007, he was assassinated by a young Turkish nationalist.

Dink's insights suggest that the word "genocide" may be the correct term but the wrong solution to the controversy. Simply put, the emotive power of the word has overpowered Armenian-Turkish dialogue. No one willingly admits to committing genocide. Faced with this accusation, many Turks (and others in their position) believe that they are being invited to compare their grandparents to the Nazis.

It may be that the word "genocide" has exhausted itself, and that the success of Lemkin's invention has also been its undoing. Lemkin probably never anticipated that coining a new standard of awfulness would set off an unfortunate global competition in which nations—from Armenia's neighbor Azerbaijan to Sudan and Tibet—vie to get the label applied to their own tragedies. As the philosopher Tzvetan Todorov has observed, even though no one wants to be a victim, the position does confer certain advantages. Groups that gain recognition as victims of past injustices obtain "a bottomless line of moral credit," he has written. "The greater the crime in the past, the more compelling the rights in the present—which are gained merely through membership in the wronged group." Conversely, the grandchildren of the alleged perpetrators aspire to absolve their ancestors of guilt and, by association, of a link to Adolf Hitler and the Holocaust.

In *A Problem From Hell,* Power chastised the international community for its timidity and failure to stop genocides even after this appalling phenomenon had been named and outlawed. But the problem can be posed the other way around: Could it be that international actors hide behind the ambiguities of genocide terminology in order to do nothing—and that the very power of the word "genocide" and the responsibilities it invokes deter action? It may be no coincidence that the first successful prosecution under the UN Genocide Convention, that of a Rwandan war criminal, came only in September 1998, nearly 50 years after the convention was adopted. In the Armenian case, the phrase "Armenian genocide" has become customary in the scholarly literature. Those who avoid it today risk putting themselves in the company of skeptics who minimize the tragedy or deny it outright. Many progressive Turkish intellectuals, too, now use the term. Among them are such brave voices as the journalist Hasan Cemal, grandson of Ahmed Cemal Pasha, one of the three Young Turkish leaders who ran the brutal Ottoman government in 1915.

But that does not mean that Meds Yeghern is an inferior and less expressive phrase. If it becomes more widely used, it might acquire the same resonance as the words "Holocaust" and "Shoah" have in describing the fate of the European Jews. There is also the legal term "crimes against humanity," first applied in 1915 specifically in reference to the Armenian massacre. This concept lacks the emotional charge and the definitional problems of the word "genocide" and covers mass atrocities not falling under its narrow definition—those in which the perpetrators may not have intended to eradicate an entire nation but have still killed an awful lot of innocent people.

The challenge for the United States, then, is not simply to find a way to once again use the term "Armenian genocide," a phrase it has employed before, but to do so while also accepting the limitations of a concept that has grown emotionally fraught and overly legalistic. The mere act of using the term, without a deeper engagement with the history of the Armenians and the Turks, would do little to resolve the bigger underlying question—namely, how to persuade Turkey to honor the losses of the Ottoman Armenians and other minorities a hundred years ago.

Having been a neutral power in 1915, the United States can assert that it bears no historical grudge against Turkey. Washington can therefore help bring about the rapprochement between the Armenians and the Turks that Dink advocated. The United States can urge Turkey to hasten the process of historical reckoning by taking steps to keep the small Armenian Turkish population from leaving the country, to conserve what little Armenian cultural heritage survives in Turkey, and to restore the place of Armenians and other ethnic minorities in Turkey's history books.

Armenians need to be able to finally bury their grandparents and receive an acknowledgment from the Turkish state of the terrible fate they suffered. These steps toward reconciliation will surely become more possible as a more open Turkey begins to confront its past as a whole. If that can be made to happen, everything else will follow.

Critical Thinking

1. Why won't the U.S. label the killing of the Armenians as genocide?

2. Why does the author argue that the term genocide is overly legalistic?

3. Why did the Ottoman Turkish government target the Armenians?

Internet References

Genocide Convention
  preventgenocide.org
Turkish Embassy, Washington DC
  vasington.be.mfa.gov.tr
U.S. Holocaust Museum
  www.ushmm.org/

THOMAS DE WAAL is a Senior Associate at the Carnegie Endowment for International Peace and the author of the forthcoming *Great Catastrophe: Armenians and Turks in the Shadow of Genocide.* Follow him on Twitter @TomdeWaalCEIP.

Article

Prepared by: Robert Weiner, *University of Massachusetts, Boston*

Race in the Modern World: The Problem of the Color Line

Kwame Anthony Appiah

Learning Outcomes

After reading this article, you will be able to:

- Explain the difficult efforts of scholars to come up with a definition of race.
- Discuss the importance of Pan-Africanism.

In 1900, in his "Address to the Nations of the World" at the first Pan-African Conference, in London, W. E. B. Du Bois proclaimed that the "problem of the twentieth century" was "the problem of the color-line, the question as to how far differences of race—which show themselves chiefly in the color of the skin and the texture of the hair—will hereafter be made the basis of denying to over half the world the right of sharing to their utmost ability the opportunities and privileges of modern civilization."

Du Bois had in mind not just race relations in the United States but also the role race played in the European colonial schemes that were then still reshaping Africa and Asia. The final British conquest of Kumasi, Ashanti's capital (and the town in Ghana where I grew up), had occurred just a week before the London conference began. The British did not defeat the Sokoto caliphate in northern Nigeria until 1903. Morocco did not become a French protectorate until 1912, Egypt did not become a British one until 1914, and Ethiopia did not lose its independence until 1936. Notions of race played a crucial role in all these events, and following the Congress of Berlin in 1878, during which the great powers began to devise a world order for the modern era, the status of the subject peoples in the Belgian, British, French, German, Spanish, and Portuguese colonies of Africa—as well as in independent South Africa—was defined explicitly in racial terms.

Du Bois was the beneficiary of the best education that North Atlantic civilization had to offer: he had studied at Fisk, one of the United States' finest black colleges; at Harvard; and at the University of Berlin. The year before his address, he had published *The Philadelphia Negro,* the first detailed sociological study of an American community. And like practically everybody else in his era, he had absorbed the notion, spread by a wide range of European and American intellectuals over the course of the nineteenth century, that race—the division of the world into distinct groups, identifiable by the new biological sciences—was central to social, cultural, and political life.

Even though he accepted the concept of race, however, Du Bois was a passionate critic of racism. He included anti-Semitism under that rubric, and after a visit to Nazi Germany in 1936, he wrote frankly in *The Pittsburgh Courier,* a leading black newspaper, that the Nazis' "campaign of race prejudice . . . surpasses in vindictive cruelty and public insult anything I have ever seen; and I have seen much." The European homeland had not been in his mind when he gave his speech on the color line, but the Holocaust certainly fit his thesis—as would many of the centuries' genocides, from the German campaign against the Hereros in Namibia in 1904 to the Hutu massacre of the Tutsis in Rwanda in 1994. Race might not necessarily have been *the* problem of the century—there were other contenders for the title—but its centrality would be hard to deny.

Violence and murder were not, of course, the only problems that Du Bois associated with the color line. Civic and economic inequality between races—whether produced by government policy, private discrimination, or complex interactions between the two—were pervasive when he spoke and remained so long after the conference was forgotten.

All around the world, people know about the civil rights movement in the United States and the antiapartheid struggle in

South Africa, but similar campaigns have been waged over the years in Australia, New Zealand, and most of the countries of the Americas, seeking justice for native peoples, or the descendants of African slaves, or East Asian or South Asian indentured laborers. As non-Europeans, including many former imperial citizens, have immigrated to Europe in increasing numbers in recent decades, questions of racial inequality there have come to the fore, too—in civic rights, education, employment, housing, and income. For Du Bois, Chinese, Japanese, and Koreans were on the same side of the color line as he was. But Japanese brutality toward Chinese and Koreans up through World War II was often racially motivated, as are the attitudes of many Chinese toward Africans and African Americans today. Racial discrimination and insult are a global phenomenon.

Of course, ethnoracial inequality is not the only social inequality that matters. In 2013, the nearly 20 million white people below the poverty line in the United States made up slightly more than 40 percent of the country's poor. Nor is racial prejudice the only significant motive for discrimination: ask Christians in Indonesia or Pakistan, Muslims in Europe, or LGBT people in Uganda. Ask women everywhere. But more than a century after his London address, Du Bois would find that when it comes to racial inequality, even as much has changed, much remains the same.

Us and Them

Du Bois' speech was an invitation to a global politics of race, one in which people of African descent could join with other people of color to end white supremacy, both in their various homelands and in the global system at large. That politics would ultimately shape the process of decolonization in Africa and the Caribbean and inform the creation of what became the African Union. It was politics that led Du Bois himself to become, by the end of his life, a citizen of a newly independent Ghana, led by Kwame Nkrumah.

But Du Bois was not simply an activist; he was even more a scholar and an intellectual, and his thinking reflected much of his age's obsession with race as a concept. In the decades preceding Du Bois' speech, thinkers throughout the academy—in classics, history, artistic and literary criticism, philology, and philosophy, as well as all the new life sciences and social sciences—had become convinced that biologists could identify, using scientific criteria, a small number of primary human races. Most would have begun the list with the black, white, and yellow races, and many would have included a Semitic race (including Jews and Arabs), an American Indian race, and more. People would have often spoken of various subgroups within these categories as races, too. Thus, the English poet Matthew Arnold considered the Anglo-Saxon and Celtic races to be the

main components of the population of the United Kingdom; the French historian Hippolyte Taine thought the Gauls were the race at the core of French history and identity; and the U.S. politician John C. Calhoun discussed conflicts not only between whites and blacks but also between Anglo-Canadians and "the French race of Lower Canada."

People thought race was important not just because it allowed one to define human groups scientifically but also because they believed that racial groups shared inherited moral and psychological tendencies that helped explain their different histories and cultures. Of course, there were always skeptics. Charles Darwin, for example, believed that his evolutionary theory demonstrated that human beings were a single stock, with local varieties produced by differences in environment, through a process that was bound to result in groups with blurred edges. But many late-nineteenth-century European and American thinkers believed deeply in the biological reality of race and thought that the natural affinity among the members of each group made races the appropriate units for social and political organization.

Essentialism—the idea that human groups have core properties in common that explain not just their shared superficial appearances but also the deep tendencies of their moral and cultural lives—was not new. In fact, it is nearly universal, because the inclination to suppose that people who look alike have deep properties in common is built into human cognition, appearing early in life without much prompting. The psychologist Susan Gelman, for example, argues that "our essentializing bias is not directly taught," although it is shaped by language and cultural cues. It can be found as far back as Herodotus' *Histories* or the Hebrew Bible, which portrayed Ethiopians, Persians, and scores of other peoples as fundamentally other. "We" have always seen "our own" as more than superficially different from "them."

What was new in the nineteenth century was the combination of two logically unrelated propositions: that races were biological and so could be identified through the scientific study of the shared properties of the bodies of their members and that they were also political, having a central place in the lives of states. In the eighteenth century, the historian David Hume had written of "national character"; by the nineteenth century, using the new scientific language, Arnold was arguing that the "Germanic genius" of his own "Saxon" race had "steadiness as its main basis, with commonness and humdrum for its defect, fidelity to nature for its excellence."

If nationalism was the view that natural social groups should come together to form states, then the ideal form of nationalism would bring together people of a single race. The eighteenth-century French American writer J. Hector St. John de Crèvecoeur's notion that in the New World, all races could be "melted into a new race of man"—so that it was the nation that made the race, not the race the nation—belonged to an older way of thinking, which racial science eclipsed.

The Other Dismal Science

In the decade after Du Bois' address, however, a second stage of modern argumentation about human groups emerged, one that placed a much greater emphasis on culture. Many things contributed to this change, but a driving force was the development of the new social science of anthropology, whose German-born leader in the United States, Franz Boas, argued vigorously (and with copious evidence from studies in the field) that the key to understanding the significant differences between peoples lay not in biology—or, at least, not in biology alone—but in culture. Indeed, this tradition of thought, which Du Bois himself soon took up vigorously, argued not only that culture was the central issue but also that the races that mattered for social life were not, in fact, biological at all.

In the United States, for example, the belief that anyone with one black grandparent or, in some states, even one black great-grandparent was also black meant that a person could be socially black but have skin that was white, hair that was straight, and eyes that were blue. As Walter White, the midcentury leader of the National Association for the Advancement of Colored People, whose name was one of his many ironic inheritances, wrote in his autobiography, "I am a Negro. My skin is white, my eyes are blue, my hair is blond. The traits of my race are nowhere visible upon me."

Strict adherence to thinking of race as biological yielded anomalies in the colonial context as well. Treating all Africans in Nigeria as "Negroes," say, would combine together people with very different biological traits. If there were interesting traits of national character, they belonged not to races but to ethnic groups. And the people of one ethnic group—Arabs from Morocco to Oman, Jews in the Diaspora—could come in a wide range of colors and hair types.

In the second phase of discussion, therefore, both of the distinctive claims of the first phase came under attack. Natural scientists denied that the races observed in social life were natural biological groupings, and social scientists proposed that the human units of moral and political significance were those based on shared culture rather than shared biology. It helped that Darwin's point had been strengthened by the development of Mendelian population genetics, which showed that the differences found between the geographic populations of the human species were statistical differences in gene frequencies rather than differences in some putative racial essence.

In the aftermath of the Holocaust, moreover, it seemed particularly important to reject the central ideas of Nazi racial "science," and so, in 1950, in the first of a series of statements on race, unesco (whose founding director was the leading biologist Sir Julian Huxley) declared that national, religious, geographic, linguistic and cultural groups do not necessarily coincide with racial groups: and the cultural traits of such groups have no demonstrated genetic connection with racial traits. . . . The scientific material available to us at present does not justify the conclusion that inherited genetic differences are a major factor in producing the differences between the cultures and cultural achievements of different peoples or groups.

Race was still taken seriously, but it was regarded as an outgrowth of sociocultural groups that had been created by historical processes in which the biological differences between human beings mattered only when human beings decided that they did. Biological traits such as skin color, facial shape, and hair color and texture could define racial boundaries if people chose to use them for that purpose. But there was no scientific reason for doing so. As the unesco statement said in its final paragraph, "Racial prejudice and discrimination in the world today arise from historical and social phenomena and falsely claim the sanction of science."

Construction Work

In the 1960s, a third stage of discussion began, with the rise of "genetic geography." Natural scientists such as the geneticist Luigi Luca Cavalli-Sforza argued that the concept of race had no place in human biology, and social scientists increasingly considered the social groups previously called "races" to be social constructions. Since the word "race" risked misleading people on this point, they began to speak more often of "ethnic" or "ethnoracial" groups, in order to stress the point that they were not aiming to use a biological system of classification.

In recent years, some philosophers and biologists have sought to reintroduce the concept of race as biological using the techniques of cladistics, a method of classification that combines genetics with broader genealogical criteria in order to identify groups of people with shared biological heritages. But this work does not undermine the basic claim that the boundaries of the social groups called "races" have been drawn based on social, rather than biological, criteria; regardless, biology does not generate its own political or moral significance. Socially constructed groups can differ statistically in biological characteristics from one another (as rural whites in the United States differ in some health measures from urban whites), but that is not a reason to suppose that these differences are caused by different group biologies. And even if statistical differences between groups exist, that does not necessarily provide a rationale for treating individuals within those groups differently. So, as Du Bois was one of the first to argue, when questions arise about the salience of race in political life, it is usually not a good idea to bring biology into the discussion.

It was plausible to think that racial inequality would be easier to eliminate once it was recognized to be a product of sociology and politics rather than biology. But it turns out that all sorts of status differences between ethnoracial groups can persist long after governments stop trying to impose them. Recognizing that institutions and social processes are at work rather than innate qualities of the populations in question has not made it any less difficult to solve the problems.

Imagined Communities

One might have hoped to see signs that racial thinking and racial hostility were vanishing—hoped, that is, that the color line would not continue to be a major problem in the twenty-first century, as it was in the twentieth. But a belief in essential differences between "us" and "them" persists widely, and many continue to think of such differences as natural and inherited. And of course, differences between groups defined by common descent can be the basis of social identity, whether or not they are believed to be based in biology. As a result, ethnoracial categories continue to be politically significant, and racial identities still shape many people's political affiliations.

Once groups have been mobilized along ethnoracial lines, inequalities between them, whatever their causes, provide bases for further mobilization. Many people now know that we are all, in fact, one species, and think that biological differences along racial lines are either illusory or meaningless. But that has not made such perceived differences irrelevant.

Around the world, people have sought and won affirmative action for their ethnoracial groups. In the United States, in part because of affirmative action, public opinion polls consistently show wide divergences on many questions along racial lines. On American university campuses, where the claim that "race is a social construct" echoes like a mantra, black, white, and Asian identities continue to shape social experience. And many people around the world simply find the concept of socially constructed races hard to accept, because it seems so alien to their psychological instincts and life experiences.

Race also continues to play a central role in international politics, in part because the politics of racial solidarity that Du Bois helped inaugurate, in co-founding the tradition of pan-Africanism, has been so successful. African Americans are particularly interested in U.S. foreign policy in Africa, and Africans take note of racial unrest in the United States: as far away as Port Harcourt, Nigeria, people protested against the killing of Michael Brown, the unarmed black teenager shot to death by a police officer last year in Missouri. Meanwhile, many black Americans have special access to Ghanaian passports,

Rastafarianism in the Caribbean celebrates Africa as the home of black people, and heritage tourism from North and South America and the Caribbean to West Africa has boomed.

Pan-Africanism is not the only movement in which a group defined by a common ancestry displays transnational solidarity. Jews around the world show an interest in Israeli politics. People in China follow the fate of the Chinese diaspora, the world's largest. Japanese follow goings-on in São Paulo, Brazil, which is home to more than 600,000 people of Japanese descent—as well as to a million people of Arab descent, who themselves follow events in the Middle East. And Russian President Vladimir Putin has put his supposed concern for ethnic Russians in neighboring countries at the center of his foreign policy.

Identities rooted in the reality or the fantasy of shared ancestry, in short, remain central in politics, both within and between nations. In this new century, as in the last, the color line and its cousins are still going strong.

Wouldn't It Be Nice?

The pan-Africanism that Du Bois helped invent created, as it was meant to, a new kind of transnational solidarity. That solidarity was put to good use in the process of decolonization, and it was one of the forces that helped bring an end to Jim Crow in the United States and apartheid in South Africa. So racial solidarity has been used not just for pernicious purposes but for righteous ones as well. A world without race consciousness, or without ethnoracial identity more broadly, would lack such positive mobilizations, as well as the negative ones. It was in this spirit, I think, that Du Bois wrote, back in 1897, that it was "the duty of the Americans of Negro descent, as a body, to maintain their race identity until . . . the ideal of human brotherhood has become a practical possibility."

But at this point, the price of trying to move beyond ethnoracial identities is worth paying, not only for moral reasons but also for the sake of intellectual hygiene. It would allow us to live and work together more harmoniously and productively, in offices, neighborhoods, towns, states, and nations. Why, after all, should we tie our fates to groups whose existence seems always to involve misunderstandings about the facts of human difference? Why rely on imaginary natural commonalities rather than build cohesion through intentional communities? Wouldn't it be better to organize our solidarities around citizenship and the shared commitments that bind political society?

Still, given the psychological difficulty of avoiding essentialism and the evident continuing power of ethnoracial identities, it would take a massive and focused effort of education, in schools and in public culture, to move into a postracial world. The dream of a world beyond race, unfortunately, is likely to be long deferred.

Critical Thinking

1. What are W. E. B. Dubois' major ideas about race?
2. What is meant by an imagined community?

Internet References

Brennan Center for Justice
www.brennancenter.org

W. E. B. Dubois papers
credo.library.umass.edu

KWAME ANTHONY APPIAH is Professor of Philosophy and Law at New York University. His most recent book is *Lines of Descent: W. E. B. Du Bois and the Emergence of Identity*. Born to a British mother and Ghanaian father, Kwame Anthony Appiah grew up traveling between his two homelands, an experience that has shaped his wideranging writing on ethnicity, identity, and culture. He is the author of numerous books, including *Lines of Descent,* and has won scores of prizes, among them the National Humanities Medal and the Arthur Ross Book Award. Now a professor at New York University, Appiah explores the past, present, and future of thinking about race in "Race in the Modern World" (page 1).

Article

Prepared by: Robert Weiner, *University of Massachusetts, Boston*

The Surveillance State and Its Discontents

ANONYMOUS

Learning Outcomes

After reading this article, you will be able to:

- Understand what is meant by the surveillance state.

- Understand the impact of the Internet on intelligence operations.

This year, leaks of classified U.S. government documents rewrote our understanding not only of the American intelligence apparatus, but of the possibilities and pitfalls of the Internet writ large. The statesmen, hackers, and activists in this category of Global Thinkers are working on the bleeding edge of the digital revolution, where a battle is being fought over who will control the defining tool of the 21st century. They represent those seeking to harness the web in the name of national security, those working to bring it under the letter of the law, and those hoping to liberate it in the name of human freedom.

Edward Snowden

For Exposing the Reach of Government Spying.

Former Contractor, National Security Agency I Russia

Perhaps the most surprising thing about the man behind the biggest story of 2013 is that we know his name. When Edward Snowden took credit for giving journalists classified documents from the U.S. National Security Agency, in the process revealing several clandestine intelligence programs, he deviated from the long-standing tradition of anonymity among leakers. Consequently, Snowden has become the public face of a raging international debate over surveillance.

Opinions on the merits of Snowden's actions are as divergent as the terms used to describe him. Patriot, whistleblower, and hero. Traitor, enemy, and defector. "I'm an American," Snowden told the South China Morning Post while in Hong Kong in June, the month that the first media accounts based on his pilfered files appeared. "I acted in good faith, but it is only right that the public form its own opinion."

From Hong Kong, Snowden flew to Moscow, seeking asylum as U.S. authorities charged him with espionage, theft, and "unauthorized communication of national defense information." Intelligence sources have pegged Snowden's cache of documents at approximately 50,000 pages, and their contents have inspired intense backlash. Leaks from the files, for instance, have compelled foreign governments targeted by U.S. spying to seek a U.N. resolution about the rights of individuals to retain their privacy on the Internet.

Snowden remains in Russia, at least temporarily, and he has reportedly given two journalists, Glenn Greenwald and Laura Poitras, full access to his documents. Insofar as, in Greenwald and Poitras's hands, leaks from the documents can go on without him—and they certainly will—it's unclear whether Snowden the man now matters as much as Snowden the symbol. Still, his actions to date have positioned him as the single most important figure in the global surveillance debate—and the most divisive world figure of 2013.

Keith Alexander

For Masterminding the Surveillance State.

Director, National Security Agency; Commander, U.S. Cyber Command | Fort Meade, MD.

He has been called "Emperor Alexander" and "Alexander the Geek" by his colleagues. But even those nicknames don't truly capture the scope of this four-star general's impact: Keith Alexander is the architect of a sweeping surveillance infrastructure that is monitoring Internet traffic in the United States and routinely scooping up Americans' phone records, emails, and text messages. Surveillance also extends to countries around the world.

Alexander made headlines in 2013 when leaked documents showed that the National Security Agency has been collecting data from the world's largest technology companies—a revelation that set off a storm of public criticism and induced heartburn among some in the intelligence community who fear Alexander is running roughshod over Americans' constitutional rights. One former intelligence official who worked with the general has said that he "tended to be a bit of a cowboy" whose philosophy was, "Let's not worry about the law. Let's just figure out how to get the job done."

But even as his work is increasingly saturated in controversy, Alexander's domain is growing. From a 350-acre headquarters in Maryland, Alexander oversees tens of thousands of employees and a budget that expanded by billions of dollars even as belt-tightening gripped most of the government. The NSA recently constructed a $2 billion data-processing center in Utah, and it is considering quadrupling the size of its Fort Meade facility.

The surveillance state Alexander has built, it seems, is here to stay, and its next move may be to take over the security of major companies threatened by cyberattack. As Alexander, who is supposed to retire in 2014, said recently, "I am concerned that this is going to break a threshold where the private sector can no longer handle it and the government is going to have to step in."

For Alexander's critics, the notion is a sinister one.

Glenn Greenwald, Ladra Poitras

For Giving Edward Snowden A Voice.

Journalists | Brazil, Germany

Glenn Greenwald and Laura Poitras are believed to be the only journalists with full access to leaker Edward Snowden's purloined trove of documents from the National Security Agency. Thus, what more we learn about the NSA's classified spying operations is largely up to Greenwald, a blogger turned columnist for the Guardian, and Poitras, a documentary filmmaker. (Both have joined a recently announced online news venture backed by eBay founder Pierre Omidyar.)

In the work they've done thus far with Snowden's documents, revealing the extent of the NSA'S spying on U.S. citizens, world leaders, and others, Greenwald has been the more public half of the duo. A frequent guest on radio and TV shows, he's working on a highly anticipated book and has engaged in more than a few Twitter fights with his critics. Poitras is a quieter force. Instead of hitting the talk-show circuit, she has opted to set up shop in Berlin and pore through Snowden's documents, writing for outlets like *Der Spiegel* and teaming up with other reporters, such as the *New York Times'* James Risen.

Poitras, the first journalist with whom Snowden made contact via email, told the *Times* of their initial correspondence, "I thought, ok, if this is true, my life just changed." Since that moment, Poitras and Greenwald have been consumed by Snowden's leaks—a situation that shows no signs of abating. As Poitras told the *Times,* she and Greenwald may never really be free of the surveillance apparatus they've worked to expose. "I don't know if I'll ever be able to live someplace and feel like I have my privacy," Poitras said. "That might be just completely gone."

But having their worlds permanently altered is a reality Greenwald and Poitras have accepted: Both say there are more surveillance stories to come.

Dilma Rousseff

For Confronting Washington and Its Spies.

President | Brazil

When Edward Snowden disclosed the extent of the National Security Agency's surveillance activities in Latin America, it awoke memories of a historical U.S. paternalism at odds with the region's growing sense of independence and strength. No leader has come to embody the resulting outrage quite like Brazilian President Dilma Rousseff. She has openly criticized the United States, including in a scathing speech at the U.N., and she even canceled a state dinner in Washington. "The right to safety of citizens of one country can never be guaranteed by violating fundamental human rights of citizens of another country," Rousseff argued before the U.N.

Her anger is informed by her background as a leftist revolutionary. The daughter of a Bulgarian ex-communist who fled his home in the 1930s, Rousseff was a militant left-wing university student by the time a military dictatorship took over Brazil.

With her husband, she smuggled guns, bombs, and money for the guerrilla group Colina. After allegedly helping to plan the 1969 theft of $2.5 million from the mistress of a former Säo Paulo governor, Rousseff was apprehended, spending 3 years in prison and enduring intense torture. She maintains her innocence.

In standing up to U.S. spying—even as Brazil admitted to following and photographing the movements of foreign diplomats in its capital a decade ago—Rousseff's anti-authoritarian impulses have aligned conveniently with her country's desire to flex its muscles and represent the interests of its region. Among Rousseff's generation of Latin American leaders, the legacy of U.S. interventions in the affairs of its southern neighbors and, more recently, a drug war that has left tens of thousands dead have fanned the flames of discontent with Uncle Sam. Amid the NSA scandal, Rousseff has had no problem reminding the United States that its era of dominance in Latin America is over.

Ron Wyden

For Insisting That the Law Should Never be Secret. Senator | Washington

"Does the NSA collect any type of data at all on millions or hundreds of millions of Americans?" When Sen. Ron Wyden asked that question of the U.S. director of national intelligence, James Clapper, in March, he already knew the answer. But his hands were tied because, as he explained in July, "under the classification rules observed by the Senate, we are not even allowed to tap the truth out in Morse code—and we tried just about everything else we could think of to warn the American people."

Edward Snowden revealed what Wyden couldn't: The National Security Agency is surveying U.S. citizens. That revelation allowed Wyden to wage his fight publicly—and wage he has.

He wants to roll back the NSA's authority to collect data on Americans and make public the government's interpretations of terrorism laws. To get there, he has pressured intelligence officials and even President Barack Obama to disclose their legal interpretations of the Patriot Act and the extent of the surveillance apparatus the law has been used to create. He has also pushed legislation to make the intelligence community more accountable to Congress.

It's not unusual for Wyden's principles to separate him from his colleagues. He was on the losing side of votes to strip wiretap provisions from the Patriot Act, and in March, he was the only Democrat to join Republican Sen. Rand Paul in protesting the administration's targeted killing policy. The NSA

revelations just gave Wyden a new platform for expressing his outrage about violations of Americans' civil liberties.

After a move to restrict the NSA'S surveillance program failed by just 12 congressional votes in July, Wyden told *Rolling Stone,* "I think Congress will come back in the fall and there will be new support for the kinds of views we're talking about." That hasn't happened yet, but expect Wyden to keep working overtime to secure that support.

Jesselyn Radack

For Championing the Rights of Whistleblowers.

National Security and Human Rights Director, Government Accountability Project | Washington

Before she became a leading defender of government whistleblowers, Jesselyn Radack was one herself. In 2002, Radack resigned from the U.S. Justice Department after John Walker Lindh, the so-called "American Taliban," was questioned without his lawyer present. Radack, who had advised prosecutors against the interrogation, gave emails about the event to a reporter.

Today, at the Government Accountability Project, Radack is a go-to defender of and public advocate for government employees who have disclosed sensitive information about intelligence and counterterrorism programs to journalists. Her most famous client is former National Security Agency official Thomas Drake, who told a *Baltimore Sun* reporter about a failed computer system at the agency that cost taxpayers millions of dollars. The government tried to prosecute Drake under the Espionage Act, but dropped the charges before going to trial.

In 2013, Radack has been a regular on talk shows and op-ed pages, advocating for whistleblower protections and criticizing the unprecedented prosecutions of leakers by the Obama administration. She's not representing the world's most famous whistleblower, Edward Snowden, but she is using her quasi-celebrity to try to shift the media story from details about the former NSA contractor to the specifics and potential impact of the surveillance programs he has revealed. (In October, she visited Snowden in Russia to present him with an "Integrity in Intelligence" award.)

"Instead of focusing on Snowden and shooting the messenger, we should really focus on the crimes of the NSA," Radack told ABC's *This Week* in June. "Because whatever laws Snowden may or may not have broken, they are infinitesimally small compared to the two major surveillance laws and the Fourth Amendment of the Constitution that the NSA's violated."

Moxie Marlinspike

For Making it Harder For The NSA—and Google—To SPY on You.

CO-Founder, Whisper Systems | San Francisco

In January 2011, as Egyptians took to the streets in revolution, Moxie Marlinspike, then working at a tiny tech start-up, designed a pair of encrypted communications services, RedPhone and TextSecure, that protect phone calls and text messages from eavesdroppers. "[RedPhone] is targeted just for Egypt, but sets the stage for worldwide support," he told *Wired* at the time.

Twitter later bought the start-up, Whisper Systems, for an undisclosed amount, and Marlinspike's creations have been released as Android apps. Versions for the iPhone are in development. He also designed a program that helps people remain anonymous when they're using Google's services. "Who knows more about citizens in their own country, North Korean leader Kim Jong II or Google? Why is Google not scary? Because we choose to use it," Marlinspike, who goes by a pseudonym, said at a 2010 computer security conference.

In 2013, revelations about mass spying by the National Security Agency have made Marlinspike seem like a surveillance soothsayer. (In May, he also exposed a Saudi telecommunications company that tried to hire him to monitor its customers.) He has had plenty to say about the NSA news: "It is possible to develop user-friendly technical solutions that would stymie this type of surveillance. . . . It's going to take all of us," he wrote on his blog. Calling for "all the opposition we can muster," he implored other techies to join him in developing ways to undermine surveillance.

For a man who says he "secretly hate[s] technology," Marlinspike's tools and message have already gone a long way toward defying states and companies that spy.

Kevin Mandia

For Identifying the Perpetrators of China's Cyber-Offensive.

Founder, Chief Executive, Mandiant | Alexandria, VA.

Since at least January 2010, when Google reported that a "highly sophisticated" attack on its corporate infrastructure

and "at least 20 other large companies" had originated in China, U.S. business executives have worried about the integrity of their secrets. But because of the difficulty of determining the source of a hack and the seeming absurdity (not to mention sensitivity) of accusing Beijing, the allegations went unproved.

That is, until a February 2013 report by the cybersecurity firm Mandiant traced individual hacks of U.S. companies to an exact address in Shanghai.

In the report, CEO Kevin Mandia and his team of corporate lawyers and cybergeeks traced more than 90 percent of the attacks they had seen to the headquarters of Unit 61398 of the People's Liberation Army. "Either they are coming from inside Unit 61398," Mandia told the *New York Times*, "or the people who run the most controlled, most monitored Internet networks in the world are clueless about thousands of people generating attacks from this one neighborhood."

In May, in a white paper titled "Chinese Motivations for Corporate Espionage: A Historical Perspective," Mandiant made a case for why Beijing would conduct large-scale corporate spying. "China's political history and popular culture is littered with examples of changing allegiances, profiteering, lies, spying, etc., in the name of victory, which, more often than not, ultimately equals moral legitimacy," the report said.

China denied the report's accusations, but they are widely seen as credible. Similar to what Edward Snowden's leaks have done to the U.S. National Security Agency, Mandiant's report has made it much more difficult for China to pretend it isn't launching attacks.

Dmitri Alperovitch

For Leveling the Cyber Playing Field.

Co-Founder, Crowdstrike | Irvine, Calif.

Amid reports in 2010 that Chinese hackers were stealing billions of dollars in trade secrets from American companies, Dmitri Alperovitch saw a business opportunity. Eschewing ineffective firewalls and virus scanners, Alperovitch decided the best way to defeat serious hackers was to use some of their own tactics against them, gaining intelligence about who they are, how they operate, and what they steal.

It's a strategy Alperovitch calls "active defense," and he cofounded a company, CrowdStrike, to help clients employ it. "Why can't you go into [a] network for the purpose of getting your data back or [to] take data off that machine to mitigate the damage?" Alperovitch said in an interview with MIT Technology Review, which named him one of 2013s top innovators under age 35.

This provocative approach isn't without controversy. Many cybersecurity experts argue that it's a terrible idea for a private business to retaliate, a practice colloquially called "hacking back," because it will only encourage hackers—who probably have more resources than the business—to become more aggressive. Alperovitch, in response, says that with active defense, he isn't advocating hacking back, but rather a framework that allows networks under attack to identify the source and use limited offensive measures, such as misinformation and malware, against it.

CrowdStrike has plenty of supporters: It has garnered some $60 million from eager investors and wowed audiences at tech conferences. This has helped Alperovitch make headway in his mission to, as he said in August, "change the existing security paradigm . . . and protect the intellectual property and trade secrets that are the crown jewels of our knowledge-based economy."

Critical Thinking

1. How can electronic surveillance by the state be reconciled with the individual liberty of citizens?

2. Why is it necessary for the US to engage in electronic surveillance of its allies?

Create Central

www.mhhe.com/createcentral

Internet References

Central Intelligence Agency
http://www.cia.gov/index.htm

Federal Bureau of Investigation
http://www.fbi.gov

National Intelligence Council
http://www.dni.gov/index.php/about

National Security Agency
http://www.nsa.gov